Intelligent and Adaptive Systems in Medicine

Series in Medical Physics and Biomedical Engineering

Series Editors: John G Webster, E Russell Ritenour, Slavik Tabakov, and Kwan-Hoong Ng

Series in Medical Physics and Biomedical Engineering

Intelligent and Adaptive Systems in Medicine

Edited by

Olivier C. L. Haas
Coventry University, UK

Keith J. Burnham
Coventry University, UK

CRC Press
Taylor & Francis Group
Boca Raton London New York

CRC Press is an imprint of the
Taylor & Francis Group, an **informa** business

MATLAB® is a trademark of The MathWorks, Inc. and is used with permission. The MathWorks does not warrant the accuracy of the text or exercises in this book. This book's use or discussion of MATLAB® software or related products does not constitute endorsement or sponsorship by The MathWorks of a particular pedagogical approach or particular use of the MATLAB® software.

CRC Press
Taylor & Francis Group
6000 Broken Sound Parkway NW, Suite 300
Boca Raton, FL 33487-2742

© 2008 by Taylor & Francis Group, LLC
CRC Press is an imprint of Taylor & Francis Group, an Informa business

No claim to original U.S. Government works

ISBN-13: 978-0-7503-0994-3 (hbk)
ISBN-13: 978-0-367-38762-4 (pbk)

Library of Congress Cataloging-in-Publication Data

Intelligent and adaptive systems in medicine / editors, Olivier C.L. Haas and Keith J. Burnham.
p. ; cm. -- (Series in medical physics and biomedical engineering)
Includes bibliographical references and index.
ISBN 978-0-7503-0994-3 (alk. paper)
1. Artificial intelligence--Medical applications--Congresses. 2. Intelligent control systems--Congresses. 3. Biomedical engineering--Computer simulation--Congresses. I. Haas, Olivier. II. Burnham, Keith J. III. Title. IV. Series.
[DNLM: 1. Expert Systems--Congresses. 2. Biomedical Technology--methods--Congresses. 3. Diagnosis, Computer-Assisted--Congresses. 4. Medical Informatics Applications--Congresses. W 26.55.A7 I607 2008]

R859.7.A72I48 2008
610.285--dc22   2007035188

Visit the Taylor & Francis Web site at
http://www.taylorandfrancis.com

and the CRC Press Web site at
http://www.crcpress.com

About the Series

The Series in Medical Physics and Biomedical Engineering describes the applications of physical sciences, engineering, and mathematics in medicine and clinical research.

The series seeks (but is not restricted to) publications in the following topics:

- Artificial organs
- Assistive technology
- Bioinformatics
- Bioinstrumentation
- Biomaterials
- Biomechanics
- Biomedical engineering
- Clinical engineering
- Imaging
- Implants
- Medical computing and mathematics
- Medical/surgical devices
- Patient monitoring
- Physiological measurement
- Prosthetics
- Radiation protection, health physics, and dosimetry
- Regulatory issues
- Rehabilitation engineering
- Sports medicine
- Systems physiology
- Telemedicine
- Tissue engineering
- Treatment

The Series in Medical Physics and Biomedical Engineering is an international series that meets the need for up-to-date texts in this rapidly developing field. Books in the series range in level from introductory graduate textbooks and practical handbooks to more advanced expositions of current research.

The Series in Medical Physics and Biomedical Engineering is the official book series of the International Organization for Medical Physics.

The International Organization for Medical Physics

The International Organization for Medical Physics (IOMP), founded in 1963, is a scientific, educational, and professional organization of 76 national adhering organizations, more than 16,500 individual members, several corporate members, and four international regional organizations.

IOMP is administered by a council, which includes delegates from each of the adhering national organizations. Regular meetings of the council are held electronically as well as every three years at the World Congress on Medical Physics and Biomedical Engineering. The president and other officers form the executive committee, and there are also committees covering the main areas of activity, including education and training, scientific, professional relations, and publications.

Objectives

- To contribute to the advancement of medical physics in all its aspects
- To organize international cooperation in medical physics, especially in developing countries
- To encourage and advise on the formation of national organizations of medical physics in those countries which lack such organizations

Activities

Official journals of the IOMP are *Physics in Medicine and Biology* and *Medical Physics and Physiological Measurement*. The IOMP publishes the bulletin *Medical Physics World* twice a year, which is distributed to all members.

A World Congress on Medical Physics and Biomedical Engineering is held every three years in cooperation with IFMBE through the International Union for Physics and Engineering Sciences in Medicine (IUPESM). A regionally based international conference on medical physics is held between world congresses. IOMP also sponsors international conferences, workshops, and courses. IOMP representatives contribute to various international committees and working groups.

The IOMP has several programs to assist medical physicists in developing countries. The joint IOMP Library Programme supports 68 active libraries in 41 developing countries, and the Used Equipment Programme coordinates equipment donations. The Travel Assistance Programme provides a limited number of grants to enable physicists to attend the world congresses. The IOMP Web site is being developed to include a scientific database of international standards in medical physics and a virtual education and resource center.

Information on the activities of the IOMP can be found on its Web site at www.iomp.org

Contents

Preface

The idea for this book stemmed from the multidisciplinary meeting on "intelligent systems in medical diagnostics and therapy," which was held at Coventry University, United Kingdom, in October 2002. The meeting was sponsored by the EUropean Network on Intelligent TEchnologies for Smart Adaptive Systems (EUNITE), which aimed to promote research and developments in adaptive and intelligent systems. It provided a unique opportunity for technical specialists in the fields of control engineering, medical imaging, and statistics to meet with medical professionals and allied sciences. The book aims to capture the genuine richness emanating from the discussions and debate, the various strands of thought, the raising of awareness, and of stimulating the ideas for future multidisciplinary lines of inquiry.

In setting out to achieve the foregoing aims, the book attempts to address, and, indeed, highlight the multidisciplinary nature supporting the discipline of "techno-medicine" and its reliance on engineering systems that are required to be intelligent and adaptive to react in the most appropriate manner to improve patient treatment, as well as their overall experience, while taking into account the needs of the clinical staff performing the procedures.

The techno-medicine era that our hospitals are currently, fundamentally a part of means that hospitals and clinics are now truly multidisciplinary organizations where patient diagnostics, treatment delivery, and follow-up involve, for example, medical professionals, engineers, scientists, and mathematicians. In the United Kingdom, the creation of "super university hospitals" linked with local universities is aimed at improving contacts between medical specialists and academics with the ultimate goal to increase the quantity, quality, and applicability of collaborative multidisciplinary medical research. Key to the success of such a national initiative is the realization that technical specialists should focus on clinically realizable solutions as opposed to "blue sky" research. Conversely, medical professionals should gain an appreciation of the working principles of the techniques used by theoretical and technical specialists to be able to provide appropriate information to further develop these intelligent and adaptive systems. Indeed, the main challenges of the techno-medical specialist are to derive the most effective means to capture, consolidate, and further develop the everevolving "knowledge." Understanding of the limitations of these challenges by all parties involved is fundamental to ensure a successful and sustainable healthcare industry that is ready to meet the challenges of the future.

Intelligent and adaptive systems have evolved and are utilized almost routinely throughout many aspects of the modern clinic, from the initial diagnosis to the planning of the treatment, its execution, its verification, culminating with patient follow up, and automated statistical records.

Traditionally, many branches of medicine are based on the experience of the clinician and can vary depending on training, demographic concentrations of illnesses, and diseases. Such experience, including the experience gained from studying the effects of various treatments, is a very important factor that needs to be captured, consolidated, and further developed. Today, computer-based intelligent and adaptive systems are very much part of this process, helping to analyze the link between treatment outcome, procedures, individual circumstances, as well as an individual's genetic makeup.

The continuing advances in medical imaging and computing technology have enabled intelligent systems to help clinicians to diagnose diseases, identify their location, and outline the structures of interest in the images. These intelligent systems are today required to be adaptive to exploit the new, so-called "four dimensional" images, which are in effect volumetric images that change with time. The image-processing work on deformable models described in this book aims to help the reader appreciate the technical solutions that are available to process and exploit the information contained in such enormous dataset.

Clinicians are required to make critical decisions on a daily basis and are ultimately responsible for the treatment they prescribe. The complexity of modern treatment has prompted the development of various software tools to help plan and predict the effect of various medical procedures in advance. Various approaches are presented in this book. Some are based on understanding the current procedures and best practice by interviewing the staff involved in the decision-making process. Others are based on deriving computer models that are able to predict the outcome of a particular set of procedures.

Following treatment planning, adaptive as well as intelligent computer systems are employed to help clinicians deliver a given treatment. Although the equipment may be operated by a medical specialist, the actual procedure is likely to be performed by the machine. To be safe and effective, these machines are controlled by computer systems that constantly monitor the machine performance as well as the demands from the user. Each time a user enters a new command, a layer of processing occurs, establishing if this command is appropriate and how to respond to it. The most common techniques and algorithms aimed to improve machine operation are reviewed in this book.

Although the increase in machine complexity may improve treatment or at least facilitate its delivery, it will always be accompanied by an increased risk of equipment failure. This book addresses such issues and discusses tools and techniques to monitor the "health" of the machine as well as that of the patients.

This book is therefore an ideal starting point to help medical specialists, scientists, engineers, and mathematicians to gain a common understanding of the potential for utilizing intelligent and adaptive systems for medical applications. It demonstrates how actual challenging medical problems can find practical solutions, which call for continuous, novel, scientific, and technical development.

Acknowledgments

The editors would like to extend their gratitude to Professor D. A. Linkens for his encouragement; first, for his assistance in setting up and supporting the EUNITE meeting at Coventry University and second, in supporting, together with R.F. Mould, the concept of archiving the materials and discussions of the meeting, which now form the bases of the chapters of this book. Finally, acknowledgments are extended to the contributors to this book.

Editors

Olivier C. L. Haas, DUT, BEng, MSc, PhD, MIEEE, graduated in electrical engineering from Joseph Fourier University, Grenoble, France, and obtained his PhD in systems modeling optimization and control applied to radiotherapy from Coventry University, United Kingdom (1997). He is currently a reader in applied control systems, leading the medical activities of the Control Theory and Applications Center at Coventry University, United Kingdom, and honorary research fellow at the University Hospitals Coventry and Warwickshire NHS Trust, Coventry, United Kingdom. Dr. Haas pioneered the use of engineering techniques in radiotherapy physics and has been a strong promoter of multidisciplinary applied research. His main research interests include systems modeling, optimization, image processing, and control engineering applied to medical systems and, in particular, to the treatment of cancer with radiotherapy. He is currently work-package 1 coordinator within the MAESTRO Integrated European Project, supervising the integration of image and signal processing together with control engineering and medical physics to develop a practical patient motion compensation system for radiotherapy. He has more than 80 publications to his credit, including journals and conference papers, book chapters, books, and edited conference proceedings.

Keith J. Burnham has been the professor of Industrial Control Systems and director of the Control Theory and Applications Center (CTAC), a multidisciplinary research center based within the Faculty of Engineering and Computing, Coventry University, United Kingdom, since 1999. The CTAC is engaged in a number of collaborative research programs with industrial organizations involving the design and implementation of adaptive control systems. Keith Burnham obtained his BSc (mathematics), MSc (control engineering), and PhD (adaptive control) at Coventry University in 1981, 1984, and 1991, respectively. He is regularly consulted by industrial organizations to provide advice in areas of advanced algorithm development for control and condition monitoring. He is currently a member of the Editorial Board of the Transactions of the Institute of Measurement and Control (InstMC), the InstMC Systems Control Technology Panel, the Informatics and Control Working Group of the Institution of Mechanical Engineers (IMechE), and the Automotive Control Technical Committee of the International Federation of Automatic Control (IFAC). He is also a chartered mathematician (CMath) and a corporate member of the InstMC, the Institute of Mathematics and its Applications (IMA) and the Institution of Engineering and Technology (IET).

Contributors

Richard Aldridge
School of Computing Sciences
University of East Anglia
Norwich, U.K.

Mark K. Bennett
Royal Victoria Infirmary
Newcastle Upon Tyne, U.K.

Gloria Bueno
School of Engineering
University of Castilla-La Mancha
Ciudad Real, Spain

Keith J. Burnham
CTAC
Engineering and Computing
Coventry University
Coventry, U.K.

Paulo de Carvalho
Department of Informatics
    Engineering
University of Coimbra
Coimbra, Portugal

Robert C. Crichton
Department of Clinical Physics
    and Bioengineering
University Hospital
Coventry, U.K.

António Dourado
Department of Informatics
    Engineering
University of Coimbra
Coimbra, Portugal

Mark Fisher
School of Computing Sciences
University of East Anglia
Norwich, U.K.

John H. Goodband
SIGMA
Engineering and Computing
Coventry University
Coventry, U.K.

Peter Groumpos
Laboratory for Automation
    and Robotics
Department of Electrical and
    Computer Engineering
University of Patras
Rion Patras, Greece

Olivier C. L. Haas
CTAC
Engineering and Computing
Coventry University
Coventry, U.K.

Jorge Henriques
Department of Informatics
    Engineering
University of Coimbra
Coimbra, Portugal

Charles Johnston
Engineering and Computing
Coventry University
Coventry, U.K.

William M. Kelly
Radiotherapy Department
University Hospital
Coventry, U.K.

Pedro Lago
Department of Applied Mathematics
    Faculty of Sciences
University of Porto
Porto, Portugal

Teresa Feio Mendonça
Department of Applied Mathematics
    Faculty of Sciences
University of Porto
Porto, Portugal

John A. Mills
Department of Clinical Physics
    and Bioengineering
University Hospital
Coventry, U.K.

Raouf Naguib
BIOCORE
Engineering and Computing
Coventry University
Coventry, U.K.

Robert Newman
School of Computing and IT
Wolverhampton University
Wolverhampton, U.K.

Catarina Nunes
Department of Applied Mathematics
    Faculty of Sciences
University of Porto
Porto, Portugal

Elpiniki Papageorgiou
Laboratory for Automation and
    Robotics
Department of Electrical
    and Computer Engineering
University of Patras
Rion Patras, Greece

Colin R. Reeves
CTAC
Engineering and Computing
Coventry University
Coventry, U.K.

Phil Sharpe
Department of Clinical Physics
    and Bioengineering
University Hospital
Coventry, U.K.

James Shuttleworth
Engineering and Computing
Coventry University
Coventry, U.K.

Chrysostomos Stylios
Laboratory of Knowledge and
    Intelligent Computing
Department of Informatics and
    Communications Technology,
TEI of Epirus,
Artas, Greece

Yu Su
School of Computing Sciences
University of East Anglia
Norwich, U.K.

Alison Todman
Engineering and Computing
Coventry University
Coventry, U.K.

1

Introduction

Olivier C. L. Haas and Keith J. Burnham

CONTENTS

1.1   Introduction to Intelligent and Adaptive Systems

Medicine is an area in constant evolution where software and hardware systems, employed to support the clinical staff decision-making process as well as the treatment delivery, are increasingly complex and numerous. Today, many of such systems can be considered as "intelligent and adaptive," relying on measured data and the information provided by medical specialists to "learn."

There are many definitions for intelligent and adaptive systems. In the context of this book, a system is considered as a combination or assemblage of interdependent, interrelated, or interacting elements which perform a set of functions. An intelligent system is defined as one that is able to learn from experience to reach an appropriate decision or react in an appropriate manner to a set of specific stimuli. Intelligent systems should be capable of learning, enabling them to change their behavior as a function of their environment. Such an ability to change, adjust, and modify their behavior to suit a new set of conditions or objectives is characteristic of adaptive systems. This means that intelligent systems are often adaptive too.

To realize the full potential of intelligent and adaptive systems in medicine, there is a need for the developers and the end users to understand both the underlying technologies and the specificities of the medical application considered. Indeed, to develop intelligent and adaptive systems successfully, medical experts should appreciate how the information they provide will be exploited by a set of software routines in an effective as well as a safe manner. Technical experts should also be aware of the limitations, constraints, and ultimately the responsibility associated with designing systems for medical applications.

This book aims to provide the medical practitioner with a clear and concise explanation of a range of intelligent and adaptive systems, highlighting their benefits and limitations with realistic medical examples. Each technical chapter is designed to include a concise explanation of techniques and algorithms exploited by the family of intelligent and adaptive systems presented and an "application section" that describes in detail the medical problems and the solutions adopted. There are many problems that remain to be solved in medical applications and even more candidate solutions to address them. In addition to raising the awareness of the medical practitioner to adaptive and intelligent systems, it is believed that this book will also help academics identify the potential for further research into bridging the gap between theory and practice associated with the new "techno-medicine" era.

The intelligent and adaptive systems reported are mainly software based. The structure of the intelligence is provided by the design of the intelligent algorithms. The system's knowledge is obtained from a set of training data such as historical data describing a particular relationship or expert opinions from which a representative model can be constructed. Typical intelligent systems include artificial neural networks (ANN) also referred to as neural networks (NN), fuzzy logic (FL), combinations and derivations of ANN and FL such as neurofuzzy (NF), adaptive neurofuzzy inference system (ANFIS), and fuzzy cognitive maps (FCM). Such intelligent systems were developed using human behavior as an inspiration. For example, ANN started from an analogy with the general behavior of neurons in the brain. Current developments attempt to include biological findings into the algorithms, for example, in the case of ANN, the need for the axons to be activated above a given level to actually transmit an electrical signal. Adaptive systems can change their internal behavior to adapt to significant changes in the system behavior by modifying their internal structure or changing their parameters of operation. By doing so, they can change the approach to solving a problem. Alternatively, adaptation or adjustment can be achieved owing to the intelligent and adaptive system structure. In the latter case, the system response is modified depending on information received. Intelligent and adaptive control systems generally include strategies based on artificial intelligence (AI) and heuristics. Although AI and heuristic control algorithms have received much academic interest, they are still in their infancy in terms of the number of practical control applications. This chapter therefore focuses on describing traditional deterministic control systems that are inherently safe and robust at the same time possessing some degree of "intelligence" and "adaptivity."

It is important to note that, in addition to the algorithm performing the intelligent and adaptive strategy, information-processing tools are also required to be present. Indeed the amount of data in medicine is generally very large and, at the same time, some of this information can be corrupt or incomplete. It is therefore important to present the "appropriate" data to the adaptive and intelligent system. The reader should therefore pay attention to the intelligence and adaptivity of the algorithms presented as well as the data preprocessing activities that also accompany the algorithms presented in this book. In particular, there is a need to have a better understanding of the statistical properties of the uncertainties and assumed process noise.

There are many medical applications where intelligent and adaptive systems are employed, for example, coronary heart disease and stroke, cancer, mental health, accident, and surgery (see Chapter 5 by Dourado et al.). The first challenge was, therefore, to restrict the focus of this book in terms of the medical applications considered by identifying an area of medicine, where these techniques had been applied from the initial diagnosis through to the patient follow-up. Cancer was selected as the main medical application to serve as a candidate to support for the detailed description of the implementation of intelligent and adaptive systems. In addition to being one of the main foci of medical research in the Western world (see, for example, the MAESTRO project in Europe [MAESTRO 2007]), oncology and in particular radiotherapy is a prime example where a significant amount of new techniques and technologies are being developed to facilitate diagnostics, treatment organization, planning, optimization, delivery, and follow-up. Medical devices involved in oncology are becoming increasingly complex, capable of performing many tasks in an automated manner. The development of these sophisticated machines and their maintenance require the adoption of techniques that have previously been employed in the aerospace and the automotive industries.

Having introduced the subject of intelligent and adaptive systems in medicine, the next section describes the structure of the book in terms of the interconnection between the different chapters.

1.2 Contents

This book on *Intelligent and Adaptive Systems in Medicine* comprises 11 chapters organized into three parts according to the type of "intelligence" and "adaptation" exploited by the techniques and algorithms described. Each part includes reviews of theories and applications together with specific studies employing various combinations of intelligent and adaptive techniques. Each technical chapter includes a detailed description of the intelligent and adaptive system, a problem statement highlighting the issues associated with the medical application considered, and a section considering the modifications added to the algorithms together with implementation issues.

The first part of this book presents some of the tools employed to automate tasks, for example, drug infusion during anesthesia (Feio-Mendonça et al., Chapter 3) and improve the safety of machine operation by monitoring their "health" (Mills et al., Chapter 4). The second part considers AI techniques applied to treatment planning, decision making, delivery, and follow-up. The third part considers medical imaging techniques for the purpose of diagnostic and task automation.

Part I starts in Chapter 2 with a brief description of systems modeling and control engineering algorithms employed in Chapters 3 and 4. It aims to provide the reader with an appreciation of the control engineering approach to problem solving.

Chapters 3 (Feio-Mendonça et al.) and 4 (Mills et al.) approach the concepts of intelligent and adaptive systems from a traditional control engineering perspective, where the knowledge gained from the system studied is not only aimed at replicating its observable behavior but also to understand its underlying principles. The advantage of such a technique is that if a mismatch between the system and its model occurs, it is possible to interrogate the model to assess if all processes have been modeled adequately. The intelligence lies, therefore, in the ability of the model to understand the process modeled and replicate its behavior. Having understood how the system operates it is then possible to derive the most appropriate "control strategy" to improve its performance and hence adapt to the environment.

Chapter 3 by Feio-Mendonça et al. entitled "Intelligent Control Systems in Anesthesia" starts with a brief description of anesthesia and in particular balanced anesthesia, where a cocktail of drugs is administered to the patient. The difficulty in the problem addressed in Chapter 3 lies in its safety critical aspect together with the equipment utilized to control the depth of anesthesia, the number of drugs used, the difficulty associated in obtaining consistently reliable information from the sensors/equipment monitoring the intravenous drug concentration, and the high interpatient variability. The latter means that each patient is likely to react differently to the same cocktail of drugs.

The second section of Chapter 3 focuses on the means to control anesthesia in a clinical environment. It follows a control engineering approach to the problem by initially explaining the importance of deriving an accurate and robust model of patient response to atracurium, a muscle relaxant drug. Models of increasing complexity and reliability are presented. As in many engineering applications, factors which are important to explain a system's behavior may not always be measurable. It is shown that in such cases it is possible to estimate these variables by relating them to measurable quantities.

Unlike traditional electrical or mechanical engineering systems, medical systems, such as the intake of drugs by the body have a wide variability. Further, the monitoring of substances in the blood stream is subject to large amounts of noise with the extreme case being temporary sensor failures. Despite these noise effects and uncertainties, the closed-loop control system reaction based on this nonideal information must perform in a robust and consistent manner to become accepted in a clinical setting. While other non-medical systems may have similar issues, they can, however, be submitted to a range of experiments to help identify their time-varying and dynamic behavior. It is, however, not "desirable" to inject the patient with drugs just to study their responses. In the particular case of anesthesia, a *bolus* dose is given to the patient at the start. The large amount of drug used in the *bolus* makes it very difficult to assess the effect of small changes, such as that required to maintain the patient relaxation.

The remaining part of the section progresses from the description of robust control strategies based on classical, adaptive, and finally switching control, and culminates with the description of "hypocrates" the robust automatic control system for the control of neuromuscular blockade on patients undergoing anesthesia. The software package developed by Feio-Mendonça et al. includes a set of tools for control as well as noise reduction and adaptation, to enable the user to evaluate different control strategies. Many of the algorithms implemented in "hypocrates" have been clinically tested.

Chapter 4 by Mills et al. entitled "Control Methods and Intelligent Systems in Quality Control of Megavoltage Treatment Machines" approaches the development of intelligent and adaptive systems from a medical physics perspective. The aim of the systems presented is to maintain increasingly demanding linear accelerator (linac) performance. Quality control aims to identify sources of error to alleviate in an effective manner the amount of machine "downtime," thus reducing the time when the machine is not available to treat patients.

The challenge faced by the medical physicists and engineers is to keep the machine downtime to a minimum. Quality control activities must be performed on a regular basis to meet stringent requirements and regulations. In addition, many radiotherapy physics departments will have their own set of test procedures to monitor the machine performance to detect and possibly preempt the problems. Mills et al. divide the monitoring process into three categories: immediate, short term, and long term. Immediate monitoring is related to tuning the equipment to improve current performance. Short term relates to the daily monitoring of the machines and assessment of their performance over a whole treatment day. Ideally, once the so-called "morning checks" have been performed, and the machine has been deemed to work satisfactorily, the machine's performance should remain satisfactory throughout the day. The problem, however, is that machine performance changes with environmental changes such as humidity, magnetic field, and temperature. Long-term monitoring is used to assess the deterioration of machine components over a longer period of operation, for example, weeks, months. The latter can be exploited to advantage in predicting routine maintenance schedules, hence reducing unplanned maintenance activities that could affect patient treatments. The first part of the chapter makes use of statistics and systems modeling techniques to monitor the health of the linac and in particular its ability to generate and steer the treatment beam. The study demonstrates the benefits of regular beam monitoring and adjustment in terms of machine performance. In addition, it highlights a typical issue with respect to a comparative study in a medical/clinical environment where medical equipment is involved. If the equipment is used clinically, it is not possible to let it deviate from optimal performance by too large a margin. This means that whilst ideally one set of tests should not have involved any machine retuning, the latter was necessary to ensure that the machine remained clinical.

The second part of Chapter 4 deals with the design, implementation, and testing of the cyber-assistant, which is a new intelligent and adaptive system designed to (i) facilitate the gathering of data, (ii) improve the consistency of the information collected by forcing the operator to follow given procedures, and (iii) give feedback to the person performing the test. Some of the tasks performed during routine maintenance require the staff to use both their hands. A voice recognition system was thus implemented to recognize a set of predefined commands and acquire numerical data. The intelligence of the system lies in its ability to guide the user through a set of actions to be performed for each procedure. Depending on the outcome of the measurement or task performed different actions may be required. The knowledge base of the system is created from historical data and updated at each measurement. The assessment of the system shows that it offers great potential, further increased by the rapid development of hand-held devices and wireless technologies.

Part II of this book comprises four chapters. Chapter 5 by Dourado et al. entitled "Neural, Fuzzy, and Neurofuzzy for Medical Applications" aims to set the scene in terms of traditional intelligent and adaptive systems applied to a wide range of medical applications. It starts with a general introduction to the topic of soft computing and AI. Having introduced the concepts of numeric and linguistic information as a means to express intelligence, Chapter 5 gives a brief overview of ANN, FL, and NF including a description of their working principles. Such an introduction to the subject should provide the reader with a general appreciation of these AI-based techniques and their applicability to medicine. The applications mentioned include medical imaging, biological signal processing and interpretation, control of biological systems, and prognosis and decision support.

Chapter 6 by Papageorgiou et al. entitled "The Soft Computing Technique of Fuzzy Cognitive Maps (FCMs) for Decision-Making in Radiotherapy" describes an application of FCM to implement decision-support systems for radiotherapy. The overall radiotherapy process is very complex and involves many factors and constraints that may affect a range of decisions that have to be made from the planning through to the delivery of the treatment. The clinical treatment simulation tool based on a fuzzy cognitive map (CTST-FCM) model aims to emulate the process followed by the clinicians and physicists to decide whether a course of action will give rise to a positive outcome. It uses experts to identify the influence of the various elements or concepts entering into the decision process, their interdependence as well as the constraints imposed by regulatory bodies and good practice. The CTST-FCM creates a dynamic model for estimating the final dose delivered to the target volume and normal tissues with the ability to evaluate the success of radiotherapy. If the solution reached from a set of initial conditions determined by experts is not acceptable, the system reoptimizes the interconnection between its different concepts/relationships to reach an acceptable solution.

The overall approach is assessed using two test cases. The first test case entails a relatively complex conformal treatment aiming to minimize the dose delivered to healthy tissues such as bladder and rectum whilst delivering a high dose to the prostate. The second case study involves the treatment of a prostate cancer using a standard treatment comprising four orthogonal beams to deliver a cubic volume of high radiation covering the region identified by the clinician. The issue with such treatments is that they also deliver a high dose of radiation in the healthy tissues present in this "cubic" volume. The advantage is that it is a well-tried and tested technique. In both instances it is shown that the CTST-FCM is able to reach a clinically acceptable solution.

Chapter 6 provides a clear example of the ability of soft computing to model expert knowledge using fuzzy concepts. This is demonstrated by using the nonlinear Hebbian learning algorithm to train the initial model resulting in the creation of an advanced sophisticated system. The main strength of FL is also its main weakness. Indeed, to be accurate an intelligent system needs to be well trained. The selection of the number of rules and their expressions rely on human experts. Issues related to the knowledge and "expertise" of the experts utilized to train the system are common. To improve the accuracy of the results, several expert opinions can be combined. Further, experts with slightly different backgrounds should also be requested to take part in the training of such a system. In addition to several experts, it may be useful to ask a group of experts to propose and agree on rules together as opposed to proposing rules independently of other experts. The weakness is that human nature may bias the decision of the group toward the expert with the most influence, even if this is not the "best" expert. Such considerations indicate the need for the fusion of knowledge based systems such as FCMs with computational intelligence techniques such as 'learning methods' to develop intelligent and adaptive systems.

Chapter 7 by Goodband and Haas entitled "Neural Networks in Radiation Therapy" starts with a brief introduction of radiotherapy and the various improvements made to the techniques, including conformal radiation therapy (CRT), intensity modulated radiation therapy (IMRT), image guided radiation therapy (IGRT), and adaptive radiation therapy (ART). The chapter deals with applications in IMRT and ART. This is followed by a brief overview of the ANN structures and training algorithms that can be adopted to obtain a relationship between a set of inputs and a set of outputs. Differences between ANN schemes used to model static or dynamic/time-varying relationships are also highlighted.

Having established sufficient theoretical bases to understand the approaches adopted to date, a review of applications in radiotherapy is then given under the following themes: classification, nonlinear regression or mapping of nonlinear input–output relationships, and finally control or time series prediction. The reader will note that the range of problems addressed using ANN is quite varied. By contrast, the structures of the ANN schemes and the training approaches adopted are very similar. This means that the ability of the ANN to be "tuned" to solve a specific problem may not have been fully exploited. Although a generic ANN can provide good results, they have not been able to match the accuracy of traditional medical physics techniques in many instances. This relative lack of success has contributed to the slow uptake of ANN in radiotherapy physics.

The remaining part of Chapter 7 aims to demonstrate that when applied correctly an ANN can be successful and "safe." Two case studies are adopted to illustrate good practice in designing and optimizing ANNs to obtain not only a solution, but an accurate as well as robust solution.

The first case study demonstrates the benefit of adopting a divide-and-conquer approach implemented through a "committee machine," as opposed to using a single ANN, to solve a challenging inverse problem in IMRT. The "intelligence and adaptation" of the "committee machine" lies in its ability to exploit and combine local knowledge by ANNs trained to model different aspects of the system. The problem solved deals with the determination of the position and depth of material within an IMRT compensator from a desired dose distribution. However, many inverse problems could also be solved using a similar approach as long as there is a forward model or a cause–effect relationship that can be measured. The data corresponding to the cause and the effect are then swapped (the "effect" becoming the "cause") and given to the ANN scheme to determine the inverse relationship.

The first case study demonstrates the applicability of an ANN to model static relationships, that is, the physics of the attenuation process within the compensator is not likely to change with time. The second case study aims to expose the ability of ANNs to solve problems where the system changes with time. The adaptive nature of the scheme lies in its ability to adjust to a changing environment. The problem considered stems from the implementation of novel IGRT and ART treatment techniques. It is shown through a comparative study between different training and regularization algorithms that a robust and safe ANN can be employed to predict the short-term evolution of patient motion tracked using external markers.

Chapter 8 by Johnston and Reeves entitled "Neural Networks for Estimating Probability Distributions for Survival Analysis," describes the applications of intelligent and adaptive systems to patient follow-up and in particular survival analysis. Chapter 8 starts with a brief review of the main types of statistical methods traditionally employed to model time to events for example, time to failure for machines and time to death for survival analysis. The three groups of survival analysis modeling methods are nonparametric, parametric, and quasi-parametric. The first and the latter are the most commonly used in medical applications. The Cox proportional analysis, a quasi-parametric modeling technique and method of choice in medical statistics, is selected in Chapter 8 as a basis for comparison with an ANN. The ANN scheme implemented is based on multiple-layer perceptron. An interesting aspect of Chapter 8 is the comparison reported by the authors between statistical terminology and ANN terminology. The chapter includes a clear mathematical description of the ANN applied to survival analysis data and a training section, which similar to Chapter 7, emphasizes the need to apply the so-called regularization techniques to ensure that the ANN does not model the noise in the data as opposed to the underlying relationship. Several regularization techniques are briefly discussed.

It is often found in the literature that ANN users do not spend sufficient time to analyze the features that are essential to model nonlinear processes. This chapter, however, makes use of a preliminary analysis of the data to select the noncorrelated/independent features through principal component analysis. This enables the number of inputs and features used to train the ANN to be reduced without reducing the underlying meaning of the features. This may surprise a few medical practitioners, as apparently relevant criteria/features are not directly used. In effect, the knowledge linked with all the features is summarized by the principal components.

The clinical data selected relate to breast cancer and includes 600 cases. It is shown that the training and in particular the selection of regularization parameters is essential in obtaining a good model fit and generalization. The ability of the ANN to model survival data appropriately is analyzed by comparing the predictions made by the model with sample estimates. It is concluded that ANNs are well suited to estimate probability distributions for survival analysis.

Part III of the book describes medical imaging applications of adaptive and intelligent systems. The availability of digital images throughout hospitals and clinics has provided researchers with a large pool of information. Numerous methods have been developed and are currently under development to improve the processing of images, the extraction of relevant and the subsequent analysis of the information.

Chapter 9 by Fisher et al. entitled "Some Applications of Intelligent Systems in Cancer Treatment: A Review" extends the review from Chapter 5 to focus on the application of intelligent and adaptive systems applied to diagnosis, planning, and delivery of cancer treatment using radiotherapy. Intelligent computer aided diagnosis (CAD) systems, which aim to help clinicians reach appropriate decisions, are reviewed for two typical cancer types: prostate and breast. Each subsection starts with a description of the means available to clinicians to diagnose the disease and establish its stage. These belong to two families: biological investigation to detect tumor-specific antigen or tumor-associated antigen and medical imaging–based investigation to find specific patterns. It is shown that the key to the success of the intelligent CAD is the information used to train the algorithms and reach a diagnosis. Current developments in medical imaging are also reported and provide a brief introduction to the work of Bueno on deformable models in Chapter 10. A further detailed description of CAD methods is given by Shuttleworth et al. in Chapter 11. The remaining part of Chapter 9 provides a brief review of the terminology, current issues, and latest development in radiotherapy treatment techniques such as IMRT and IGRT.

Chapter 10 by Bueno entitled "Fuzzy Systems and Deformable Models" provides a description of two adaptive and intelligent techniques: fuzzy C-means clustering (FCMC) and deformable models. Despite continuous improvement in medical imaging technology, processing and analysis of images remains a challenging problem. This is due to the relatively high inter- and intra-patient variability in shape and densities of organs and anatomical body structures. In addition, whilst multiple imaging modalities are combined to facilitate diagnosis and planning of treatment, there are still some uncertainties with the geometrical localization of the cancerous tissues. Therefore the information available to partially automate the analysis of medical images is often incomplete.

The first method presented in Chapter 10 exploits the ability to differentiate tissues using imprecise and incomplete information. It aims to classify image pixels into different structures. The classification is performed according to the degree of membership to the different classes or structures identified during the training stage. The intelligence of the FCMC is in the mathematical expression of the objective aimed to be attained through training. It results in the identification of a number of classes, for example, healthy, malignant; and their corresponding centroids. The centroids characterize the difference between the classes. The smaller the distance between a particular centroid and a pixel the more likely it is that this particular pixel belongs to the structure corresponding to that centroid. Although, ideally a pixel should belong to a single structure, its characteristics may be similar to several structures. The degree of similarity is usually referred to as the membership value. In the work presented, the resulting pixel class is determined by taking the class for which its degree of membership is the highest. The key to the success of the algorithm is the identification of the appropriate number of structures, or clusters, and the position of the centroids.

FCMC is applied to the classification of mammogram images which include seven classes that are identified based on tissue abnormalities and severity of the disease. The tissues encountered in the images are divided into different clusters. It is shown that using unsupervised learning leads to a higher number of clusters than with supervised learning where the number of clusters is initialized based on histogram information. Having demonstrated the applicability of the method using a published database, the approach is then implemented within a computer-aided design system and tested using clinical data.

The second family of intelligent and adaptive techniques described in Chapter 10 is that of a deformable model. Three types of deformable models are described: energy minimizing deformable models, geodesic active contour models, and statistical physics–based deformable models (SDMs).

In the case of original deformable models based on energy minimization, the intelligence is encoded into an objective function, referred to as an energy function. The energy function aims to fit an "elastic body," the shape of which can be modified to fit a particular region of interest (ROI) in an image. The challenge lies in finding an appropriate formulation of the objective to balance the importance of the information from the image with the general knowledge of the shape of the object from anatomical or medical information. The need for such a method arises from the difficulty of being able to identify the contour of a ROI based solely on visual information from the image. The energy function can thus be viewed as the combination of expert knowledge with image information. The intelligence of the method lies in the definition and combination into an objective function of a set of objectives that represents knowledge about the image, for example, pixel gray levels, gradient of image, and knowledge about the shape of the object that is being segmented, for example, smoothness of the contour, concavity, or convexity of the ROI. Additional objectives related to the dynamic behavior or the statistical properties of the ROI can also be added. The latter enables the algorithm to adapt to changes in ROI shape in time and space. The difficulty associated with the implementation of the method lies in the determination of the most appropriate trade-off between relying on image and shape information.

Geodesic active contour models aim to achieve the same objectives, hence solve an energy function. However, a level set representation is used to improve the ability of the algorithm to adapt to changes in shape and surface with time, and reduce the algorithm sensitivity to initial conditions, thus improving its numerical stability. Level set is a numerical analysis tool that models the boundaries between different parts of the image in terms of their spatial and temporal evolutions using partial differential equations. The advantage of level set is that it is possible to change the object topology, that is, its geometric properties that are independent of, for example, size or shape by changing a constant corresponding to the level plane dividing them.

SDMs exploit the matrix/vector discrete formulation of the energy function to facilitate the description of the relationship among different structures taking into account the "statistical" variability of specific anatomical structures in three dimensions. This last family of deformable models is particularly relevant to medical applications due to their ability to take into account the intra- and inter-variability of anatomical regions of interest as well as the spatial relationship among multiple-anatomical structures. Following a detailed description of the technique, a SDM is applied to three-dimensional (3D) brain structures modeling. Three-dimensional magnetic resonance images are employed due to their ability to differentiate between soft tissues. A training set consisting of 24 patients was adopted to provide initial statistical information to be included in the model. The SDM was then used to extract 24 3D structures including head, brain, ventricles, and cerebellum. Using this relatively small data set, it was shown that the agreement between manual and computer based segmentation of the ROI within the brain was 93%. This could however be further improved by increasing the number of patients in the database. Indeed, the advantage of these models is that it is relatively easy to update them by adding more images to their data set, hence improving the statistical description of the model.

Chapter 11 by Shuttleworth et al. entitled "Texture Analysis and Classification Techniques for Cancer Diagnosis" describes the criteria that can be formulated to express knowledge about the image being analyzed, and their use with a view to diagnose the severity of colon cancer. The previous chapters have described in detail intelligent and adaptive techniques that can be exploited to help clinicians make decisions. Chapter 11 focuses on the exploitation and understanding of the information required by the intelligent and adaptive technique to create the knowledge required to classify tissue samples based on texture.

To be intelligent a system requires knowledge. This knowledge should be formulated such that it can be processed by computers. A common language to express knowledge is mathematics. Mathematics is therefore adopted to translate subjective terminology and information into measurable and reproducible quantities. This results in the formulation of a set of mathematical criteria that aim to describe unambiguously properties of the image.

The specific properties investigated in Chapter 11 include smoothness and coarseness of the textures. The first type of texture feature formulated exploits statistical information from the spatial distribution of pixel intensities (e.g., gray level or color) in an image. Three types of statistical techniques are described. The first, based on image histogram information relies only on pixel intensities. Such a method is fast to compute but may lead to inappropriate characterization of observable features. Co-occurrence matrices are then described in detail concerning their ability to incorporate spatial information in an image. Commonly used features that can be extracted from co-occurrence matrices are formulated mathematically and issues connected with the direction and rotation dependence highlighted. Gradient analysis is then briefly described as the means to exploit contrast information to assess the coarseness of texture.

The second type of texture analysis is based on the modeling or the detection of structures or pattern of shape detected in an image. The model-based approach links together, through a set of rules, texture primitives that represent subpatterns in the image. Mathematical morphology detects the shape directly by implementing them as structuring elements for an erosion operation. A problem associated with shape-based techniques is that the shape of biological structures can vary significantly. An alternative is to employ frequency-based techniques such as Fourier analysis, wavelets, and Gabor filters. The relative advantages of these techniques are briefly described before considering color texture, where images are analyzed based on color mixing or chrominance and luminance.

The first section of Chapter 11 describes the means to identify the information contained in the image as a set of mathematical expressions, referred to as a feature vector in the image processing community. The second section considers the rationalization and exploitation of such information or knowledge. The intelligence in such systems lies in their ability to exploit the most appropriate subset or combination of features, to be able to identify clinically relevant differences between medical images of tissues that are required to be classified. The section starts with a description of the selection mechanisms that can be adopted with a view to reduce the number of features considered, and hence the complexity of the decision-making process. Having selected a set of features the next stage is to determine how these features can be employed or combined to classify the tissues analyzed as being normal or cancerous, for example. Three classification methods: regression, ANN, and discriminant analysis are briefly described, together with logistic regression and classification trees.

Having described the means to calculate features associated with an image as well as the methods required to rationalize and then to integrate these features into an intelligent and adaptive system, Chapter 11 concludes with a medical imaging application. The application deals with the use of color textures analysis for the identification of dysplastic severity in colon cancer tissue biopsies. The authors take the reader through successive refinement of color-based texture analysis to highlight the benefits of color texture in the processing of cytological and histological microscopy images. Accuracy, sensitivity, and specificity close to 100% are achieved to automatically classify colon tissue samples into three different stages of cancer progression.

Having presented the outline together with a personal view of the intelligent and adaptive systems described in this book, the next Chapter introduces the subject of systems modeling and adaptive control to prepare the reader for the material presented in Chapters 3 and 4.

1.3 Conclusions

This chapter has attempted to present the topics covered in this book on *Intelligent and Adaptive Systems in Medicine*. The chapter has highlighted and has presented the organization of each chapter in terms of theoretical content as well as practical applications in the medical field. The chapters are grouped into three parts according to the intelligent and adaptive techniques reported and the general application domain. Part I presents medical applications of systems modeling and control engineering. Part II presents medical applications of AI and Part III presents medical imaging applications involving both AI and traditional image processing. Oncology was selected as the main medical application considered in this book as a means to convey the applicability of intelligent and adaptive systems to diagnosis, treatment planning, delivery, follow-up, as well as medical equipment maintenance. However, the scope of the techniques presented in this book is far reaching and not limited to oncology, see, for example, Chapter 3 deals with anesthesia and Chapter 5 for a general review of a range of applications.

Reference

MAESTRO (2007). Methods and Advanced Equipment for Simulation and Treatment in Radio Oncology. http://www.maestro-research.org/index.htm (22-10-2007).

Part I

Systems Modeling, Identification and Control Theory Applied to Medicine

2

Systems Modeling and Control Applied to Medicine

Olivier C. L. Haas and Keith J. Burnham

CONTENTS

2.1 Introduction

In a book of this nature it would be inappropriate to attempt to provide an exhaustive review of the essential theory and, therefore, only a brief outline is given. The first part of this chapter addresses issues connected with system modeling and identification theory. The second part illustrates how the knowledge of the system gained through the modeling activity can be exploited to advantage when designing control systems. The control algorithms presented include proportional-integral-derivative (PID) controllers as well as an introduction to "modern" self-tuning and adaptive control strategies.

2.1.1 System Modeling and Identification for Control

This section begins with a review of model identification and parameter estimation considerations that are commonly employed. Some of the key issues connected with the methods adopted by the authors of the chapters in this book are highlighted and alternative solutions are given. A brief overview of the typical forms used for representing systems is also given.

2.1.1.1 Model Identification

The desire to obtain mathematical models of physical systems has witnessed an increasing interest spanning over many generations, and dates back at least as far as 3000 years to the ancient Egyptians. Models are essentially built for a purpose; their complexity will depend on the purpose for which they have been developed, for example, for simulation of system behavior, for prediction and forecasting of events, for monitoring the condition of a system, for detecting faults, and, last but not least, for the purpose of control.

While in reality almost all physical systems exhibit some form of nonlinearity, it is common practice, at least initially, to obtain linear models. Such linear system models approximate the system behavior over a range where local linearity is assumed to hold. Linear system theory has the advantage that it is well developed and understood. However, depending on the severity of the nonlinearity, the use of these linear models is limited to relatively small operating regions. Extrapolation beyond the range often leads to poor results due to the increasing inappropriateness of the model. When models are required to replicate system behavior over an increased range, where the assumptions on linearity no longer hold, then alternative approaches need to be deployed, including the use of multiple linear models (switched or blended) (see Chapter 3), time-varying and operating condition, or state-dependent models retaining a linear structure, and directly dealing with nonlinear models, e.g. bilinear models.

Because linear systems theory is well developed and widely applicable, with due recognition of the above limitations, consideration is restricted here to linear system models only. For a good introduction to linear systems theory, see, for example, Balmer (1991). For the reader who wishes to pursue a nonlinear systems modeling approach, there is a wealth of existing literature and continuously emerging developments in terms of new research (see, for example, Atherton, 1982; Mohler, 1991; Cook, 1994).

Today, many applications require the development of discrete-time models to facilitate their computer simulation and subsequent control system design and implementation. When obtaining a discrete-time model for a continuous system, a fundamental step is that of choosing an appropriate sampling interval. This step itself suggests that prior knowledge of the system is necessary/

desirable, and in practice the final choice of this crucial parameter often requires refinement through iteration. It is again stressed that it is particularly important to appreciate, at the outset, the intended purpose of the model; in the context of this book, the ultimate aim is to develop models for use in specific model–based scenarios, for example, control, monitoring, prediction, and simulation. In either case, the sampling interval should be fast enough to capture the salient dynamics of the system, for the intended purpose of the model, yet slow enough to obtain a meaningful parameterization.

Irrespective of the application domain or the modeling approach adopted, the following key issues need to be addressed:

- How to choose a particular model or system representation
- How to parameterize such a model
- How to validate the chosen model

The purpose of this section is to consider the theoretical bases upon which to provide an insight to the answers to these fundamental questions. Most of the approaches considered utilize discrete-time model representations of systems. An overview can be found in Fnaiech and Ljung (1987) and the references therein and an excellent review of the most common identification techniques is provided in the authoritative texts of Young (1984) and Ljung (1999). Assuming that an appropriate sampling interval has been found, the model identification problem generally involves three steps: structure identification or selection based on empirical knowledge and experience, parameter identification or estimation and finally model validation.

2.1.1.2 System Representations

There are, in general, three main approaches to represent a mathematical model of continuous-time system, namely, differential equation, transfer function, and state space representation, all of which have discrete-time counterparts. The principle exploited by these system representations is that the current behavior or output of a system depends on

  i. Its past behavior (or previous output, denoted by y).
 ii. Its previous stimuli (or input denoted by u) that caused its past and current behavior.
iii. Other causes and factors not directly taken into account to model the systems intrinsic response but having an influence on the system behavior. These other factors are usually amalgamated into unmeasurable disturbances and noise (denoted by ε_1 and ε_2) and can often be added conveniently to the noise free and disturbance free model (see Figure 2.1). While the precise nature of the noise and disturbances are unknown or uncertain, it is normally assumed that some *a priori* knowledge of the statistical properties are available.

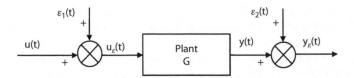

FIGURE 2.1
Illustrating a system or plant G in which input $u(t)$ and output $y(t)$ are corrupted by unmeasurable disturbance and additive noise denoted by $\varepsilon_1(t)$ and $\varepsilon_2(t)$, respectively.

The advantage of such model representations over the alternative AI approaches, such as NNs or FL is the reduced number of variables that are required to tune a model. Another benefit is that these rather more traditional approaches allow the incorporation of *a priori* knowledge in a structured and systematic manner. In addition, a relatively small number of parameters means that these models are able to capture the most important and relevant system behavior. These representations can be viewed as the "structure" of the intelligence. They are capable of summarizing and most importantly exploiting the knowledge about the system being modeled. A common system representation is the differential equation as it enables the modeling of a system from first principles. To facilitate the analysis of the model, differential equations are usually transformed into their equivalent transfer function forms in terms of the complex Laplace variable. State space representations offer the additional advantage of being able to define system state variables that correspond to entities within the system that sometimes are inconvenient or not possible to measure but are known to have an influence on the system behavior.

2.1.1.2.1 Differential Equation

Models that may be derived from the physical laws that describe a process or the individual components within the process are usually expressed in linear differential equation form. A general single-input–single-output nth-order linear differential equation representation is given by

$$\alpha_n \frac{d^n y}{dt^n} + \alpha_{n-1} \frac{d^{n-1} y}{dt^{n-1}} + \cdots + \alpha_0 y = \beta_m \frac{d^m u}{dt^m} + \cdots + \beta_0 u \quad (m < n) \quad (2.1)$$

where the α_i and β_i are normalized constant coefficients such that $\alpha_n = 1$, with u and y representing the continuous system input and output signals, respectively.

2.1.1.2.2 Transfer Function

A convenient and well-established description of a linear system is the so-called transfer function representation. This is defined to be the ratio of the Laplace transform of the system output to the Laplace transform of the system input, under the assumption that all the initial conditions are zero.

Taking the Laplace transform throughout the linear differential equation (2.1) leads to the transfer function for the plant, denoted G(s):

$$G(s) = \frac{Y(s)}{U(s)} = \frac{\beta_m s^m + \beta_{m-1} s^{m-1} + \cdots + \beta_1 s + \beta_0}{s^n + \alpha_{n-1} s^{n-1} + \cdots + \alpha_1 s + \alpha_0} \tag{2.2}$$

where s denotes the Laplace variable and Y(s) and U(s) denote the Laplace transforms of y and u, respectively.

2.1.1.2.3 State Space

When utilizing linear model–based control, the system can also be represented in state space form. Such a representation transforms the nth-order differential equation (Equation 2.2) into n simultaneous first-order differential equations. This necessitates the introduction of additional variables, namely the state variables. Choice of these state variables is not unique and, therefore, a number of state-space representations (Equation 2.3) of a given differential equation of the form (Equation 2.2) are possible (Friedland, 1987; Ogata, 2002). It is possible, but not necessary, to relate the state variables (states) to physical quantities within the system. A general state space form for a single-input–single-output system is given by

$$\begin{cases} \dot{x} = Ax + bu \\ y = c^T x + du \end{cases} \tag{2.3}$$

where $x \in \mathbb{R}^n$ is the vector of state variables; $A \in \mathbb{R}^{n \times n}$ a matrix of real constants; and $b,c,d \in \mathbb{R}^n$ are the input, output, and feed forward vectors of real constants, respectively.

In the case of a continuous linear system expressed in the phase-variable canonical form (Friedland, 1987; Ogata, 2002), d is usually equal to zero and the matrix A and vectors b and c are of the following forms:

$$A = \begin{bmatrix} 0 & 1 & . & . & . & 0 & 0 \\ 0 & 0 & . & . & . & 0 & 0 \\ . & . & & & & . & . \\ . & . & & & & . & . \\ . & . & & & & 1 & 0 \\ 0 & 0 & . & . & . & 0 & 1 \\ -\alpha_0 & -\alpha_1 & . & . & . & -\alpha_{n-2} & -\alpha_{n-1} \end{bmatrix} \quad b = \begin{bmatrix} 0 \\ 0 \\ . \\ . \\ . \\ 0 \\ 1 \end{bmatrix} \quad c = \begin{bmatrix} \beta_0 \\ \beta_1 \\ . \\ \beta_m \\ 0 \\ . \\ 0 \end{bmatrix} \tag{2.4}$$

The state space representation offers an insight into properties of system behavior including features such as controllability and observability that are not so readily apparent from the differential equation and transfer function representations.

When utilizing computer control it becomes necessary to make use of a discrete-time representation of the plant. Discrete-time control involves the incorporation of a zero-order-hold, to keep the system input at a constant value between samples. This is required to be taken into account within the resulting discrete-time system model representation. In the design of discrete-time model-based controllers, the continuous differential equation representation is replaced by a discrete-time difference equation. Similarly, discrete-time state space representations are also employed.

2.1.1.2.4   *Discrete-Time Difference Equation Representation*

A common discrete-time representation is the so-called extended autoregressive moving average with exogeneous inputs (ARMAX) model structure. It takes the form:

$$A(q^{-1})y(t) = q^{-k}B(q^{-1})u(t) + C(q^{-1})\varepsilon(t) \tag{2.5}$$

where the polynomials $A(q^{-1})$, $B(q^{-1})$, and $C(q^{-1})$ are defined by the general polynomial:

$$L(q^{-1}) = l_0 + l_1 q^{-1} + l_2 q^{-2} + \cdots + l_{n_l} q^{-n_l} \tag{2.6}$$

with $n_a \geq n_b$, $a_0 = c_0 = 1$, $b_0 \neq 0$, $k \geq 1$ is the system time delay expressed as an integer multiple of the adopted sampling interval, q^{-1} is the backward shift operator defined by $q^{-i}y(t) \equiv y(t-i)$, where $u(t)$, $y(t)$, and $\varepsilon(t)$ represent the sampled input, output, and white noise signals, respectively. It is common to assume that the noise-coloring polynomial is unity, such that $C(q^{-1}) = 1$, that is, the output noise is assumed to be white, so that the system is modeled as

$$A(q^{-1})y(t) = q^{-k}B(q^{-1})u(t) + \varepsilon(t) \tag{2.7}$$

As for the continuous case, there are many possible state space representations depending on the choice of the state variables. One common discrete-time representation that is equivalent to the model representations (Equation 2.3) is the so-called implicit delay observable canonical form:

$$x(t + 1) = \mathbf{P}\,x(t) + \mathbf{Q}\,u(t) + \mathbf{R}\,\varepsilon(t)$$
$$\tag{2.8}$$
$$y(t) = \mathbf{H}\,x(t) + \varepsilon(t)$$

where $x(t)$ is the vector of state variables and, assuming white noise, the matrices $\mathbf{P}$, $\mathbf{Q}$, $\mathbf{R}$, and $\mathbf{H}$ are in the observable canonical form, where the time delay k is implicit in the structure and creates an inflated dimension as shown:

$$
P = \begin{bmatrix} 0 & \cdots & \cdots & \cdots & \cdots & 0 \\ 1 & \ddots & & & & \vdots \\ & \ddots & \ddots & & & 0 \\ & & \ddots & \ddots & & -a_{n_a} \\ & & & \ddots & 0 & \vdots \\ 0 & \cdots & \cdots & 0 & 1 & -a_1 \end{bmatrix}, \; Q = \begin{bmatrix} b_{n_b} \\ \vdots \\ b_0 \\ 0 \\ \vdots \\ 0 \end{bmatrix}, \; R = \begin{bmatrix} 0 \\ \vdots \\ 0 \\ -a_{n_a} \\ \vdots \\ -a_1 \end{bmatrix}, \; H = \begin{bmatrix} 0 & \cdots & 0 & 1 \end{bmatrix}
$$

denoting n = max (n_a, n_b), it follows that initial dim (state space) $n_i = n_a + k$. Note that the initial dimension may need to be reduced to ensure controllability of the pair **PQ**.

2.1.1.3   Structure Identification

An important first step in model identification is that of model structure determination; that is, for a single model, the model order and the magnitude of any pure time delays are required. In the case of a multiple-model approach, such as that presented in Chapter 3 by Mendonça et al., this task is more demanding. It requires the number of submodels to be identified and for each, their associated structure, as well as the variables (e.g., inputs, outputs, states, and auxiliary variables), often called premise variables in the literature, to be chosen for the scheduling of the models. It should not be a surprise that the problem of structure determination is automatically increased when dealing with multiple models, as the individual submodels could potentially be of different structures.

Numerous methods have been developed to assess the prediction performance of single models. Criteria have been developed to assess the models in terms of accuracy, at the same time penalizing complexity to avoid modeling spurious effects including noise and uncertainties; an aim to keep the model parsimonious, i.e. of low order for the purpose of control. These criteria include final prediction error (FPE) (Akaike, 1969), Akaike information criterion (AIC) (Akaike, 1974), Young information criterion (YIC) (Young and Beven, 1994), minimum description length (MDL) (Rissanen, 1978), Bayesian criteria (Kashyap, 1977), cross-validation (Stoica et al., 1986), the unbiased criterion (Ivakhnenko and Yurachkovsky, 1988), as well as the simplest and perhaps most popular criterion, namely, the use of separate validation data.

While the MDL and the Bayesian criteria are consistent, in the sense of convergence to the simplest model structure that contains the true system, this is not the case for the AIC. With the only drawback being a longer data sequence required, compared to the other procedures, the use of separate validation data gives asymptotically a model with the best possible expected prediction performance within the given model set. It would also be worth using a separate validation set in conjunction with either one of the remaining criteria or all of the remaining criteria. While all these methods are realizable for single-model configuration, the number of cases to be computed significantly increases with a multiple-model approach, as it would ideally be required to optimize the structure for each submodel (order) as well as the optimal number of submodels.

2.1.1.4 Parameter Estimation

The parameter identification (or estimation) problem is defined by the cost function to be minimized (Ljung, 1999). Re-arranging (1.7) it is possible to obtain $y(t) = \mathbf{x}^T(t)\theta + \varepsilon(t)$ where $\mathbf{x}(t)$ is the observation vector and θ is a vector containing the model parameters i.e.

$$\mathbf{x}^T(t) = ([-y(t-1)) \cdots -y(t-n_a); \ u(t-k) \cdots u(t-k-n_b)]$$

and

$$\theta^T = [a_1 \cdots a_{n_a}; \ b_0 \cdots b_{n_b}]$$

It is normal to estimate the parameters of an assumed model; hence the structure is required to be known in advance. It is very common in engineering problems to choose a quadratic cost function for its convex property. A commonly used cost function is the mean square error (MSE) between the measured and the h-step-ahead prediction of the output of the system; the MSE of the prediction error is defined as

$$\text{MSE}_{\text{pe}}(\hat{\theta}) := \frac{1}{N}\sum_t \|y(t) - \hat{y}(t \mid t - h, \hat{\theta})\|_2^2 \tag{2.9}$$

where h is typically chosen as unity. Minimizing such a cost criterion leads to a parameter identification framework, which is referred to as prediction error minimization (PEM). One particular subset of PEM, known as output error minimization, is extensively used to obtain models for the purpose of simulation. It provides a basis for the design and evaluation of model-based control systems through simulation. In general, it can be interpreted as having a horizon h tending toward infinity or at least to the number N of samples measured. The corresponding minimized-cost function defines the MSE in terms of simulation error, and is given by

$$\text{MSE}_{\text{sim}}(\hat{\theta}) := \frac{1}{N}\|y(t) - \hat{y}(t)\|_2^2 \tag{2.10}$$

where MSE_{sim} is the MSE between the measured output y of the system and the simulated output $\hat{y}$ of the model. Additionally, in the case where recursive estimation techniques are applied to obtain a time-varying model and when measuring some statistical performances based on Monte Carlo analysis, the variance of the estimated parameters, denoted by $\text{var}(\hat{\theta}(t))$ is used.

The sum of the squares of the error (SSE) is the criterion employed within the recursive least squares (RLS) scheme. RLS is one of the most widely and successfully used estimation techniques that minimizes the error between actual output and estimated output $\|y - X\hat{\theta}\|_\Lambda^2$ where the observation matrix X is constructed of observation vectors as its rows. The RLS algorithm is given by (Hsia, 1977)

$$\phi(t) = \Phi(t - 1)x(t)[1 + x(t)^T \Phi(t - 1)x(t)]^{-1}$$

$$\Phi(t) = [1 + \phi(t)x(t)^T]\Phi(t - 1)/\lambda(t) \tag{2.11}$$

$$\hat{\theta}(t) = \hat{\theta}(t - 1) + \phi(t)[y(t) - x(t)^T \hat{\theta}(t - 1)]$$

where $[y(t) - x(t)^T\hat{\theta}(t-1)]$, referred to as the estimation prediction error, is a sequence of fitting errors that includes both measurement errors (noise) and estimation errors. Upon convergence of RLS, the estimation error tends to zero and $[y(t) - x(t)^T\hat{\theta}(t-1)]$ tends to the noise. In the algorithm (see Equation 2.11), $\phi(t)$ is the gain vector, $\Phi(t)$ is the error covariance matrix with $0 < \lambda(t) \le 1$ being a forgetting factor used to inflate elements of the covariance matrix at each time step; thus keeping the algorithm alert and assisting adaptation (Hsia, 1977). Choice of the value of the forgetting factor is a compromise between algorithm alertness and noise sensitivity (Burnham et al., 1985). To alleviate this problem, use may be made of a variable forgetting factor $\lambda(t)$, which is automatically adjusted as a function of the estimation prediction error, to retain the information content within the algorithm (Fortescue et al., 1981; Wellstead and Sanoff, 1981). While use of a forgetting factor facilitates the tracking of slow variation in parameters, a technique that facilitates the tracking of rapid parameter variation is that of covariance matrix reset. Such a scheme, which can be operated in conjunction with forgetting factors, may trigger reset on set point change periodically or on detection of large errors in estimation.

It should be noted that unbiased parameter estimates can only be obtained from RLS, if it may be assumed that the observation vector of past input and output values and the noise sequence are uncorrelated (Young, 1974); this being true only in the case of a white noise sequence. If this is not the case then the problem of biased estimates may be alleviated using a modified algorithm such as extended least squares (ELS) (Hsia, 1977), recursive maximum likelihood (RML) (Hsia, 1977), or recursive instrumental variables (RIV) (Young, 1970). In addition, if poor parameter estimates are obtained due to insufficient input signal excitation, cautious least squares (CLS) may be employed (Randall and Burnham, 1994) in which the algorithm is kept alert without disturbing the plant. The method of CLS is also of value when attempting to constrain the estimated parameters to remain within sensible regions obtained from both operational experience and knowledge of the plant.

The reader should note the subtle yet important differing concepts between modeling and simulation as a basis for control system design and evaluation, as well as models for realizing online implementation of the developed model-based control algorithms in the field. The former are required to be of a higher level of sophistication, with the latter being as simple as can be practically possible.

2.1.1.5 Model Validation

Model validation usually requires the measured data obtained from a system to be divided into two distinct sets, namely, the identification set, which is utilized, as its name implies, to identify the model parameters and the validation set, which allows the model to be tested on unseen data, that is, data not used for the identification. The conceptual idea behind such a process is that the model should aim to replicate the system behavior (the identification set[s]) as best as possible, but not at the detriment of the generalization to other data sets (the validation set[s]) issued from the same system. These potentially conflicting objectives can traditionally be combined into a weighted sum, where a weighting factor or hyperparameter specifies a user-defined trade-off between accuracy of the model on the identification data set and its generalizability to others, that is, the validation data set. Another alternative to a weighted combination of the objectives is to use Pareto ranking to assess the correlation of the objectives and select *a posteriori* the best compromise solution from the set of nondominated solutions (see Rodriguez-Vazquez and Fleming, 1998; Haas, 1999).

2.1.1.6 Regularization Techniques

The "blind" application of system identification techniques can often lead to identified models that may be numerically difficult to compute, or worse, which may not actually offer an appropriate mathematical representation of the system being considered. The aim of regularization techniques is to guide or steer the identification algorithms to converge toward a solution, which is actually a representative of the system being modeled.

From a theoretical perspective, regularization techniques can be broadly used to solve or alleviate three types of problems: ill-posed problems (Polson et al., 1992), well-posed but ill-conditioned problems (Tarantola, 1987), and well-posed–well-conditioned problems (Johansen, 1996). For well-posed–well-conditioned problems, a class of composite criteria–based regularization techniques may be used to add *a priori* knowledge.

There are broadly two different types of regularization techniques: (i) through control of dimension, for example, truncated singular value decomposition (TSVD), choice of a parsimonious parameterization and (ii) through minimization of a composite criterion. Regularization is essentially an approach, which aims to improve the robustness of numerical solutions by recasting the estimation problem and introducing additional information, to either reduce the problem dimension (truncation) or reduce the solution space of the original problem. If a stable solution cannot be obtained, the original problem is replaced by a class of "equivalent" problems with constraints imposed to limit the solution space to that of a subset in agreement with *a priori* knowledge. Application of regularization to the parameter estimation problem may reduce the variance of the estimates and the MSE prediction at the expense of a possible bias (Johansen, 1997). Hence, use of regularization leads to more accurate estimates, provided the bias is smaller than the reduction in variance. An overview of regularization can be found in references Johansen (1996), Neumaier (1994), Sjöberg et al. (1993), and Demoment (1989).

The *a priori* knowledge and empirical data can be used to formulate an optimization criterion that penalizes

a. Mismatch between model prediction data and empirical data,

b. Nonsmoothness of model predictions,

c. Mismatch between the predictions of an identified model and a default model, and

d. Violation of certain soft constraints.

In addition, it is possible to allow hard equality or inequality constraints to be incorporated within the optimization problem.

Tikhonov regularization (TR), which was made widely known through the publication of (Tikhonov and Arsenin (1977), is probably the best known and most widely applied regularization scheme. It is characterized by the filter factors:

$$f_i(\alpha) = \frac{\sigma_i^2}{\sigma_i^2 + \alpha} \tag{2.12}$$

which is equivalent to minimizing the cost function (Kirsch, 1996):

$$J_T(\hat{\theta}) = \|y - \mathbf{X}\hat{\theta}\|^2 + \alpha\|\hat{\theta}\|^2 \tag{2.13}$$

This compound criterion exhibits an additional term $\alpha|\hat{\theta}|^2$ that penalizes large values of the estimated solution. This is an intuitive approach in which the resulting algorithm aims to contain the solution within a reasonable region of the parameter space (Johansen, 1997). It is not surprising that this idea arose simultaneously in different branches of engineering and science. Probably, the most popular example is ridge regression (Hoerl and Kennard, 1970), the notation used in the statistical community for TR, which was developed simultaneously and independently to the work of Tikhonov. When *a priori* knowledge about the solution in the form of an approximation of the expected solution θ^* is available, a more generalizable form of TR is characterized by

$$\tilde{J}_T(\hat{\theta}) = \|y - \mathbf{X}\hat{\theta}\|^2 + \alpha\|\theta^* - \hat{\theta}\|^2 \tag{2.14}$$

where deviations from the prior value θ^* to some $\hat{\theta}$ is penalized. In another form of representation, as used in Golub and Van Loan (1996), Equation 2.14 may be expressed equivalently as

$$\tilde{J}_T(\hat{\theta}) = \left\| \begin{pmatrix} \mathbf{X} \\ \sqrt{\alpha}I \end{pmatrix}\hat{\theta} - \begin{pmatrix} y \\ \sqrt{\alpha}\theta^* \end{pmatrix} \right\|_2^2 \tag{2.15}$$

Some recent work (Linden et al., 2005; Linden, 2005) has also shown the link between regularization techniques and CLS, a method developed in earlier work within the Control Theory and Applications Centre at Coventry

University, U.K. It has been shown that CLS is a subset of the regularization techniques. A recent development of this work has led to the creation of an online Tikhonov regularization (OTR) procedure.

Inspired by research in cautious control (see, for example, Aström and Helmersson, 1986) the CLS algorithm arose in the field of system identification in 1986, with the aim being to stabilize the online estimation process for bilinear systems in self-tuning control (STC) (Burnham and James, 1986). In addition to the minimization of the residuals:

$$(y-\hat{y})^{\mathrm{T}}\Lambda(y-\hat{y}) = \left\| y - \mathbf{X}\hat{\theta} \right\|_{\Lambda}^{2} \tag{2.16}$$

this approach also aims to minimize a penalty term, which is given by

$$(\theta^{*}-\hat{\theta})^{\mathrm{T}}\Psi(\theta^{*}-\hat{\theta}) = \left\| \theta^{*} - \hat{\theta} \right\|_{\Psi}^{2} \tag{2.17}$$

where Λ and Ψ are diagonal weighting matrices and θ^{*} denotes a "safe set" of parameters. The similarities to TR are immediately obvious; however, the use of adaptivity through implementation of exponential forgetting, performed by Λ, and the individual weighting of each parameter in the penalty term, performed by Ψ, distinguishes this approach from standard TR. Moreover, the most significant difference between CLS and TR is its realization as an online estimation scheme, which allows repeated regularization to be implemented in the estimation algorithm.

2.1.1.6.1 *Realization of Online Tikhonov Regularization*

The minimization of the residuals and the penalty term presented in Equations 2.16 and 2.17 are not realized directly, but split into two parts leading to a two-step algorithm. This tandem solution consists of two interacting RLS algorithms, thus facilitating a straightforward implementation.

In the first step, the residuals $\left\| y - \mathbf{X}\hat{\theta} \right\|_{\Lambda}^{2}$ are minimized by applying the standard adaptive RLS algorithm with a forgetting factor. Then, at a certain stage in time, say $t = i$, the RLS algorithm is "interrupted" and p additional iterations are performed by a so-called nested recursive least squares (NRLS) algorithm, which minimizes the OTR penalty term $\left\| \theta^{*} - \hat{\theta} \right\|_{\Psi}^{2}$. The NRLS algorithm can be represented by (cf. Linden, 2005; Burnham, 1991)

$$\tilde{\phi}_{j} = \tilde{\Phi}_{j-1}\sqrt{\Psi}_{j}\, e_{j}\big[1+\Psi_{j}\, e_{j}^{\mathrm{T}}\tilde{\Phi}_{j-1}e_{j}\big]^{-1}$$

$$\tilde{\Phi}_{j} = \big[I-\tilde{\phi}_{j}\sqrt{\Psi}_{j}\, e_{j}^{\mathrm{T}}\big]\tilde{\Phi}_{j-1} \tag{2.18}$$

$$\tilde{\theta}_{j} = \tilde{\theta}_{j-1} + \tilde{\phi}_{j}\big[\sqrt{\Psi}_{j}\, e_{j}^{\mathrm{T}}(\theta^{*} - \tilde{\theta}_{j-1})\big]$$

where $e_{j}^{\mathrm{T}} = [\delta_{1j}\,\delta_{2j}\cdots\delta_{pj}]$, δ_{ij} the Kronecker delta and Ψ_{j} are the diagonal elements of Ψ, that is, $\Psi = \mathrm{diag}\,(\Psi_{1}, \Psi_{2}, \ldots, \Psi_{p})$. At each activation of the NRLS subalgorithm the covariance matrices Φ and $\tilde{\Phi}$, as well as the current estimates $\hat{\theta}$ and $\tilde{\theta}$, are interchanged.

This section concludes the brief overview of parameter identification and regularization techniques enabling a designer to create a mathematical model of a system. Traditionally such models have been used within numerical simulation environments to experiment with different system designs, develop software that interacts with the modeled systems or with hardware components that represent subsystems, i.e. hardware in the loop design in simulation. The next section presents the second stage in the control design process, namely, the exploitation of the model information to design a control system.

2.1.2 Intelligence and Adaptivity of Traditional Control Schemes

Having introduced the model structures as well as the parameter identification techniques necessary to allow appropriate parameterization of the models, this section focuses on the control strategies that are utilized and extended within Chapter 3. Attention is limited here to the PID controller and a brief view of the self-tuning pole-placement controllers is also presented.

Control engineering aims to automate and improve the performance of a system by manipulating some of the factors influencing its behavior. The system being controlled is usually referred to as the "plant." Most industrial plants are controlled using closed-loop strategies whose basic principle is illustrated in Figure 2.2. The plant output $y(t)$, also referred to as process variable (PV), is measured and compared with the desired value, usually referred to as the set point or reference signal, denoted $r(t)$, giving rise to an error signal $e(t) = r(t) - y(t)$. The controller C then continually attempts to reduce this error by adjusting the control signal $u(t)$, which drives the system G, that is, the plant, through an actuator such that its output $y(t)$ matches the reference signal $r(t)$. The variables influenced by the controller to alter the plant response through actuation devices are referred to as control variables (CV) or referred to as the control action, generally denoted by $u_a(t)$. The effect of the actuator on the control signal results in an action $u_a(t)$ which is physically applied to the plant. Other factors that may influence the plant output but cannot be controlled are referred to as disturbances and measurement noise. It is common to 'lump' these effects into $d(t)$, an unknown additive quantity at the system output, as shown in Figure 2.2. The resulting plant output with an additive disturbance can then be denoted $y_d(t)$. These uncertainties are

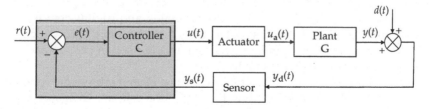

FIGURE 2.2
Illustrating closed-loop control system where the controller aims to compensate for deviation of the process variable from a reference signal by adjusting the control variables.

unknown, partially known, measurable, or predictable, for example, temperature variation during the day, patient weight, and respiratory motion. In the latter case, the solution is to monitor the PV and correct any deviation from the desired behavior. This can be achieved by acting on the error between the PV and the reference. Ideally the plant should be the system studied, for example, a medical device or the response of a person to drug perfusion. In practice, however, the plant may include "external" systems linked with the monitoring devices employed, that is, the sensor resulting in a signal $y_s(t)$ and the actuator(s) enabling the controller to act on the plant. Actions taken by an external system to improve the quality of signals, such as filtering, will also affect the performance of the plant system being controlled. All these factors should be taken into account in the system model by the designer at the outset. This will lead to a suitably sophisticated model for the purpose of control system design through simulation. However, once designed the controller should be reengineered to be as simple as possible. Bearing in mind that there will be an inevitable mismatch between the actual plant and the model, the control system must be designed either to be robust to acknowledge the mismatch or to be able to adapt to reduce the effect of the mismatch.

2.1.2.1 Analog and Digital Control Systems

Most systems considered for control engineering purposes are dynamic in nature, with their behavior changing with time. A control system should therefore regularly monitor and act on the plant. When a controller is implemented using analog electronics, it will act continuously on the plant and its speed of reaction will only be limited by the bandwidth of the electronic circuits. Today, the increased computational power combined with low cost and high versatility of microprocessors has led to many control systems being based on digital techniques. Digital controllers have two main differences with their analog counterparts. First, they require the analog (or continuous) signals from the plant to be digitized by analog to digital converters (ADC) and the control signals/variables to be converted from digital to analog using digital to analog converters (DAC). This conversion limits the resolution of sensors and actuators to a finite number of discrete steps. The step size, referred to as quantization, depends both on the ADC and DAC resolution expressed in bits, for example, 12 bits or a resolution of 1 in 4096, and, on the range of the signals. Today most ADCs/DACs have 12-bit resolution with an increasing number adopting 16 bits. The second major difference between analog and digital devices is the signal sampling mechanism that determines the frequency with which both the information from the plant is obtained and the action calculated by the controller is sent to the plant actuators.

If the sampling is too slow, aliasing could occur, causing the signal frequency to appear at a lower frequency than the continuous signal. To ensure that its perception of reality is accurate, it is necessary to sample the signal sufficiently fast. Shannon's sampling theorem states that the signals should be sampled at least twice the bandwidth of interest. In real life, the systems to be controlled rarely have well-defined bandwidths and sharp cutoff points to the noise in the signals (Parr, 1998). Noise is characterized by high-frequency components, resulting in measured signals having high-frequency components too. Before sampling, a signal should therefore be passed through a low-pass antialiasing filter to ensure that the bandwidth of interest is sampled appropriately. In medical applications such as biological systems, the bandwidth of the system is relatively low (a few Hertz). Similarly, most medical robotic devices such as medical imaging scanners, linacs, patient support systems (PSSs) have bandwidths in the order of 2–10 Hz. Hence, in practice, a sampling frequency of 5–10 times the system frequency is used, that is, 20–50 Hz. It may sometimes be difficult to appreciate, however, that unlike analog control systems, where the components are physically fixed onto a circuit board; a digital controller could dramatically change behavior if the sampling frequency is changed. Thus, a digital control system is designed to operate at a specific sampling frequency. As stated in Section 2.1.1.1, it is a crucial parameter to select, as it not only affects the models but also affects the controllers, which are ultimately based on the validity of the models in the first instance.

The majority of commercial controllers currently in use in industry are realized in discrete form. In the case of controllers, such as the PID, originally implemented using analog circuits, there may be a tendency to attempt to emulate the continuous controller action. The high-computational power of microprocessors and microcontroller chips enables rapid sampling and frequent control actions. However, the need to ensure that computational delay, or computational time, denoted by T_c, is kept to a minimum with respect to the sampling interval T, can become a potential obstacle. In a similar manner to the system bandwidth, the computational cost together with the data communication time can change due to the traffic on the bus system used and the microprocessor load. To run on a digital system, the software needs to communicate with some form of operating system. If the operating system is not "real time" then the time at which the calculations are made cannot be guaranteed. This means that the computational time will change and may impact on the time at which the control action is sent to the system, resulting in reduced performance. Indeed discrete-time control assumes that the control actions are sent at regular and identical time intervals. If the interval duration changes, then the theoretical basis that justified the calculation of the next control action becomes invalid. A number of real-time operating systems exist and are currently used to ensure that the program execution is performed in a timely manner. However, even real-time operating systems have limitations and exceeding these limitations will cause problems. As a rule of thumb the designer should ensure that $T_c \leq 0.1T$. Further, fast sampling could lead to the closed-loop system being controlled to move from a very responsive mode to an unstable mode due to the presence of noise and delays in the system. While the main limitation of early digital control systems was associated with processing speed, today the speed of response of the actuator, for example, valve response time, is starting to become the limiting factor when selecting a sampling frequency. Indeed, it is not possible to make the system act faster than it is physically capable. Hence the actuator bandwidth should be taken into account in setting up the sampling time. Further, while frequent actuator actions may theoretically give rise to better closed-loop plant performance, it will also reduce the life expectancy of the actuator device and may increase the risk of failure. In summary, the sampling frequency should be selected such that it is greater than 5 to 10 times the system's frequency and its period should be smaller than 10 times the processing time.

Having reviewed general aspects of discrete-time control, the next section focuses on the most widely used controller (the three-term PID) as well as pole placement model based control systems, which are referred to in the control engineering literature as "modern" control systems.

2.1.2.2 Proportional-Integral-Derivative Controller

Originally introduced in the 1940s, the three-term PID controller remains prevalent among the numerous control schemes available today (Bueno and Favier, 1991; Bobál et al., 2005). For a wide variety of applications, where satisfactory set point tracking can be achieved with a relatively simple cascade controller, the PID strategy is likely to retain its prominence in industry. PID controllers provide satisfactory performance when operating in a region "near" to a point of tuning. Such a controller assumes that the plant has linear characteristics about the operating point and its dynamic behavior is predominantly of first or second order.

2.1.2.2.1 *Continuous Proportional-Integral-Derivative Control Formulation*

The PID control action $u(t)$ is obtained through three adjustable parameters, or gains, which operate on the error signal $e(t)$ such that the overall unity feedback closed-loop system behaves in some satisfactory manner. The PID performs three basic actions on the error signal: multiplication, integration, and differentiation. In continuous form, the PID algorithm may be expressed as

$$u(t) = K_p\left(e(t) + \frac{1}{T_i}\int_0^t e(\tau)d\tau + T_d\frac{de(t)}{dt}\right) \qquad (2.19)$$

where K_p is the proportional gain, T_i the integral time constant, and T_d the derivative time constant. An alternative implementation expression is given by

$$u(t) = K_p e(t) + K_i\int_0^t e(\tau)d\tau + K_d\frac{de(t)}{dt} \qquad (2.20)$$

where the three independent gains are K_p the proportional gain, K_i the integral gain equivalent to $K_i = K_p/T_i$, and K_d the derivative gain $K_d = K_p T_d$.

Considering Equation 2.20, the multiplicative action is implemented through the product of the proportional gain with the error. One of the issues with proportional action is that it disregards history and current trends and thus cannot change its behavior even if the same error condition keeps occurring. For example, if there is a steady-state error, that is, a constant error that remains once the system transient dynamics have disappeared, the proportional action alone will not be able to remove it. Indeed a constant error value will lead to an identical proportional control action being calculated. The proportional action can thus be regarded as a purely reactive mode.

The integral action has, however, the ability to "remember" the past. As long as the error is not zero, the integral and thus the response of the integral control action will keep increasing. When the error becomes zero the integral action will stop increasing and become constant. However, unlike the proportional action, it will not become null. This characteristic is also the cause of a detrimental effect referred to as wind-up, where the integral contribution keeps increasing beyond a reasonable amount. Wind-up can typically occur if the plant actuator is saturated and cannot provide a sufficiently high-control action as required by the integrator. Typical schemes to alleviate this problem involve "informing" the integrator about the system by limiting its maximal contribution, if no observable change in the system behavior is noted despite applying increasingly high integral control action. The larger the gain associated with the integral action the faster the system will be able to react to set point changes or disturbances. The price to pay for a more responsive system is that it may also become less stable.

Differential action tends to be associated with the ability of the controller to "anticipate" changes by monitoring the rate of change of the error or of the PV. Hence, when the error is increasing, the differential control action will tend to increase the controller output, while it will reduce if the error is decreasing. If the error is constant, the differential action will be zero and hence have no effect. The differential action is thus able to anticipate changes and dampens the system being controlled, thereby reducing the amount of oscillation that may have been introduced by the integral action. If large changes in the signals occur, such as sudden demand change or noise, then the resulting differential control action will be undesirably high. To avoid such high-control action, the differential control action is often implemented using the PV (i.e., the plant output) as opposed to the error signal. To reduce the effect of noise, a low-pass filter is also applied to the signal before being acted upon by the differentiator.

Although it is known as a three-term control system, the PID controller may be regarded as a two-degree of freedom scheme, because the ratio between the integral and derivative gains, or equivalently their time constants, is fixed. The ratio is typically of the order of 4 to 1 (even, if some other schools of thought advocate the use of 6 to 1). There are occasions, however, where it is inadvisable to use derivative or integral action, leading to PI, PD, and simple P control.

- On many plants, care is needed when tuning the derivative gain, due to the measurement noise at the system output. The latter can become amplified and inappropriately transferred to the control action. In such circumstances, the derivative control action can be disabled, thus leading to a PI controller.

- In the case of systems exhibiting inherent integral action, use of an integral term is unnecessary and if PID is used blindly on such systems, the resulting control action can become oscillatory leading to a degradation of performance. Such a controller, employing P and D terms only, can be regarded as a pole-placement scheme, see Section 2.1.2.3.2.

- When faced with a system with both problems outlined above (e.g., inherent integral action and noisy measurements), one can be reduced to a simple P controller. While this may provide a satisfactory solution, the degree of freedom offered to the plant engineer/designer is reduced, and this may be seen as a trade-off.

To increase the flexibility of PID control schemes numerous refinements have been proposed. The next section describes some of the most common.

2.1.2.2.2   Enhancements to Conventional Proportional-Integral-Derivative Control

Although the PID controller has had considerable success among the process industries, particularly where plant characteristics are predominantly first or second order, it is limited somewhat due to its simplicity and when applied to more complex plant the tuning of the controller gains to provide the "best" performance, in some sense, is no longer as straightforward. It is common to find on many practical plants that PID controllers are poorly tuned; often they are rarely retuned from commissioning. It is recognition of this, which is in part due to lack of understanding of plant operators, that has led to the development of automatic loop tuning software to enable automatic adjustment of the controller tuning parameters.

Many proposals for improved solutions consist of auto-tuning algorithms, most of which assume the process model to be linear and of first or second order in nature. There are basically three types of auto-tuning methods: (i) those based on identification of parameters of an assumed low-order model structure, (ii) those based on identification of the critical point in terms of a frequency response, and (iii) those based on identification of the dominant first or second order characteristics of a system using methods of pattern recognition (Bristol, 1977). Having identified the salient features of the system in the sense of a first- or second-order model, common to all of the above methods are empirical design rules to obtain the appropriate controller gains. Many of these design methodologies are based on extensions or derivatives of the early proposals of Ziegler and Nichols (1942) and Cohen and Coon (1953). A good summary of these tuning methods may be found in Ziegler and Nichols (1942), Zhuang and Atherton (1993), Åström et al. (1993), Åström and Hägglund (1995), Menani and Koivo (1996), Atherton (1999), Atherton (2000), and Bobál et al. (2005). However, once such auto-tuning controller has been tuned, it suffers from the same problems as the manually tuned PID schemes in that it remains unable to adapt to changes in operating conditions and unplanned disturbances on the plant.

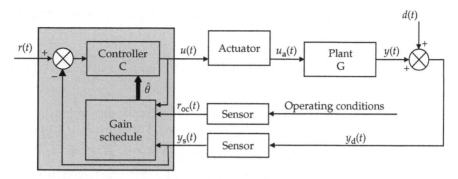

FIGURE 2.3
Illustrating gain-scheduling control strategy where the gains of the controller are updated based on the plant operating condition as well as the plant environment.

Other attempts to improve the overall PID performance include gain scheduling (see Chapter 3) and simple self-tuning techniques, see for example Vega et al. (1991). The gain-scheduling approach, illustrated in Figure 2.3 is in essence an attempt to produce a nonlinear PID scheme and may be obtained by carrying out an auto-tuning procedure at different operating points. Similar approaches include FL-based schemes. Such approaches are particularly useful for controlling systems whose dynamics change with operating conditions. Gain scheduling assumes that the variations observed in a system are consistent with both operating point and time. If this is not the case, then self-tuning PID controllers, whose parameters are continually adjusted or tuned online, may be used to advantage.

Developments in PID control continue to emerge (Atherton, 2000; Bobál et al., 2005). In parallel with these developments, sophisticated distributed control systems are becoming available whereby dedicated controllers for specific loops make use of PID strategies. In such schemes, the PID set points are derived from more advanced model–based control strategies, which operate at a higher supervisory level within the overall scheme.

2.1.2.2.3 Discrete Proportional-Integral-Derivative Control Formulation

Discrete versions of the PID controller can be obtained by different numerical methods (rectangular or trapezoidal rule) for integrating a continuous signal in terms of its sampled values (see, for example, Warwick and Rees, 1986). When the sampling frequency is sufficiently fast, there is little difference among "the backward rectangular, the forward rectangular, and the trapezoidal" numerical integration methods. Having replaced the continuous integral by its equivalent discrete form, it is then necessary to implement a recursive method to calculate the error to minimize the amount of information to be stored. The latter involves storing the previous value of the integral control action and updating it by taking into account the last value of the integral component. For example, making use of rectangular integration, it may be deduced that the incremental form of the discrete PID controller is given by

$$u(t) = u(t-1) + \sum_{j=1}^{3} K_j e(t-j+1) \tag{2.21}$$

where t denotes the discrete-time step index, with $u(t)$ and $e(t) = r(t)-y(t)$ being the sampled control and error signals, respectively. The gains K_1, K_2, and K_3 relate to the gains in Equation 2.14, with $K_1 = K_p + K_i T + \frac{K_d}{T}$, $K_2 = \frac{-2K_d}{T} - K_p$, and $K_3 = \frac{K_d}{T}$; with T being the sampling interval. Readers interested in a summary of the various representations and derivations of discrete PID controllers should consult Bobál et al. (2005).

Discrete-time computer control requires the introduction of a zero-order-hold device into the loop to maintain the signal generated by the microprocessor/microcontroller to the same value for the duration of the sampling period. Unfortunately, however, the introduction of such a sampling action may destabilize the resulting closed-loop system (Rohrs et al., 1984) and, to achieve a similar degree of stability to the continuous case, the controller gains may be required to be decreased; this being at the expense of an overall reduction in system performance.

2.1.2.3 Linear Model–Based Control

One of the main advantages of the PID controller described in Section 2.1.2.2 is that a mathematical model of the process is not a necessary requirement. However, if knowledge of a process is available then it is possible to derive a mathematical model that attempts to describe the relationships between the system inputs and outputs. Such knowledge can be exploited to advantage within a model-based control strategy.

2.1.2.3.1 Linear Self-Tuning Control

Ideally, a controller should be able to automatically adjust its parameters in real time to achieve the required control objective. This leads to the idea of an STC that essentially involves the online implementation of two algorithms: one for parameter estimation and the other for calculating an appropriate control action. The concept of STC first emerged in the late 1950s (Gregory, 1959), but the lack of computing power as well as theoretical foundation at that time rendered the proposed idea impractical. Early attempts to realize the self-tuning concept include Kalman (1958) and Young (1966), in which laboratory-based experiments were conducted. A major step was made when it was demonstrated that the parameter estimates did not necessarily have to converge to the true values for a linear self-tuning minimum variance controller to be optimal (Astrom and Wittenmark, 1973), thus constituting the so-called self-tuning property (Wellstead et al., 1979). This effectively made STC a feasible solution for industrial control problems. An authoritative account of the developments

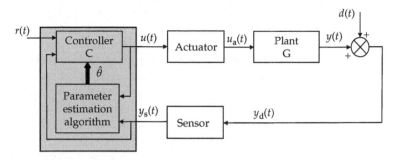

FIGURE 2.4
Illustrating explicit self-tuning control strategy where the controller parameters are calculated and then updated based on an estimated model of the plant.

in linear self-tuning and adaptive control may be found in the texts of Harris and Billings (1985) Wellstead and Zarrop (1991) and Bobál et al. (2005).

STC may be classified as either implicit (direct) or explicit (indirect) depending on whether the controller parameters are estimated directly from within the estimation algorithm or calculated indirectly based on estimated values of the model parameters (Lam, 1980). A further classification is possible by grouping STCs in terms of dual and nondual abilities. A dual controller generates a control action that is considered as being ideal for both estimation and control, whereas a nondual scheme generates an ideal control signal only. The rapid developments in microcomputer technology since the 1990s has meant that these approaches can now be practically implemented.

A distinctive feature of an STC is the need to estimate the parameters of an assumed mathematical model of the plant; the structure of the model (i.e., n_a, n_b, and k) normally being identified before the application of STC. The operation of an explicit STC in which the control action is calculated based on estimated model parameter values, which are assumed to be correct, is illustrated schematically in Figure 2.4.

The model parameters that have to be estimated or updated at each time step can be obtained from an estimation algorithm of the form described in Section 2.1.1.2.4, i.e., recalling that Equation 2.7 is equivalently

$$y(t) = \sum_{i=1}^{n_a} -a_i y(t-i) + \sum_{i=0}^{n_b} b_i u(t-k-i) + \varepsilon(t)$$

$$(2.22)$$

So that

$$y(t) = \mathbf{x}^{\mathrm{T}}(t)\,\theta + \varepsilon(t)$$

where $\mathbf{x}(t)$ and θ are, respectively, the observation and parameter vectors

$$\mathbf{x}^{\mathrm{T}}(t) = [-y(t-1) \cdots -y(t-n_a); u(t-k) \cdots u(t-k-n_b)]$$

and

$$\theta^{\mathrm{T}} = [a_1 \cdots a_{n_a}; b_0 \cdots b_{n_b}]$$

At each time step the sampled values of the system input $u(t)$ and output $y(t)$ are fed into the parameter estimation algorithm. Normally, RLS, or a derivative thereof, is used to find or an estimate of the model parameter vector, denoted $\hat{\theta}$. The control law design algorithm makes use of the updated model parameter values to generate an appropriate control action $u(t)$, which satisfies some particular control law policy. The control action $u(t)$ is then applied to the system and the cycle repeats. It is important to ensure that the two algorithms are operated as efficiently as possible to reduce the detrimental effects due to computational delay.

While there are many possible choices available for the parameter estimation scheme, one of the most popular approaches is that of RLS (see Section 2.1.1). This is mainly due to ease of implementation and its ability to provide for a satisfactory compromise between tracking accuracy and simplicity. In addition, RLS forms the basis of a number of enhanced estimation schemes (see Section 2.1.1). It should be noted that the need for enhanced estimation algorithms in conventional linear STC arises when estimated model parameter values are required to vary in an attempt to accommodate the unmodeled (or inappropriately modeled) plant nonlinearities. Indeed, should the system be nonlinear then the estimated parameter values of an assumed linear model structure may lead to a reduced performance of the controller, or even to instability. It is believed that a more appropriate nonlinear model, *albeit* itself an approximation, should give rise to reduced variation in the model parameters and therefore lead to improved overall controller performance. In fact the use of a more appropriate nonlinear model allows tighter constraints to be placed on the estimated parameters and this is considered to be a distinct advantage over the standard linear STC approach.

In principle, any online parameter estimation algorithm can be combined with any suitable control law design algorithm to provide an overall STC. Irrespective of the control strategy employed, the overall result will normally involve a relocation of the closed-loop poles, giving rise to a satisfactory compromise between transient performance and set point tracking accuracy. For example, the suboptimal pole-placement strategies (Wellstead et al., 1979; Warwick, 1981), which are based on the dyadic form (Young and Willems, 1972), are able to achieve directly the specified transient performance, but generally are required to be tailored to meet steady state requirements. Conversely the optimal tracking strategies, for example, minimum variance (Astrom and Wittenmark, 1973), generalized minimum variance (Clarke and Gawthrop, 1975), and generalized predictive control (Clarke et al., 1987a, b) are able to achieve the required steady-state performance but may need to be tailored through an adjustment of controller cost weightings to meet the transient requirements (Hang et al., 1991) as well as incremental forms with inherent integral action (Clarke and Gawthrop, 1979). While successful implementation of linear STC to industrial plant has been achieved, it has, however, been somewhat limited.

In the context of intelligent and adaptive systems, the STC intelligence is provided both by the model of the system (which summarizes the system knowledge) it is using to calculate the control action as well as the criteria involved in deciding the most appropriate control action. In a similar manner to many application domains, the decision that has to be made by the controller is a trade-off between attempting to achieve the most accurate response and the cost associated with achieving this response. The adaptivity of STC lies in the ability to adapt their responses to environmental changes. Their adaptation can be performed due to the strategies built into the control system without changing the controller parameters, or by changing the controller parameters online to account for significant behavioral changes in the system.

2.1.2.3.2 Polynomial Pole–Placement Control Law Design

Pole placement, sometimes referred to as pole assignment or eigenvalue assignment, is used to specify the desired characteristic equation for the closed-loop system. It relocates, that is, "places," the poles of the closed-loop system directly. The notion of pole placement introduced in this section as the placement of poles is also indirectly achieved using any other form of controller. Note that while pole placement is able to change the pole position, it also affects the overall gain of the closed-loop system, hence pole placement is traditionally combined with some form of gain compensation. This steady-state gain compensation can be implemented through a pure integrator or a feedforward term to give an overall unity gain.

The polynomial pole–placement control law proposed by Wellstead et al. (1979) is given by

$$F(q^{-1})\, u(t) = G(q^{-1})\, y(t) \tag{2.23}$$

where the controller polynomials are:

$$F(q^{-1}) = f_0 + f_1 q^{-1} + f_2 q^{-2} + \cdots + f_{n_f} q^{-n_f}; \quad f_0 = 1$$

$$G(q^{-1}) = g_0 + g_1 q^{-1} + g_2 q^{-2} + \cdots + g_{n_g} q^{-n_g}$$

and their respective orders are:

$$n_f = n_b + k{-}1$$

$$n_g = n_a {-} 1$$

The coefficients of $F(q^{-1})$ and $G(q^{-1})$ are obtained from the polynomial identity:

$$A(q^{-1})\, F(q^{-1}) - q^{-k} B(q^{-1})\, G(q^{-1}) = \Gamma(q^{-1}) \tag{2.24}$$

where

$$\Gamma(q^{-1}) = \gamma_0 + \gamma_1 q^{-1} + \gamma_2 q^{-2} + \cdots + \gamma_{n_\gamma} q^{-n_\gamma}$$

is the closed loop characteristic polynomial having the desired closed-loop characteristics, i.e. poles.

The pole-placement strategy is implemented by

i. Determining the orders of the controller polynomials
ii. Equating the desired closed-loop characteristics with the actual polynomial identify.
iii. Solving for the controller polynomials g_i, $i = 0 \ldots n_g$ and f_j, $j = 1 \ldots n_f$
iv. Calculating the steady-state gain as well as the steady-state error and compensating for this by specifying the feed forward gain as the inverse of the steady-state gain and/or by adding an integrator in the forward loop.

Illustrative Example 1

Consider a system in which $n_a = 2$, $n_b = 1$, and $k = 1$

It follows that $n_f = 1$ and $n_g = 1$. Substituting in Equation 2.24, yields

$$(1 + a_1 q^{-1} + a_2 q^{-2})(1 + f_1 q^{-1}) - (b_0 q^{-1} + b_1 q^{-2})(g_0 + g_1 q^{-1}) = 1 + \gamma_1 q^{-1} + \gamma_2 q^{-2}$$

Equating coefficients and solving for the controller parameters gives

$$f_1 = b_1(b_0 s_2 - b_1 s_1)/\Delta$$
$$g_0 = b_1 s_2 - (a_1 b_1 - a_2 b_0) s_1/\Delta$$
$$g_1 = a_2(b_0 s_2 - b_1 s_1)/\Delta$$

where

$$\Delta = a_1 b_0 b_1 - a_2 b_0^2 - b_1^2$$
$$s_2 = (\gamma_2 - a_2)$$
$$s_1 = (\gamma_1 - a_1)$$

It is readily seen that this can be formulated as a matrix equation

$$\begin{bmatrix} 1 & -b_0 & 0 \\ a_1 & -b_1 & -b_0 \\ a_2 & 0 & -b_1 \end{bmatrix} \begin{bmatrix} f_1 \\ g_0 \\ g_1 \end{bmatrix} = \begin{bmatrix} s_1 \\ s_2 \\ 0 \end{bmatrix}$$

which may be written in the form

$$\mathbf{A} \, \mathbf{c} = \mathbf{s_p}$$

where $\mathbf{c}$ is the vector of controller coefficients, which is updated at each time step as a function of both estimated model parameters and coefficients of the desired closed-loop pole polynomial. The matrix formulation provides a straightforward means of obtaining $\mathbf{c}$ and may be accomplished in a variety of ways all requiring either explicit or implicit matrix inversion. This may be solved by various means, for example,

   i. Matrix inversion and multiplication,
  ii. Gauss–Jordan elimination,
 iii. Gaussian elimination, and
 iv. Triangular factorization.

The preferred approach, which is readily programmable utilizes the triangular factorization method in the controller calculation. The method is to factorize the matrix **A** into the product of upper and lower triangular matrices U and L, respectively, so that $\mathbf{A} = \mathbf{L\,U}$. Nonzero pivots in row operations can be ensured by premultiplying by a permutation matrix **P** so that $\mathbf{P\,A} = \mathbf{L\,U}$.

The permutation matrix ensures that the larger elements are "brought to the top" of the system of linear equations. This has the effect of providing greater accuracy in the computation. Essentially, the problem is reduced from one of solving $\mathbf{c} = \mathbf{A}^{-1}\mathbf{s}_p$ to one of solving $\mathbf{L\,v} = \mathbf{s}_p$ for **v** and substituting in $\mathbf{U\,c} = \mathbf{v}$. This is a much simpler problem due to the triangular properties of the matrices **U** and **L**.

Illustrative Example 2

Consider again the system in which $n_a = 2$, $n_b = 1$, and $k = 1$. After row reduction and factorization

$$\mathbf{U} = \begin{bmatrix} 1 & -b_0 & 0 \\ 0 & a_1 b_0 - b_1 & -b_0 \\ 0 & 0 & \frac{a_2 b_0^2 - a_1 b_0 b_1 + b_1^2}{a_1 b_0 - b_1} \end{bmatrix}$$

and

$$\mathbf{L} = \begin{bmatrix} 1 & 0 & 0 \\ a_1 & 1 & 0 \\ a_2 & \frac{a_2 b_0}{a_1 b_0 - b_1} & 1 \end{bmatrix}$$

The polynomial pole–placement controller can thus force a system to respond according to dynamics specified by the user. Theoretically, pole-placement control strategy is able to cancel the system behavior and impose an entirely new behavior. In practice, however, it requires an accurate model of the system to be controlled. The problem is then in the presence of noise and uncertainties the identification process may lead to an inaccurate model and thus an ineffective control strategy. The state space formulation of pole placement enables the control engineer to compensate to some extent for noisy measurements.

2.1.2.3.3 State-Space Pole-Placement Control Law Design

The polynomial control law in the previous STC makes direct use of the noisy measurements of the current system output in the control input calculation. This noise-contaminated measurement will possibly degrade the control. The state-space control law uses a steady-state Kalman filter (SKF) that reconstructs a noise-free estimate of the system output, which is subsequently used in the control input calculation. This filtering action is inherent in the method and does not require extra filtering of the measured output. This noise-free filtered estimate of the system output should provide a smoother control action.

2.1.2.3.4 Discrete State-Space Model

Substituting the output equation into the state equation (see Equation 2.8), eliminating the noise term, and making use of the backward shift operator give rise to the SKF

$$\hat{x}(t) = \left[I - q^{-1}\overline{P}\right]^{-1}\left[Qu(t-1) + Ry(t-1)\right]$$

in which $x(t)$ has been replaced by the state estimate $\hat{x}(t)$ and $\overline{P} = P - RH$. It is interesting to note that $\left[I - q^{-1}\overline{P}\right]^{-1}$ is a lower triangular matrix of shift operators and, due to its known structure, $\hat{x}(t)$ is readily obtained without the need for matrix inversion. The state estimate thus generated is used to update the appropriate control action in the familiar state variable feedback control law (see Figure 2.5).

Structure of the matrix: $\left[I - q^{-1}\overline{P}\right]^{-1}$.

Consider again the system in which $n_a = 2$, $n_{b=1}$, and $k = 1$. Following a reduction in the dimension of the initial state-space to ensure the controllability of the pair PQ, the matrix $\left[I - q^{-1}\overline{P}\right]^{-1}$ may be readily deduced to be

$$\left[I - q^{-1}\overline{P}\right]^{-1} = \begin{bmatrix} 1 & 0 \\ q^{-1} & 1 \end{bmatrix}$$

that is, a lower triangular matrix having a diagonal comprised of unity elements and subdiagonals comprising of backward shift operators as required.

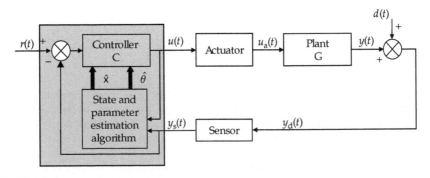

FIGURE 2.5
Illustrating pole-placement control strategy where the control strategy is based on the estimated state and parameters of the plant G.

Note that the initial dimension $n_i = n_a = +k$ has been reduced to $n_r = n_b + k$, where the subscript i and r denote initial and reduced, respectively.

2.1.2.3.5   State Variable Feedback

State variable feedback exploits the state space representation to generate a control action given by $u(t) = \mathbf{F}\hat{x}(t)$, where $\mathbf{F}$ is a feedback vector chosen such that the closed-loop system has the desired eigenvalues. Equivalently,

$$|\mathbf{I} - q^{-1}\mathbf{\Psi}| = \Gamma(q^{-1})$$

where $\mathbf{\Psi} = \mathbf{P} + \mathbf{QF}$ is the desired closed-loop system matrix.

The feedback vector is then obtained from the dyadic form:

$$\mathbf{F}^T = \mathbf{W}^{-1}\mathbf{s_s}$$

where, similar to $\mathbf{s_p}$ in the polynomial pole placement approach, $\mathbf{s_s}$ is a vector containing the coefficients corresponding to the difference between the desired closed- and open-loop polynomial coefficients:

$$\mathbf{s_s} = \begin{bmatrix} 0 \\ \vdots \\ 0 \\ \gamma_{n_a} - a_{n_a} \\ \vdots \\ \gamma_1 - a_1 \end{bmatrix}$$

and $\mathbf{W}$ is the symmetric matrix given by $\mathbf{W} = \mathbf{K\,M}$, where $\mathbf{K}$ is the Kalman controllability test matrix given by $\mathbf{K} = [\mathbf{P}^{n_r - 1}\mathbf{Q} \cdots \mathbf{PQ\ Q}]$ and $\mathbf{M}$ the lower-triangular matrix:

$$\mathbf{M} = -\begin{bmatrix} 1 & 0 & 0 \\ a_1 & & \\ & & 0 \\ a_{n-1} & a_1 & 1 \end{bmatrix}$$

Note:
For the feedback vector $\mathbf{F}$ to exist the matrix $\mathbf{W}$ must have an inverse. $\mathbf{M}$, being lower triangular will always satisfy this, therefore the only condition for $\mathbf{F}$ to exist is that $\mathbf{K}$ must be of full rank (i.e., the Kalman controllability test must be satisfied). Therefore a reduction in the dimension of the initial state space representation may be necessary. If rank $(\mathbf{K}) \neq n_i$ then the dimension of state-space should be reduced from n_i to n_r, until $\mathbf{PQ}$ becomes controllable pair. It is also interesting to note that det $(\mathbf{W}) = -\Delta$.

An advantage of state variable feedback is that the states of the system can be controlled, even if it is not possible to measure them. In practice this enables the exploitation of known behavior (but difficult to access) of the system to calculate the control action. The resulting state variable feedback controller is found to be relatively straightforward to compute and to implement.

2.1.3 Numerical Simulation Software Environments

Sections 2.1.1 and 2.1.2 have introduced the theoretical basis for systems modeling identification and control. Performing the work actually involves making use of numerical simulation software environments enabling the users to simulate responses of models, study the effects of design changes and interact with the environment. Listing all the tools available to industry and academia for this purpose is outside the scope of this book, therefore only the most widely available are listed here. Today, the most popular set of tools, in a position of quasi monopoly, in research and development are MATLAB®, Simulink®, and Stateflow® from The MathWorks, (The MathWorks, 2007). Scilab/Scicos developed by INRIA in France (Scilab, 2007), while having a user interface less developed than The MathWorks, products, offers similar performance and is free to use for research and teaching. Other simulation environments include Maple, Mathematica. More recently, LabVIEW (LabVIEW, 2007) from National Instruments Corporation has been entering into competition with The MathWorks, offering simulation tools similar to MATLAB, Simulink, and Stateflow. The main advantage of LabVIEW is however not for simulation but for communicating using standard PCs or dedicated hardware with a wide range of hardware for monitoring and control purposes. The alternative to LabVIEW to work with The MathWorks products is dSPACE. Specialized tools such as CATIA Composite Product Design (CPD), which now includes support for Dymola technologies (Dymola, 2007), and Modelica, an alternative to MATLAB and Simulink (CATIA, 2006), covers both the preliminary and detailed design phases while taking into account, even at the concept stage, the product's requirements and its manufacturability.

The combination of design tools and software that communicate with hardware enables rapid prototyping and a general speedup of the creation and development of new products. While these tools are widely used in the automotive and aerospace industries, the medical industry is only just starting to realize their immense potential.

2.1.4 Recent Development in Control Engineering for Radiotherapy

Since its inception, radiotherapy has mainly been a forward process, or open-loop-feedforward according to common control engineering terminology. Traditionally, the knowledge about the effect of a course of radiation and the different means to focus the radiation to the cancerous tissues was exploited to devise an appropriate treatment schedule and method of delivery. It has long been recognized that the localization of the cancerous volume to treat is one of the main bottlenecks for treatment improvement. Today, the coming of age of various imaging and visualization techniques, whether based on x-ray imaging technology, video imaging, or electromagnetic principles provide a facility to locate cancerous tissues and also to track, in real time, the displacement of markers and anatomical structures during the delivery of radiation. Such technology has led to the development of a new set of methods that exploit this information within closed-loop feedback mechanisms. Currently, techniques based on portal imaging can be used by staff to adjust the patient setup before the delivery of radiation. More modern techniques based on taking a computed tomography (CT) image or a cone beam CT are now starting to be implemented to improve the patient localization due to better tissue differentiation than megavoltage portal imaging. The ability to adapt the treatment to the patient response was coined "ART." However, ART was originally only involved with adaptation between treatment fractions, where the radiation delivery was replanned to account for physiological changes, for example, weight loss. Such an approach was still very much a feedforward process, where a traditional planning procedure was repeated several times during the course of the treatment. Today adaptation can take place online during the treatment delivery itself. The latest research in this area aims to develop systems to take corrective actions to ensure that radiation delivery adapts in an intelligent manner to the observable changes in shape and position of the regions of interest (Court et al., 2006; Webb, 2006). The development of these techniques is realized as a truly multidisciplinary field of research including image and signal processing, control engineering, mechanical engineering, radiotherapy physics, and biology.

Chapter 7 by Goodband and Haas show that standard systems modeling techniques as well as ANN-based techniques can be developed to facilitate the tracking of body structures. Similar recent work includes Putra et al. (2006) and Murphy and Dieterich (2006). These tracking predictions have been combined with control strategies developed to direct the motion of robotic systems, multiple-leaf collimators (MLC) (Keall et al., 2007) and PSSs (D'Souza et al., 2005; Putra et al., 2007; Skworcow et al., 2007a, b). The main commercial solution today is based on a robotic system, where an industrial robotic arm is used to move a device producing a narrow radiation beam. While the application is medical, there is "little" difference between this robotic system and that used to assemble cars. By contrast, traditional linac manufacturers are looking for solutions where they can employ their specialized equipment without having to go through a major conceptual redesign. Modern radiotherapy treatment suites consist of a linac, used to produce radiation beams; an MLC composed of several (between 80 and 240) tungsten leaves that can be moved in and out of the radiation field to attenuate the radiation; a PSS that locates the patient with respect to the treatment beams isocenter(s); and a gantry system able to rotate the beam around the patient. As opposed to industrial robotic systems, linac-based systems have relatively small motors to move the different parts of the machine. This limitation was originally introduced to prevent harming human operators around these devices and now poses a new challenge in terms of control.

Today most linac systems are still controlled using (at most) standard PID schemes. There are, therefore, many opportunities for control engineers to design new control strategies based on modern control systems (see Section 2.1.2.3). Furthermore, the need to move the MLC or PSS during the treatment delivery knowing in advance the desired position through the patient motion prediction algorithms that are currently under development, makes it an ideal application for adaptive or STC such as state variable feedback (Stewart and Davidson, 2005, 2006) or model predictive control (MPC) (Skworcow et al., 2007a, b). Note that MPC has been proposed for a hyperthermia application too (Arora et al., 2002) and can be applied whenever the effect of a particular action can be predicted, for example, tissue temperature increase. To date, however, the focus of the research is more on the so-called prediction algorithms that are required to provide a set point to the control system than on the control systems themselves. Making use of these predicted trajectories and feeding them to the existing traditional control strategies has been shown to reduce the tracking error (Keall et al., 2006; Court et al., 2006). The control system performances are, however, not improved due to the reliance on existing control systems, some of which are still analog. Control engineering research in radiotherapy treatment machines is at the current time being limited to simulation studies and laboratory experiments.

Stewart and Davidson (2005, 2006) have reported simulation studies where a feedback control scheme is used to compensate for patient setup errors and periodic tumor motion. The delay introduced by the video imaging system is simulated as being a pure time delay. A model consisting of a first-order discrete filter combined with a bias is assumed to be able to relate the respiratory breathing flow to the tumor position. The MLC leaves are also assumed to be modeled as first-order discrete filter devices. Observers are algorithms used to estimate the states, i.e. the internal and/or external behavior of a system. In this particular case, they are employed to estimate the tumor motion and the model parameters of the MLC leaves. A state variable feedback controller is derived to control a single leaf motion. In Stewart and Davidson (2006) the original feedback strategy is extended to the actual dose delivered. It is shown, using simulation, that the approach could potentially be employed to provide a means to control the dose delivered in real time.

The MPC strategy developed in Skworcow et al. (2007a, b) and extended in Putra et al. (2007) is illustrated in Figure 2.6. The feedback scheme requires an imaging system to provide tumor position measurement $z(t)$ and a sensor to measure the PSS position $y(t)$. The feedback scheme itself consists of a tumor motion predictor, an observer, and a model predictive control (MPC) system. Both, an NN (Goodband, 2006; Sahih et al., 2007) and a Kalman filter/ interactive multiple-model filter (Putra et al., 2006) were used for the motion prediction stage and it was shown that while the NN was slightly better, the prediction errors were more consistent with the Kalman-filter approach. The observer is used to estimate a simplified linear model of the PSS along the PSS longitudinal axis. The longitudinal axis corresponds to the cranio–caudal (also termed the superior–inferior) axis along which the respiratory-induced tumor motion is the most significant. Similar to many control system designs, a simplified linear model is adopted for realizing the controller and observer structures. However, the designed feedback controller

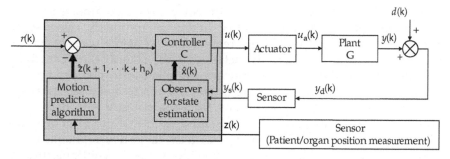

FIGURE 2.6

Illustrating model-predictive control strategy. The motion prediction algorithm predicts h_p steps ahead based on historical position $z(k)$ and an observer provides the estimated state $\hat{x}(k)$ of the plant G, for example, position or velocity, from the noisy position measurement $y_s(k)$. Based on the plant model, the estimated/predicted positions $\hat{z}(k)$ and estimated states $\hat{x}(k)$, the MPC computes the h_p optimal input sequence to move the plant to the positions given by $r(k) - \hat{z}(k)$. This strategy is called the receding horizon MPC strategy when only the first input $u(k)$ is applied to the PSS.

is subsequently applied to the full nonlinear model implemented using Simulink6.5/SimMechanics (The MathWorks, 2006). The nonlinear model offers a realistic representation of the actual system; it was validated against one of the PSS in use at the University Hospitals Coventry and Warwickshire NHS Trust, Coventry, U.K. Note the subtle differences in models used for design through simulation and those used to realize practical implementation. The latest work is now focusing on practical aspects linked with the implementation using dSPACE and LabVIEW of these model predictive control algorithms on existing clinically available systems, such as the PSS.

2.2 Conclusions

This chapter has given a brief introduction to the tools employed to approach a problem from a control engineering perspective. The first stage is to understand the problem through modeling the system being studied and parameterizing it using a range of system identification techniques such as least squares and its derivatives. The latter was presented as it is used in Chapters 3 and 4 and forms the basis for many estimation techniques. A general review of regularization techniques was presented to make the reader aware that while it is straightforward to obtain parameter values for a particular model structure, it is actually difficult to obtain the appropriate values without exploiting the vast amount of medical and engineering knowledge about most of the systems considered. The main control algorithms were then presented, including the PID family of controllers that is currently used in more than 70% of applications worldwide. "Modern control" systems exploiting discrete system representations such as self-tuning strategies were also presented. State variable feedback and pole-placement schemes were described. These algorithms were selected among many, as each control system effectively, either directly or indirectly, relocates the system poles (i.e., specifying the desired dynamic system response) through proportional, integral, or derivative actions. Finally, some of the recent developments in the field of control engineering applied to respiratory tracking for radiotherapy were briefly reported with the aim to make technical specialists aware of opportunities arising from the richness afforded by the emerging techno-medicine applications. It is hoped that this brief overview will have brought the reader up to date with the latest developments in this exciting field, and will have provided the motivation to continue through the remainder of the book.

References

Akaike, H. (1969). Fitting autoregressive models for prediction. *Ann. Inst. Stat. Math.,* **21**, 243–247.

Akaike, H. (1974). A new look at the statistical model identification. *IEEE Trans. Autom. Control,* **19**, 716–723.

Arora, D., Skliar, M. and Roemer, R. B. (2002). Model-predictive control of hyperthermia treatments. *IEEE Trans. Biomed. Eng.,* **49**(7), 629–639.

Åström, K. and Hägglund, T. (1995). *PID Controllers: Theory and Design and Tuning.* 2nd edn. Instrument Society of America, Research Triangle Park, NC.

Astrom, K. J. and Wittenmark, B. (1973). On self-tuning regulators. *Automatica,* **9**, 185–199.

Åström, K. J., Hägglund, T., Hang, C. C. and Ho, W. K. (1993). Automatic tuning and adaption for PID controllers—a survey. *Control Eng. Pract.,* **1**(4), 699–714.

Astrom, K. J. and A. Helmersson (1986). Dual control of an integrator with unknown gain. *Comput. Math. Appl.,* **12**(6, Part 1), 653–662.

Atherton, D. P. (1982). *Nonlinear Control Engineering, Van Nostrand Reinhold Company.* Student Edition (December). ISBN 978-0442304867. p. 470.

Atherton, D. P. (1999). PID controller tuning. *IEE J. Comput. Control Eng.,* **10**(2), 44–50.

Atherton, D. P. (2000). Relay autotuning: a use of old ideas in a new setting. *Trans. Inst. Meas. Control,* **22**(1), 103–122.

Balmer, L. (1991). *Signals and Systems—An Introduction.* Prentice Hall, Englewood Cliffs, NJ.

Bobal V., BohmJ., Fessl J. and Machacek J. (2005). *Digital Self-tuning Controllers, Algorithms, Implementation and Applications, Advanced Textbooks in Control and Signal Processing.* Springer-Verlag, London, ISBN: 1-85233-980-2, 317p.

Bristol, E. H. (1977). Pattern recognition: an alternative to parameter identification in adaptive control. *Automatica,* **13**, 197–202.

Bueno, S. S. and Favier, G. (1991). Self-tuning PID controllers: a review. *1st IFAC Symp. on Design Method of Control Systems.* Pre-prints, Zurich, Switzerland, pp. 459–468.

Burnham, K. J. (1991). *Self-Tuning Control for Bilinear Systems.* PhD thesis, Coventry Polytechnic Coventry, U.K.

Burnham, K. J. and James, D. J. G. (1986). Use of cautious estimation in self-tuning control of bilinear systems. *Proc. RAI/IPAR*, 1, Toulouse, France, pp. 419–421.

Burnham, K. J., James, D. J. G. and Shields, D. N. (1985). Choice of forgetting factor for self-tuning control. *J. Syst. Sci.*, **11**(2), 65–73.

CATIA (2007). CATIA Dassault Systèmes' flagship product development solution. http://www.3ds.com/corporate/about-us/brands/catia/ (22-12-2006).

Clarke, D. W. and Gawthrop, P. J. (1975). Self-tuning controller. *Proc. IEE*, **122**(9), 929–934.

Clarke, D. W. and Gawthrop, P. J. (1979). Self-tuning control. *IEE Proc. Part D: Control Theory Appl.*, **126**(6), 633–640.

Clarke, D. W., Mohtadi, C. and Tuffs, P. S. (1987a). Generalized predictive control—part 1: the basic algorithm. *Automatica*, **23**(2): 137–148.

Clarke, D. W., Mohtadi, C. and Tuffs, P. S. (1987b). Generalized predictive control—part 2: extensions and interpretations. *Automatica*, **23**(2): 149–160.

Cohen, K.H. and Coon, G.A., (1953). Theoretical consideration of retarded control. *Transact. ASME* **75**, pp. 827–834.

Cook, P. A. (1994). *Nonlinear Dynamical Systems*. Prentice Hall International Series in Systems and Control Engineering. 2nd edn (June). ISBN-13: 978-0136251613, 152p.

Court, L. E., Tishler, R. B., Petit, J., Cormack, R. and Chin, L. (2006). Automatic online adaptive radiation therapy techniques for targets with significant shape change: a feasibility study. *Phys. Med. Biol.*, **21**,**51**(10), 2493–2501. Epub 2006, April 26.

Demoment, G. (1989). Image reconstruction and restoration: overview of common estimation structures and problems. *IEEE Trans. Acoust. Speech Signal Process.*, **37**(12), 2024–2036.

D'Souza, W. D., Naqvi, S. A. and Yu, C. X. (2005). Real-time intra-fraction-motion tracking using the treatment couch: a feasibility study. *Phys. Med. Biol.*, **50**, 4021–4033.

Dymola (2007). Dynamic modeling laboratory with Modelica Dynasim (AB). http://www.dynasim.com/ (22-12-2006).

Fnaiech, F. and Ljung, L. (1987). Recursive identification of bilinear systems. *Int. J. Control*, **45**(2), 453–470.

Fortescue, T. R., Kershenbaum, L. S. and Ydstie, B. E. (1981). Implementation of self-tuning regulators with variable forgetting factors. *Automatica*, **17**, 831–835.

Friedland, B. (1987). *Control System Design*. McGraw-Hill, New York.

Golub, G. H. and Van Loan, C. F. (1996). *Matrix Computations*. 3rd edn. Johns Hopkins University Press, Baltimore.

Goodband, J. (2006). *Novel Applications Using Neural Networks and Liquid Metals in Radiation Therapy*, PhD thesis, Coventry University, Coventry, U.K.

Gregory, P. C. (ed.) (1959). *Proc. Self-Adaptive Flight Control Symposium*. Wright Air Development Center, Wright-Patterson Air Force Base, OH.

Haas, O. C. L. (1999). Radiotherapy treatment planning: new system approaches. *Advances in Industrial Control Monograph*. Springer Verlag, London. ISBN 1-85233-063-5.220p. Monograph.

Hang, C. C., Lim, K. W. and Ho, W. K. (1991) Generalised minimum variance stochastic self-tuning controller with pole restriction, *IEE Proc. D*, **138**(1), 25–32.

Harris, C. J. and Billings, S. A. (eds.) (1985). *Self-Tuning and Adaptive Control: Theory and Applications*. 2nd edn. Peter Peregrinus, London. IEE Control Engineering Series, 15.

Hoerl, A. E. and Kennard, R. W. (1970). Ridge regression: biased estimation for non-orthogonal problems. *Technometrics*, **12**, 55–62.

Hsia , T. C. (1977). *System Identification: Least-squares Methods*. Lexington Books, Lexington, MA.

Ivakhnenko, A. G. and Yurachkovsky, Y. P. (1988). System structure identification by sets of observation data on the base of unbiasness principles. *IFAC Symposium on Identification and System Parameter Estimation*, pp. 953–965.

Johansen, T. A. (1996). Identification of non-linear systems using empirical data and prior knowledge: an optimization approach. *Automatica*, **32**(3), 337–356.

Johansen, T. A. (1997). On Tikhonov regularization, bias and variance in nonlinear system identification. *Automatica*, **33**(3), 441–446.

Kalman, R. E. (1958). Design of self-optimizing control system. *Trans. ASME*, **80**, 468–478.

Kashyap, R. L. (1977). A Bayesian comparison of different classes of dynamic models using empirical models. *IEEE Trans. Autom. Control*, **5**, 715–727.

Keall, P., Cattell, H., Pokhrel, D., Dieterich, S., Wong, K., Murphy, M., Vedam, S., Wijesooriya, K. and Mohan, R. (2006). Geometric accuracy of a real-time target tracking system with dynamic multileaf collimator tracking system. *Int. J. Radiat. Oncol. Biol. Phys.*, **65**(5), 1579–1584.

Kirsch A. (1996) *An Introduction to the Mathematical Theory of Inverse Problems*. Springer-Verlag, New York.

LabVIEW (2007). The software that powers virtual instruments. http: // www.ni.com/ labview/ (22-12-2006).

Lam, K. P. (1980). *Implicit and Explicit Self-tuning Controllers*. DPhil thesis, Oxford University, U.K.

Linden, J. G. (2005). *Regularisation Techniques and Cautious Least Squares in Parameter Estimation for Model Based Control*. Master's thesis, CTAC, Coventry University, Coventry, U.K.

Linden, J., Vinsonneau, B. and Burnham, K. J. (2005). Review and enhancement of cautious parameter estimation for model based control: a specific realisation of regularisation. *Proceedings of the 17th International Conference on Systems Engineering*. Las Vegas, pp. 112–117.

Ljung, L. (1999). *System Identification: Theory for the User*. 2nd edn. Prentice Hall PTR, Upper Saddle River, NJ. ISBN 0-13-656695-2.

Menani, S. and Koivo, H. N. (1996). Relay tuning of multivariable PI controllers. *13th IFAC Triennial World Congress*, vol. K, San Francisco, July, pp. 139–144.

Mohler, R. R. (1991). *Nonlinear Systems: Applications to Bilinear Control*, Prentice Hall, Englewood Cliffs, NJ. (January) ISBN-13: 978-0136235217, 192p.

Murphy, M. J. and Dieterich, S. (2006). Comparative performance of linear and nonlinear neural networks to predict irregular breathing. *Phys. Med. Biol.*, **21,51**(22), 5903–5914. Epub 2006, October 26.

Neumaier, A. (1994). Solving ill-conditioned and singular linear systems: a tutorial on regularization. *SIAM*, **40**, 636–666.

Ogata, K. (2002). *Modern Control Engineering*. 4th edn. Prentice Hall, New York.

Parr, A. (1998). *Industrial Control Handbook*. 3rd edn. Newes, Oxford, ISBN 0 7506 3934 2.

Polson, N. G., Carlin, B. P. and Stoffer, D. S. (1992). A Monte-Carlo approach to non-normal and non-linear state-space modeling. *J. Am. Statist. Assoc.*, **87**, 493–500.

Putra, D., Haas, O. C. L., Mills, J. A. and Burnham, K. J. (2006). Prediction of tumour motion using interacting multiple model filter. *Proceedings of the 3rd IEE International Conference on Medical Signal and Information Processing (MEDSIP)*, Glasgow.

Putra, D., Skworcow, P., Haas, O. C. L., Burnham, K. J. and Mills, J. A. (2007). Output-feedback tracking for tumour motion compensation in adaptive radiotherapy. *American Control Conference*, New York, July 11–13, pp. 3414–3419.

Randall, A. and Burnham, K. J. (1994). Cautious identification in self-tuning control—an information filtering alternative. *J. Syst. Sci.*, **20**(2), 55–69.

Rissanen, J. (1978). Modeling by shortest data description. *Automatica*, **14**: 465–471.

Rodriguez-Vazquez, K. and Fleming, P. J. (1998). Multi-objective genetic programming for nonlinear system identification, *Electronics Letters*, **34**(9), pp. 930–931, April 30.

Rohrs, C. E., Athans, M. and Stein, G. (1984). Some design guidelines for discretetime adaptive controllers. *Automatica*, **20**(5), 653–660.

Sahih A., Haas O., Mills J. and Burnham, K. (2007), A new bilinear model for respiratory motion. *Proceedings of the 15th International Conference on the Use of Computer in Radiation Therapy*, vol. 1, June 4–7, Toronto, Canada, pp. 133–137

Scilab. (2007a). Scilab Home Page, http://www.scilab.org/ (22-12-2006).

Sjoberg, J., T. McKelvey, and L. Ljung (1993). On the use of regularization in system identication. *12th IFAC World Congress*, vol. 7, Sydney, pp. 381–386.

Skworcow, P., Putra, D., Sahih, A., Goodband, J., Haas, O. C. L., Burnham, K. J. and Mills, J. A. (2007a). Predictive tracking for respiratory—induced motion compensation in adaptive radiotherapy. *Meas. Control*, **40/1**, 16–19.

Skworcow P., Putra, D., Haas, O. C. L., Burnham, K. J., Mills J. A. (2007b). Compensation of system latency in tumour tracking using prediction: comparison of feedforward and model predictive strategies. *Proceedings of the 15th International Conference on the Use of Computer in Radiation Therapy*, vol. 1, June 4–7, Toronto, Canada, pp. 113–117

Stewart J., and Davison D. E., (2005). Conformal radiotherapy cancer treatment with multileaf collimators: improving performance with realtime feedback. *Proceedings of the IEEE Conference on Control Applications*, August, Toronto, Canada, pp. 125–130.

Stewart, J. Davison, D. E. (2006). Dose control in radiotherapy cancer treatment: improving dose coverage with estimation and feedback. *Proceedings of the 2006 American Control Conference*, Minneapolis, MN, June 14–16, pp. 4806–4811.

Stoica, P., Eykhoff, P., Jansen, P. and Söderström, T. (1986). Model-structure selection by cross-validation. *Int. J. Control*, **43**, 1841–1878.

Tarantola, A. (1987). *Inverse Problem Theory; Methods for Data Fitting and Model Parameter Estimation*. Elsevier Science Publ. Co. Inc., Amsterdam, The Netherlands, ISBN 0-444-42765-1.

The MathWorks (2007). MATLAB and Simulink for technical computing. http://www.mathworks.com/ (22-12-2006).

Tikhonov, A. N. and Arsenin, V. Y. (1977). *Solution of Ill-posed Problems*. (F. John, translation editor), Winston & Sons distributed by Wiley, New York, xiii, 258 pp.

Vega, P., Prada, C. and Aleixandre, V. (1991). Self-tuning predictive PID controller. *IEE Proc. Part D*, **138**(3), 303–311.

Warwick, K. (1981). Self-tuning regulators—a state-space approach. *Int. J. Control*, **33**(5), 839–858.

Warwick, K. and Rees, D. (eds) (1986). *Industrial Digital Control Systems*. IEE Control Engineering Series 29, Peter Peregrinus, London.

Webb, S. (2006). Motion effects in (intensity modulated) radiation therapy: a review. *Phys. Med. Biol.*, 51, R403–R425.

Wellstead, P. E., Edmunds, J. M., Prager, D. and Zanker, P. (1979). Self-tuning pole/zero assignment regulators. *Int. J. Control*, **30**(1), 1–26.

Wellstead, P. E. and Sanoff, S. P. (1981). Extended self-tuning algorithm. *Int. J. Control*, **34**(3), 433–455.

Wellstead, P. E. and Zarrop, M. B. (1991). *Self-tuning Systems, Control and Signal Processing*. Wiley, New York.

Young, P. C. (1966). Process parameter estimation and self adaptive control. In: *Theory of Self-Adaptive Control System*. P. H. Hammond (ed), Plenum Press, New York, pp. 118–140.

Young, P. C. (1970). An instrumental variable method for real-time identification of a noisy process. *Automatica*, **6**, 271–287.

Young, P. C. (1974). Recursive approaches to time series analysis. *Bull. IMA*, **10**, 209–224.

Young, 1984 P.C. (1984). *Recursive Estimation and Time-series Analysis*. Springer, Berlin.

Young, P. C. and Beven, K. J. (1994). Data-based modeling and the rainfall-flow non-linearity. *Environmetrics*, **5**, 335–363.

Young, P. C. and Willems, J. C. (1972). An approach to the linear multivariable servomechanism problem. *Int. J. Control*, **15**(5), 961–979.

Zhuang, M. and Atherton, D. P. (1993). Automatic tuning of optimum PID controllers. *IEE Proc. Control Theory Appl.*, **140**(3), 216–224.

Ziegler, J. G. and Nichols, N. B. (1942). Optimum settings for automatic controllers. *IEEE Trans. Autom. Control*, **64**, 759–768.

3

Intelligent Control Systems in Anesthesia

Teresa Feio Mendonça, Catarina Nunes, and Pedro Lago

CONTENTS

3.1   Introduction

Anesthesia can be defined as the lack of response or recall to noxious stimuli. The word is derived from Greek and means "without feeling"; it was first used by the Greek philosopher Dioscorides in the first century AD to describe the narcotic effect of the plant *Mandragora*.

General balanced anesthesia includes paralysis (neuromuscular blockade), unconsciousness (depth of anesthesia [DOA]), and analgesia (pain relief). The first two are concentrated in the operating theater, whereas the third is also related to postoperative conditions. In addition, different surgical procedures require different proportions of the three components.

The assessment of DOA during surgery under general anesthesia has become a very difficult process since the introduction of balanced anesthesia. In the early days of anesthesia, when a single agent (e.g., ether) was used to control all the three components of anesthesia, signs of inadequate anesthesia could be obtained relatively easily from clinical measurements and from patient movement. The main concern was that muscular relaxation could only be provided at deep levels of anesthesia. Therefore, the hypnotic and analgesic components were in excess of those required, resulting in long recoveries and other side effects. Postoperative respiratory and venous thrombotic complications were frequent. The point requiring most skill and care in the administration of anesthetics was to determine when it has been carried far enough, so as to avoid a deeper stage of anesthesia. The introduction of intravenous drugs as part of a balanced anesthesia, (i.e., the use of a muscle relaxant, an analgesic, and an anesthetic) made anesthesia safer for the patient. The three components of anesthesia could be more easily adjusted to individual requirements, which improve the patient's operative and postoperative well-being. However, this also meant that the measures of anesthetic depth became obscure or even completely ablated, making the task of measuring DOA more difficult.

This research searches for a solution to the balanced anesthesia problems to provide a robust/reliable control system that could determine the optimal infusion rate of the drugs (muscle relaxant, anesthetic, and analgesic) simultaneously, and titrate each drug in accordance to its effects and interactions. Such a system would be a valuable assistant to the anesthetist in the operating theater. This chapter is divided into four sections covering the relevant results of this research, the main concepts, terminology, and practical issues of clinical implementation. Following are brief explanations of the contents of each part.

Section 3.2 explains the measurement and control techniques of neuromuscular blockade used in the clinical environment. The evolution of control strategies and methods for achieving and maintaining an adequate level of neuromuscular blockade is presented in detail.

Section 3.3 presents the monitoring and control techniques used in the operating theater concerning DOA (anesthetic and analgesic drugs). The concepts and problems of drugs interactions are also discussed. In addition, the research objectives, modeling results, and control strategy are presented.

Finally, Section 3.4 presents the final remarks. The work presented in this chapter is based on the research carried out by this group in the clinical environment of hospital operating theaters.

3.2 Control of Neuromuscular Blockade

Muscle relaxant drugs are frequently given during surgical operations. The nondepolarizing types of drugs act by blocking the neuromuscular transmission (NMT), therefore producing muscle paralysis. The neuromuscular blockade level is measured from an evoked electromyography (EMG) obtained at the hand by electrical stimulation. The control of the neuromuscular blockade by the continuous infusion of a muscle relaxant provides a good illustration of the main features and inherent constraints associated with the control of physiological variables for optimal therapy. The situation is characterized by a large uncertainty of the dynamic behavior of the system under control and the need for a very high degree of reliability and robustness. Hence, the control system must present a reliable adaptation to the individual characteristics and requirements of a patient.

3.2.1 Measurement and Control in Clinical Environment

Figure 3.1 represents a standard block diagram for the automatic control of neuromuscular blockade by continuous infusion. The theoretical advances in the tuning of controllers characterized by the presence of nonlinearities and large uncertainty have led to the development, over the last 15 years, of

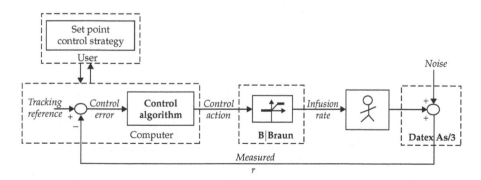

FIGURE 3.1
Automatic control scheme for a neuromuscular blockade delivery system by continuous drug infusion.

a number of control schemes [1–18]. These range from simple on–off type controllers to intelligent control schemes based on a variety of methods, i.e., using adaptive, model-based, fuzzy, and robust techniques.

3.2.2 Control of the NMT Based on Pharmacokinetic/ Pharmacodynamic Models

The dynamic response of the neuromuscular blockade may be modeled by a Wiener structure (Figure 3.2) [19]. It is composed of a linear compartmental pharmacokinetic model relating the drug infusion rate $u(t)$ (µg kg^{-1} min^{-1}) to the plasma concentration $c_p(t)$ (µg ml^{-1}), and a nonlinear dynamic model relating $c_p(t)$ to the induced pharmacodynamic response, $r(t)$ (%). The variable $r(t)$, normalized between 0 and 100, measures the level of the neuromuscular blockade, 0 corresponding to full paralysis and 100 to full muscular activity. In this study, the muscle relaxation drug used is atracurium [19,20]. The pharmacokinetic model may be described by the state equations:

$$\begin{cases} \dot{x}_1(t) = -\lambda_1 x_1(t) + a_1 u(t) \\ \dot{x}_2(t) = -\lambda_2 x_2(t) + a_2 u(t) \\ c_p(t) = \sum_{i=1}^{2} x_i(t) \end{cases} \tag{3.1}$$

where x_i ($i = 1, 2$) are state variables and a_i ($i = 1, 2$) (kg ml^{-1}), λ_i (min^{-1}) are patient-dependent parameters. The pharmacodynamic effect for atracurium may be modeled by the Hill equation:

$$r(t) = \frac{100 C_{50}^s}{C_{50}^s + c_e^s(t)} \tag{3.2}$$

where the effect concentration $c_e(t)$ [µg ml^{-1}] is related to $c_p(t)$ by

$$\dot{c}_e(t) = -\lambda c_e(t) + \lambda c_p(t) \tag{3.3}$$

where C_{50} (µg ml^{-1}), s, and λ [min^{-1}] are also patient-dependent parameters. Figures 3.3a and 3.3b illustrate the observed and simulated responses

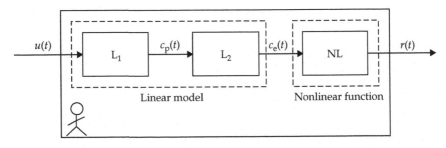

FIGURE 3.2
Structure of the neuromuscular blockade pharmacokinetic/pharmacodynamic model.

induced by a *bolus* of 500 μg kg⁻¹ of atracurium, respectively. The simulation results were obtained assuming a multidimensional log-normal distribution for the pharmacokinetic and pharmacodynamic parameters. The expected values and the covariances for the model parameters have been deduced from the data given in Refs. 19 and 20.

The assumption made on the probability distribution is not critical. Other distributions have been examined (such us truncated-normal and uniform) and the same lack of reproducibility of the observed patient's responses has been found. Besides, major alterations on the expected values and on the covariances were unable to produce a model that replicates the observed *bolus* responses.

It is clear from these results that the search for an alternative model, which mimics the patients' responses, becomes essential for the design of a robust controller for use in a clinical environment.

3.2.3 Tuning the Model to Clinical Data

The most obvious differences between the results of Figures 3.3a and 3.3b are the larger lag and variability of the observed responses. This quite different behavior can be described by the time when $r(t)$ is equal to 50%, denoted, *T*50. The median and the interquartile interval for the observed *T*50 are 1.8 and 1.0 min, respectively, whereas the corresponding values for the simulated responses are 0.7 and 0.3 min, respectively.

An empirical model for atracurium to accommodate the clinical data has been developed. The alteration has been made on the linear part of the system by the inclusion of a first-order system:

$$g(s) = \frac{1/\tau}{s + 1/\tau} \tag{3.4}$$

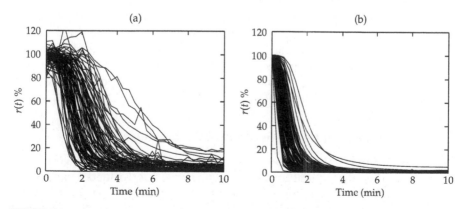

FIGURE 3.3
The responses induced by a *bolus* of 500 μg kg⁻¹ min⁻¹ atracurium on (a) 100 patients undergoing surgery and (b) simulated responses (100 models) obtained with the data given.

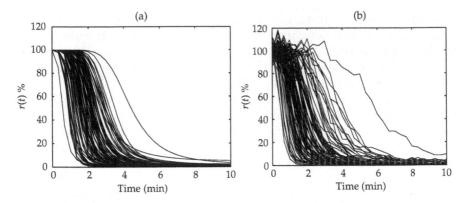

FIGURE 3.4
One hundred simulated responses induced by a *bolus* of 500 μg kg^{-1} of atracurium without noise (a) and with added measurement noise (b). Results obtained with the empirical model: uniform distribution for τ on the interval 0 and 3.5 min.

in a series connection. The time constant τ is assumed to be a random variable independent of the remaining pharmacokinetic/pharmacodynamic parameters. Figure 3.4 shows the results obtained assuming a uniform distribution for τ on the interval 0 and 3.5 min, and a multidimensional log-normal distribution for the remaining parameters; the median and the interquartile interval for $T50$ become 1.9 and 0.9 min, respectively, in close agreement with the clinical data.

3.2.4   Methods for Autocalibration of the Controller

Methods incorporating online adaptation to individual patient's characteristics extracted from the initial *bolus* given in the beginning of anesthesia have been described in some detail [9,10,14,17]. The parameters that have been found to perform the best for the characterization of the *bolus* response are based on shape parameters and on principal component analysis (PCA) [14,17]. Figure 3.5 shows a set of shape parameters easily obtained in real time. $T80$, $T50$, and $T10$ are elapsed times between the *bolus* administration and the time the response $r(t)$ becomes less than 80%, 50%, and 10%, respectively. S is a slope parameter and P is a persistence parameter, because it describes the duration of the *bolus* effect on the patient.

3.2.5   Regression Model for the Controller Parameters

The tuning of the digital proportional-integral-derivative (PID) controller to the dynamics of a patient undergoing surgery is performed by adjusting multiple linear regression models of L and R^{-1} on explanatory or predictor variables extracted from the observed *bolus* response. Consider a set of N independent observations $(\phi_i, \psi_{i1}, ..., \psi_{ip})$, $i = 1, ..., N$, where ϕ_i represents the value of the controller parameter, either L or R^{-1}, and ψ_{ij} the observed

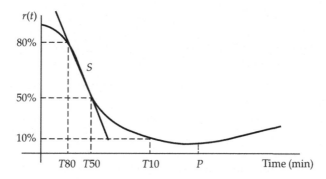

FIGURE 3.5
Shape parameters used to characterize the neuromuscular blockade response $r(t)$ induced by a *bolus* administered at $t = 0$.

values of the predictor variables. In this study, the predictor variables considered are the shape parameters and the principal components. Preliminary data analysis indicated that a multiple linear regression model was adequate [17]. Let $\Phi(N \times 1)$ represent the vector of the controller parameter variables, assumed uncorrelated. Let Ψ be an $N \times (p + 1)$ matrix of observed constants extracted from the *bolus* response:

$$
\Psi = \begin{bmatrix} 1 & \psi_{11} & \cdots & \psi_{1p} \\ 1 & \psi_{21} & \cdots & \psi_{2p} \\ \cdots & \cdots & \cdots & \cdots \\ 1 & \psi_{N1} & \cdots & \psi_{Np} \end{bmatrix} \tag{3.5}
$$

and let α denote a $([p + 1] \times 1)$ vector of unknown parameters. Then, the controller parameters and the *bolus* response are related by the equation:

$$
\Phi = \Psi\alpha + \epsilon \tag{3.6}
$$

where ϵ is a vector of uncorrelated random variables, normally distributed with mean 0 and variance σ^2. The observations on Φ and Ψ are obtained from a set of $N = 500$ simulated models for the neuromuscular blockade response. The multiple regression models are then fitted by least squares.

3.2.6 Online Estimation of Controller Parameters

Linear regression models of L and R^{-1}, on T50, T10, T80, S, and P, and on the principal components a_k ($k = 1:10$) were computed with the corresponding mean square error (MSE) and R^2, the percentage of variation in the data explained by the model [17].

The online autotuning of the controller is a simple procedure. It requires

- The registration of the response $r(t)$ induced by the bolus given in the beginning of anesthesia
- The computation of the p predictor variables using the observed response r
- The estimation of the tuned controller parameters by the regression Equation 3.6

Figures 3.6a and 3.6b show the regression plots for L and for R^{-1}, using $T50$ as predictor ($\hat{L}$ = 1.171 $T50$ + 1.199; $\hat{R}^{-1}$ = 37.03 $T50$ + 135.5). As given in Table 3.1, the parameter $T50$ explains 87% of the variation of L and 42% of the variation of R^{-1}.

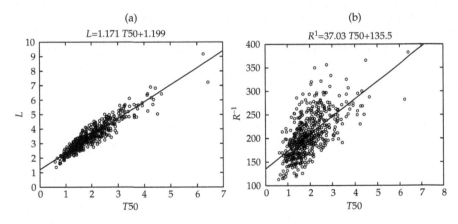

FIGURE 3.6
(a) Regression plot for $T50$ (min) versus L (min). (b) Regression plot for $T50$ (min) versus R^{-1} (μg^{-1} min ml).

TABLE 3.1

Quality of Predictors. Percentage of Variation Explained by the Linear Regression Model

	Without Noise		With Noise	
Predictors	**L**	**1/R**	**L**	**1/R**
$T50$	87	42	87	42
$T50 + P$	94	70	94	69
$a_1, \ldots, a_3$	85	47	85	47
$a_1, \ldots, a_3 + P$	95	71	94	70
$a_1, \ldots, a_{10}$	92	65	86	50
$a_1, \ldots, a_{10} + P$	95	71	95	70

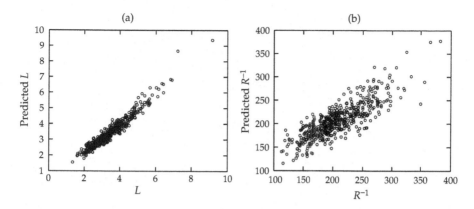

FIGURE 3.7
(a) Scatterplot of L (min) versus the predicted value of L (min). (b) Scatterplot of R^{-1} (μg^{-1} min ml) versus the predicted value of R^{-1} (μg^{-1} min ml) estimated from the *bolus* response parameters $T50$ and P.

With the inclusion of parameter P as predictor (together with $T50$), the quality of the prediction of the controller parameters L and R improves considerably. Figures 3.7a and 3.7b illustrate the scatterplots of L and R^{-1} versus the predicted values of L and R^{-1} ($\hat{L} = -0.184 + 1.303\ T50 + 0.039\ P$; $\hat{R}^{-1} = 263.978 + 24.786\ T50 - 3.582\ P$). As shown in Table 3.1, this set of predictors ($T50$ and P) explains 94% of the variation of L and 70% of the variation of R^{-1}.

The results obtained with the principal components of the *bolus* response (first 10 min) are given in Table 3.1. The predictive power of the 10 principal components (together with P) of the *bolus* response is the highest, because they explain 95% of the variation of L and 71% of the variation of R^{-1}. Because in clinical cases the P parameter determination is not always straightforward (e.g., in some cases, the *bolus* response may not even reach 5%), the principal components-based method can be considered the best predictors for L and R^{-1}. The robustness of the controller prediction parameters, in the presence of the noise in the measurement of the *bolus* response, has been investigated in detail. All the predictors have been found to be insensitive to the presence of noise, the principal components achieved the best results. Therefore, it can be concluded that the online prediction of the controller parameters from the patient *bolus* response is a robust technique that can be used in a clinical environment. Figures 3.8a and 3.8b illustrate such robustness with the scatterplots of the predicted values for L and R^{-1} deduced from the *bolus* responses without noise (illustrated in Figure 3.4a) versus the predicted values for L and R^{-1} deduced from the *bolus* responses with added noise (illustrated in Figure 3.4b).

3.2.7 Decision Support Methods

The high level of noise that in some circumstances contaminates the neuromuscular blockade measurement forced the introduction of a nonlinear filter comprising a cascade of a three-point median filter with a fifth-order

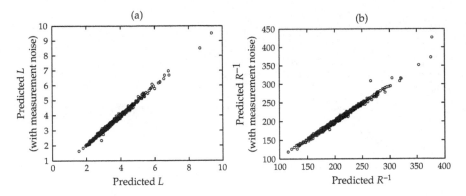

FIGURE 3.8
Scatterplot of predicted L (min) from *bolus* response without measurement noise versus with measurement noise (a) and scatterplot of predicted R^{-1} (μg^{-1} min ml) from *bolus* response without measurement noise versus with measurement noise (b). Predictor variables $T50$ and P.

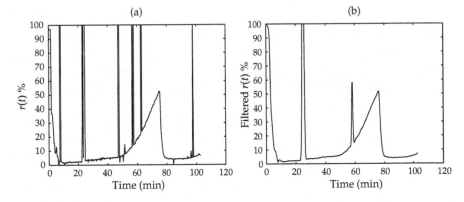

FIGURE 3.9
Faults in the measurement of neuromuscular blockade $r(t)$ (clinical case—manual control): (a) neuromuscular blockade $r(t)$ and (b) filtered $r(t)$ with the nonlinear filter comprising a cascade of a three-point median filter with a fifth-order low-pass FIR filter.

low-pass FIR filter [10,21]. However, clinical results obtained recently clearly show the presence of sensor faults ("outliers") that cannot be removed by the three-point median filter (Figures 3.9 and 3.10).

Therefore, these results motivated the development of a new filtering technique, based on Bayesian methods, more appropriate when sensor faults last for extended periods [22]. The model generating the faults is shown in Figure 3.11. Let $y(t)$ be the observed value of $x(t)$ at time t. It is assumed that $x_{min} \leq x(t) \leq x_{max}$. Under hypothesis H_0 (no fault), occurring with probability p_0, close to 1, $y(t)$ is given by

$$y(t) = x(t) + e(t) \tag{3.7}$$

where $e(t)$ is Gaussian noise of (constant) variance σ_e^2.

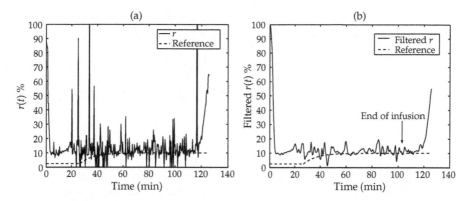

FIGURE 3.10
Faults in the measurement of neuromuscular blockade $r(t)$ (clinical case—automatic control):
(a) neuromuscular blockade $r(t)$ and (b) filtered $r(t)$ with the nonlinear filter comprising a cascade of a three-point median filter with a fifth-order low-pass FIR filter.

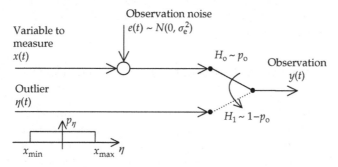

FIGURE 3.11
Model of measurement with interrupted observations.

Under hypothesis H_1 (fault), which occurs with probability $1 - p_0$, close
to zero, a measure interruption occurs. If the signal $x(t)$ to be measured is
confined between the values x_{min} and x_{max}, it is reasonable to assume that the
p.d.f. of η is uniform between these two values. This is the case of neuromuscular blockade where $r(t)$ takes values between 0% (full paralysis) and 100%
(normal muscular activity). Let Y^{t-1} be the set of observations up to $t - 1$.
According to a Bayesian approach, to detect that a given observation is actu-
ally noise, the probability $P(H_i|y(t), Y^{t-1})$ $(i = 0, 1)$ of both hypotheses, given
the observations, is computed. For H_0 this is

$$P(H_0|y(t), Y^{t-1}) = C \cdot p(y(t)|H_0, Y^{t-1})p_0 \qquad (3.8)$$

where C is a normalizing constant. Given the model of the observations
when H_0 holds,

$$P(H_0|y(t), Y^{t-1}) = C\frac{1}{\sigma_e\sqrt{2\pi}} e^{-\frac{(y(t)-x(t))^2}{2\sigma_e^2}} p_0 \qquad (3.9)$$

For computing Equation 3.9, the value of $x(t)$ is needed. Because $x(t)$ is unknown, it is replaced by an ARMAX estimate $\hat{x}(t)$. For H_1,

$$P(H_1|y(t), Y^{t-1}) = C \frac{1}{x_{max} - x_{min}} (1 - p_0) \qquad (3.10)$$

with C the same constant as in Equation 3.9. Both probabilities $P(H_0|y(t), Y^{t-1})$ and $P(H_1|y(t), Y^{t-1})$ are then compared. If

$$\frac{P(H_1|y(t), Y^{t-1})}{P(H_0|y(t), Y^{t-1})} > 1 \qquad (3.11)$$

it is accepted that a sensor fault has occurred and the observation $y(t)$ is discarded and replaced by the ARMAX estimate $\hat{x}(t)$ of the true value $x(t)$. Hence, in practice, when $y(t)$ is replaced by $\hat{x}(t)$, the set Y^{t-1} may include not only observations but also estimations.

Figure 3.12 shows clinical results obtained with the combination of the Bayesian filter (using a sixth order ARX model [22]) with the nonlinear filter. The parameters of the Bayesian filter used are $p_0 = 0.95$ and $\sigma_e^2 = 3$. As can be seen in Figure 3.12, the "outliers" have been removed and the resulting performance of this strategy is good.

3.2.8   A Guided Tour to Different Control Approaches

For clinical reasons, the patient must undergo an initial *bolus* dose to induce total muscle relaxation in a very short period of time (usually shorter than 5 min). For control purposes, the value of the reference is initially fixed at a low level during the first 30 min, being gradually raised to the set point to avoid sudden changes. The use of such a reference profile is a much better alternative to a constant one, providing a substantial improvement on the control performance. It also reflects a compromise between the variability of

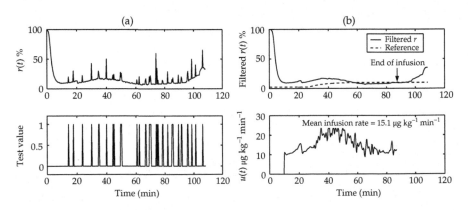

FIGURE 3.12
Clinical results obtained with the combination of the Bayesian filter (using an ARX(6) model [22]) with the nonlinear filter: (a) neuromuscular blockade $r(t)$ and test value (Equation 3.11) and (b) filtered $r(t)$ and control action $u(t)$.

the responses of the initial *bolus*, the expected noise level in the measurement of neuromuscular blockade and clinical requirements.

A digital PID controller incorporating several modifications to accommodate the characteristics of the neuromuscular blockade has been described in Ref. 10. The parameters of the PID controller (namely, the proportional gain, the derivative gain, and the integral time constant) have been obtained from the *L* and *R* parameters deduced from the Ziegler–Nichols step response method, applied to the pharmacokinetic/pharmacodynamic model [19] for the muscle relaxant. Besides, PID control parameters are adjusted to the specific clinical target by a gain scheduling technique. A simulation study indicated that the minimum value for *L* and the maximum value for *R*, deduced from the individual models for the atracurium [19,20], gave the best overall results. The high level of noise that in some circumstances contaminates the neuromuscular blockade measurement forced the introduction of a nonlinear filter comprising a cascade of a three-point median filter with a fifth-order low-pass FIR filter [10,21]. Figure 3.13 shows a typical clinical result obtained with the PID controller.

In all clinical trials of PID controller, the neuromuscular blockade level was maintained with high stability near the target value at fairly different doses, thus showing the system's adaptation to different individual requirements and circumstances. Although the controller behavior was clinically satisfactory, some results (e.g., Figure 3.14) indicated that the models on which the synthesis was based did not provide an adequate coverage of the range of patient dynamics. The responses of the patients to the initial *bolus* were fairly different (Figure 3.3a). Therefore, the information on the patient dynamics, possibly hidden in the initial *bolus* response, could be used to improve the individual tuning of the controller.

Figures 3.15 and 3.16 show the results obtained in a clinical environment with the PID controller autocalibrated using the value for *T*50 value and three

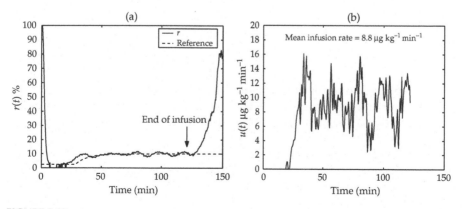

FIGURE 3.13
Clinical results obtained with the PID controller: (a) neuromuscular blockade *r*(*t*) and (b) control action *u*(*t*).

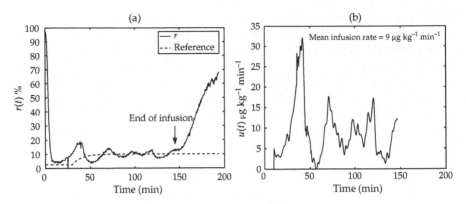

FIGURE 3.14
Clinical results obtained with the PID controller: (a) neuromuscular blockade $r(t)$ and (b) control action $u(t)$.

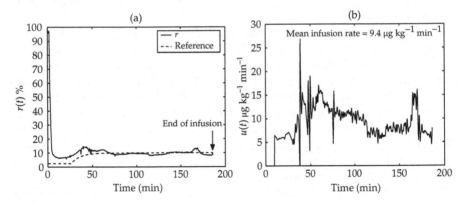

FIGURE 3.15
Clinical results obtained with the PID controller autocalibrated with the $T50$ value extracted from the initial *bolus* response: (a) neuromuscular blockade $r(t)$ and (b) control action $u(t)$.

principal components $a_1, \ldots, a_3$, respectively. The results obtained with these novel approaches have been clinically satisfactory. However, the number of cases that have been collected so far is insufficient to draw a final conclusion on the expected superior performance of the online autocalibrated methods in clinical practice, when compared to *a priori* fixed PID controller strategies. Besides, a comparison between the results of the two alternative sets of predictors would also be desirable.

As mentioned earlier, the inclusion of P as a predictor improves the quality of prediction. However, since the P parameter cannot be extracted from the *bolus* response in the majority of the cases, the relationship between P and the Walsh–Fourier periods was studied [17]. As a conclusion, it is demonstrated that P is highly correlated with $r(t^*)$, for $t^* = 14$ min. Furthermore, it is shown that P can be replaced by $r(t^*)$, for $t^* = 14$ min, in the multiple

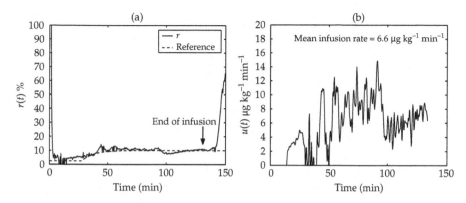

FIGURE 3.16
Clinical results obtained with the PID controller autocalibrated with the three principal components $a_1, \ldots, a_3$ extracted from the initial *bolus* response: (a) neuromuscular blockade $r(t)$ and (b) control action $u(t)$.

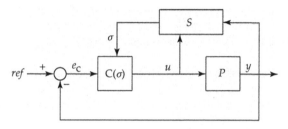

FIGURE 3.17
Switching control system.

linear regression model without a significant loss of the prediction. The use of this alternative parameter in clinical cases is clearly suggested by the simulation study carried out. The benefits of the inclusion of $r(t^*)$, for $t^* = 14$ min in the set of the predictor variables in a clinical situation are yet to be established.

Switching multiple model control has been proposed in the control of neuromuscular blockade by Neves et al. [23]. The multicontroller switching scheme is shown in Figure 3.17, as described in Refs. 24 and 25. In Figure 3.17, *ref* denotes the reference, P the process to be controlled, and $C(\sigma)$ a time-varying controller constructed from a bank of controllers $C = \{C_j, j \in J\}$. The bank of controllers is finite, i.e., $J = \{1, \ldots, N\}$. Each of the controllers, Cj, solves the tracking problem for a time-invariant model, M_j, which is viewed as a candidate for the approximate model of the process, P. The new value of σ at time t_i is computed by the selection procedure switching strategy as depicted in Figure 3.18, and corresponds to the choice of controller for which $f(e_j) = \int_0^{t_i} |e_j|^2 \, dt$ is minimum, where e_j is the identification error $e_j(t) = y(t) - y_j(t)$. The switching signal $\sigma(\cdot)$ is only allowed to change at certain time instants, t_i, such that $\inf_i \{t_{i+1} - t_i\} = \Delta > 0$, being kept constant otherwise.

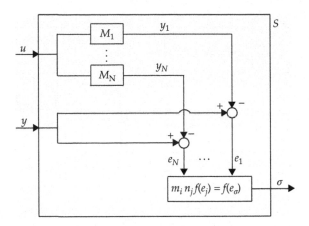

FIGURE 3.18
The switching strategy.

Switching is based on the identification error $e_j(t)$. This criterion may lead to the choice of a model–controller pair (M, C), such that C does not suitably control P, despite the proximity between P and M. This is clearly related to robustness issues.

To overcome this problem, two different approaches can be adopted:

- To use a model–controller bank such that the plant is sufficiently close to one of the models [24]
- To remove from the initial bank the model–controller pairs for which the controller is not, in some sense, "sufficiently robust"

Because stabilization is a necessary condition for reference tracking, it seems reasonable to eliminate, from the bank, the controllers that are not guaranteed to stabilize the plant P.

A controller C (with transfer function h^C) that stabilizes a nominal process (model) M (with transfer function h^M) also stabilizes the family P of processes with transfer functions of the form:

$$h = h^M\left(1 + \frac{\alpha\Delta}{\|T\|_\infty}\right), \quad 0 < \alpha < 1, \quad \|\Delta\|_\infty < 1 \tag{3.12}$$

where

$$T = \frac{h^M h^C}{1 + h^C h^M} \tag{3.13}$$

is the nominal closed-loop transfer function. It can be easily shown that a process P belongs to the family if and only if the corresponding transfer function h satisfies

$$\left\|\frac{h - h^M}{h^M}\right\|_\infty < \frac{1}{\|T\|_\infty} \tag{3.14}$$

Thus, if

$$\left\| \frac{h - h^M}{h^M} \right\|_\infty \geq \frac{1}{\|T\|_\infty} \tag{3.15}$$

the plant does not belong to the family P. Therefore, there is no guarantee that it is stabilized by the controller C. Consequently, if every pair (M, C) such that (3.15) holds (for the process under consideration) are removed from the model–controller bank, then all the remaining controllers are guaranteed to stabilize P.

Consider the family of 100 nonlinear models M_j, $j = 1 \ldots 100$ presented in Section 3.2.2 and the corresponding PID controllers C_j, $j = 1 \ldots 100$. Figures 3.19 and 3.20 illustrate the application of the switching control scheme to the processes $P = M_{27}$ and $P = M_{85}$, using the full model–controller bank (excluding P in both cases). Figure 3.19 ($P = M_{27}$) shows a typical behavior characterized by a very good overall performance. Figure 3.20 shows the results obtained for $P = M_{85}$; the performance is poor because the final pair (M_{85}, C_{21}) is "unsuitable," leading to an increased oscillatory behavior around the target value. Figure 3.21 shows the results obtained for $P = M_{85}$ using the restricted bank of controllers, C_R, satisfying the robustness condition:

$$C_R = \left\{ C_j, j \in J \backslash 85 : \left\| \frac{h - h^{Mj}}{h^{Mj}} \right\|_\infty < \frac{1}{\|T\|_\infty} \right\} \tag{3.16}$$

where h and h^{Mj} are the transfer functions of the plant and model M_j, respectively, linearized around the value of c_e corresponding to the final reference value ref $\equiv 10$. The performance achieved with the restricted bank is clearly superior as shown in Figure 3.21.

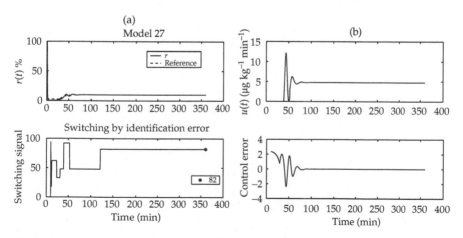

FIGURE 3.19
Results obtained for $P = M_{27}$. The pair (M_{27}, C_{27}) has been removed from the model–controller bank. The overall performance of the switching control scheme is very good. (a) Neuromuscular blockade $r(t)$ and switching signal $\sigma(t)$ (a), control action $u(t)$ and control error (b).

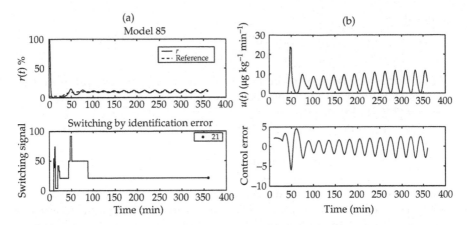

FIGURE 3.20
Results obtained for $P = M_{85}$. The pair (M_{85}, C_{85}) has been removed from the model–controller bank. The overall performance of the switching control scheme is poor. Neuromuscular blockade $r(t)$ and switching signal $\sigma(t)$ (a), control action $u(t)$ and control error (b).

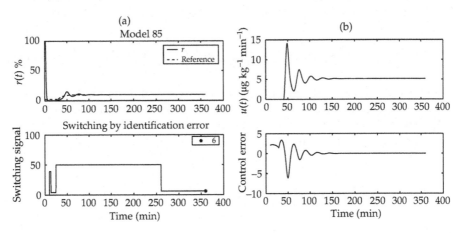

FIGURE 3.21
Results obtained for $P = M_{85}$ using the bank of controllers C_R. The pair (M_{85}, C_{85}) has been removed from the model–controller bank. The overall performance of the switching control scheme is very good. Neuromuscular blockade $r(t)$ and switching signal $\sigma(t)$ (a), control action $u(t)$ and control error (b).

As shown in Figure 3.22, the improvement is a consequence of the restriction imposed, leading to a smaller bank of model–controller pairs, by the rejection of the models M_j with a large value for

$$d(85, j) = \left\| \frac{h - h^{M_j}}{h^{M_j}} \right\|_\infty, \quad j \neq 85 \qquad (3.17)$$

The model–controller bank has to be extensive enough to accommodate the very high interindividual variability of patient dynamics to the infusion

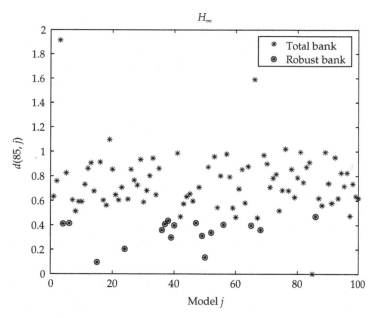

FIGURE 3.22
Norms $d(85, j)$ for $j \in J\backslash 85$.

of atracurium, and that implies that unsuitable model–controller pairs will always be present in the bank. Restrictions on the model bank to guarantee a robustness condition, as described in the earlier section, are not amenable of practical implementation. In this section, an alternative method is described and the performance achieved is evaluated.

The basic idea is simple: the unknown plant P is approximated by one of the models in the bank, M_k, and the robustness constraint assuming $P = M_k$ is applied

$$C_{\tilde{R}} = \left\{ C_j, J \in J\backslash j : d(k, j) < \frac{1}{\|T\|_\infty} \right\} \tag{3.18}$$

The selection of the model M_k, which best describes the plant dynamics, is based on the PCA of the response induced by the *bolus* given at the beginning of anesthesia [14,17]. The model in the bank, M_k, which is considered to be the best descriptor of the plant dynamics, is the nearest in the sense of the Mahalanobis distance:

$$\tilde{d}(P, k) = (a_p - a_k) \cdot \Sigma^{-1} \cdot (a_p - a_k)^t \tag{3.19}$$

where a_k is the principal components referred to in Section 3.2.7, Σ is the correlation matrix, and a_p is deduced from the patient *bolus* response (the plant response) [14,17]. A simulation study [15] indicated that results obtained with the modified robustness condition are very similar to those obtained with

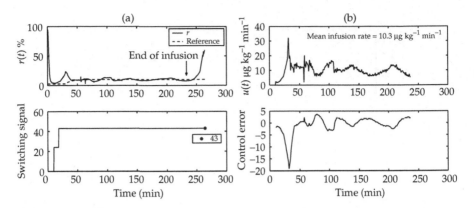

FIGURE 3.23
Clinical results obtained with the multiple model switching controller. (a) Neuromuscular blockade $r(t)$ and switching signal $\sigma(t)$, (b) control action $u(t)$ and control error.

the exact condition. Furthermore, the application of the modified robustness condition relies solely on information that can be obtained experimentally, making the practical implementation of the method feasible. Figure 3.23 shows experimental results obtained in a clinical environment using the control strategy described.

3.2.9 Robust Automatic Control System (Hipocrates)

Hipocrates [26] is an advanced software tool developed in MATLAB® for the control of neuromuscular blockade to be used on patients undergoing anesthesia. It offers a friendly graphical user interface and is easy to set up in a clinical environment. The robust control system is being used in Hospital Geral de Santo Antonio, Porto and it consists of

- *A sensor.* The Datex AS/3 neuromuscular transmission monitor to measure the neuromuscular blockade level (control is based on the T1 response, i.e., the first EMG response)
- *A delivery system.* Perfusion compact B-BRAUN with Dianet interface
- *A computer.* PC compatible computer with two serial ports for connecting the NMT sensor and the perfusion pump

At present, the package Hipocrates incorporates robust control strategies based on classical, adaptive, and switching control as well as a wide range of noise-reduction techniques and online adaptation to patient-specific characteristics, described in the earlier sections. Besides, it can be used for the control of neuromuscular blockade by continuous infusion of the nondepolarizing types of muscle relaxant drugs presently used in anesthesia, namely, atracurium, cisatracurium, vecuronium, and rocuronium. Hipocrates (software package available on request) may also be used as an advanced simulation tool for the test and comparison of different control strategies, under a wide range of situations. Therefore, it provides an excellent environment for education and training purposes.

3.3 Depth of Anesthesia: Model and Control Systems

The introduction of intravenous drugs as part of a balanced anesthesia (i.e., the use of a muscle relaxant, an analgesic, and an anesthetic) made anesthesia safer for the patient. The three components of anesthesia could be more easily adjusted to individual requirements, which improve the patient's operative and postoperative well-being.

The use of muscle relaxants in general anesthesia meant that excellent relaxation could be obtained while the patient was only lightly anesthetized. The degree of neuromuscular block can be monitored and controlled as explained in the earlier section. However, analgesia and unconsciousness (DOA) are not easy to measure. When a patient is unconscious, it is not always clear how much analgesia an agent is providing or what is the patient's DOA level. In this section, the problem of assessing and controlling the level of DOA is addressed.

3.3.1 Measurement and Control in Clinical Practice

DOA can be considered as a balance between the depression of the central nervous system (CNS) by the anesthetic drug and the stimulation of surgery. A reliable monitor of DOA is of great importance to establish effective control in the operating theater.

Different indirect monitoring methods have been investigated for DOA, such as clinical signs, electroencephalogram (EEG), and evoked potentials. Clinical signs of DOA are derived directly from the patient. Hemodynamic responses such as heart rate (HR), systolic arterial pressure (SAP), mean arterial pressure (MAP), and diastolic arterial pressure (DAP) are widely used in current anesthetic practice for DOA assessment; it is therefore likely that they carry useful information. However, the usefulness of clinical signs is reduced by drugs such as β-blockers or opioids. It is acceptable that hemodynamic parameters should be used as an extra information to assess DOA, but to rely only on them will not reflect DOA in general anesthesia [27]. Therefore, attention has turned to signals generated from within the CNS. The EEG and the auditory evoked potentials (AEP) have received a lot of attention as a credible measure for DOA, as a signal that shows similar graded changes with anesthetic concentration for different agents, as well as appropriate changes with surgical stimulus; it also indicates awareness and light anesthesia.

3.3.1.1   Bispectral Index of the Electroencephalogram

At the core of brain monitoring technology is the surface EEG. This complex physiologic signal is a waveform that represents the sum of all brain activities produced by the cerebral cortex. It has been known for decades that the EEG changes in response to the effects of anesthetic and sedative/hypnotic agents [28]. Although individual drugs can induce some unique effects on the EEG, the overall pattern of changes is quite similar for many of these agents generated from within the CNS; the EEG is not affected by neuromuscular blocking agents. The Bispectral Index (BIS) is a numerically processed, clinically validated EEG parameter. Unlike traditionally processed EEG parameters derived from power spectral analysis, the BIS is derived utilizing a composite of multiple advanced EEG signal processing techniques including bispectral analysis, power spectral analysis, and time domain analysis. These components were combined to optimize the correlation between the EEG and the clinical effects of anesthesia. Empirical and statistically derived, the BIS algorithm is the element within the BIS monitoring system that integrates and combines the EEG features, ensuring accurate interpretation of the EEG signal.

The BIS is a number between 0 and 100 scaled to correlate with important clinical endpoints during administration of an anesthetic agent. BIS values near 100 represent an awake clinical state while 0 denotes the maximal EEG effect possible (i.e., an isoelectric EEG). As the BIS value goes below 70, the probability of explicit recall decreases dramatically. At a BIS value of less than 60, a patient has an extremely low probability of consciousness. BIS values lower than 40 signify a greater effect of the anesthetic on the EEG. At low BIS values, the degree of EEG suppression is the primary determinant of the BIS value (Figure 3.24). Prospective clinical trials have demonstrated that maintaining BIS values in the range of 40–60 ensures adequate hypnotic effect during general anesthesia while improving the recovery process [29]. The Aspect Medical Systems BIS Monitor® is widely implemented in the operating theaters for monitoring the DOA level. Figure 3.25 shows the BIS monitor.

3.3.1.2   Control in Anesthesia

A reliable monitor of DOA should present similar answers with different anesthetic drugs administered to equal potency. Furthermore, it should have the ability to control anesthesia in a closed-loop system. A closed-loop control system for anesthesia allows more frequent and more accurate adjustments to the DOA. However, in clinical practice, the anesthetist still manually controls the anesthetic drugs and decides on the level of DOA by analyzing the monitored signals.

Target-controlled infusion (TCI) is an infusion system, which allows the anesthetist to select the target blood concentration required for a particular effect, and then control DOA by adjusting the requested target concentration.

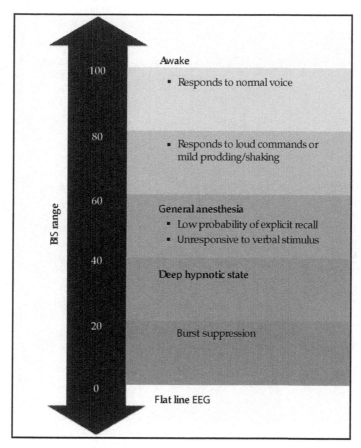

FIGURE 3.24
BIS range guidelines.

TCI systems are very much in use by anesthetists, specially because some infusion pumps incorporate TCI systems for propofol and some opioids. These pumps can be used to induce and maintain anesthesia using a specific drug. If using such a system, the anesthetist simply needs to set the initial target blood concentration required for an intravenous drug in a similar way to set the percentage concentration of an inhalational agent with a vaporizer. The target concentration is achieved and maintained with no further intervention required by the anesthetist. However, the anesthetist can make changes to the target concentration at any time. The rational administration of TCI requires an appropriate pharmacokinetic data set. A software is required to achieve and maintain a target blood concentration of an anesthetic by balancing the rate of infusion with the process of distribution and elimination. Therefore, information about the pharmacokinetic properties of the drug in appropriate patients is required. The choice of pharmacokinetic

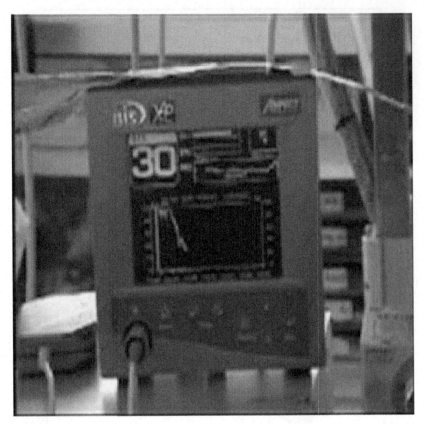

FIGURE 3.25
Aspect A2000XP BIS monitor.

model and infusion control algorithm are major determinants of the performance of a TCI system. The TCI systems are capable of creating a stable blood concentration. However, when the target concentration is changed, the resulting effect correlates better with a theoretical effect-site concentration. The efficacy of the TCI systems that can perform effect-site steering is still a process under investigation [30].

Figure 3.26 shows a data recording system in the operating room. At the moment data from the anesthesia monitor (Datex AS/3) and BIS monitor are recorded every 5 s during the surgery, and the infusion pumps for remifentanil and propofol are target controlled by the computer so that all data are synchronized. The anesthetist sets the required concentration drug target and the algorithms determine the adequate infusion rate. The drug concentrations are set independently and there is no pharmacodynamic modeling (no closed-loop software) [31–33]. This set-up is not a day-to-day practice in clinical anesthesia, in fact this is one of the few used in the world and only for research purposes.

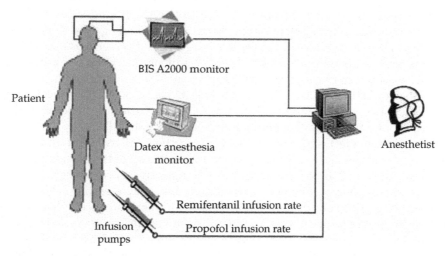

FIGURE 3.26
Diagram of the monitoring and recording system in the operating room.

3.3.2 Drug Interactions

The analgesic drug is of a high importance since it affects the pharmacody-namics of the anesthetic drug and there is no clear indicator of the degree of pain. The analgesic and anesthetic drugs are interconnected, since they interact with each other to achieve an adequate level of DOA and analgesia.

General anesthesia consists of both loss of consciousness through the action of anesthetic drugs and the inhibition of noxious stimuli reaching the brain through the acting of the analgesics. However, when two or more drugs are given simultaneously, the response may be greater or smaller than the sum of the effects of the two drugs given separately. Furthermore, one drug may antagonize or potentiate the effects of the other and in some cases, there may also be qualitative differences in response. These drug interactions provide an insight into the mechanism of general anesthesia and a practical guide-line for the optimal drug dosing during anesthesia [34,35].

The intravenous anesthetic propofol is the most used intravenous anes-thetic agent. The drug's fast distribution and rapid metabolism result in a short plasma half-life. This explains the fast and clear-headed recovery after its use and makes it an ideal drug for use in continuous infusion. One of the analgesics that is commonly combined with propofol is the new rapid-acting μ-opioid remifentanil that can be titrated to a patient's needs. Propofol and remifentanil have a synergistic relationship. The effect of the combination of these two drugs is greater than that expected as based on the concentration–effect relationships of the individual agents. In other words, a smaller amount of both drugs is needed to produce a certain effect compared to when propofol is given as a single agent. The use of remifentanil as the analgesic drug requires more attention than other analgesics. Remifentanil is a potent, short-acting opioid, and the optimal propofol concentration is much lower when combined with remifentanil compared to other analgesics. The unique characteristics of remifentanil make it a "forgiving" drug, because the return to consciousness is less postponed by higher remifentanil concentration because of the rapid decay in the effect site. This decay is much steeper than that of the other opioids and that of propofol [36]. Because of its pharmacokinetic characteristics, remifentanil may be the ideal opioid to use in combination with propofol for continuous intravenous administration [37].

Intravenous anesthetic agents affect each other's distribution and elimination. Vuyk [38] states that for propofol and the opioids, the relationship between dose and blood concentration changes by approximately 10–20% due to these pharmacokinetic interactions. He concludes that, this small variability is unlikely to be recognizable in clinical practice, because the interindividual pharmacokinetic variability of simple agents is on the order of 70–80% and interindividual pharmacodynamic variability is 300–400%. The large pharmacodynamic variability means that such interactions are of much greater importance from a clinical point of view compared with pharmacokinetic interactions.

The hypnotic should be administered to concentrations that at a minimum equal the concentration producing loss of consciousness. To inhibit somatic and autonomic responses to noxious stimuli of surgery, an opiate should be added, thereby lowering the concentration of the hypnotic. However, a ceiling effect on the reduction of the hypnotic by the opioid is reached, thus showing that opioids cannot be used as a sole anesthetic agent [34].

In general anesthesia, it is clinically important to recognize the pharmacokinetic and pharmacodynamic interaction between propofol (hypnotic) and the analgesic. The improved knowledge of these interactions can be used to optimize the quality of intravenous anesthesia and to develop practical guidelines for optimal drug dosing.

3.3.3 Research Objectives

The aim is to develop a multivariable control system to titrate simultaneously the infusion rate of the anesthetic and analgesic drugs during general anesthesia, so as to achieve and maintain an adequate level of unconsciousness and pain relief.

The analysis of the BIS signal and the hemodynamic parameters to distinguish between unconsciousness and analgesia is an important step. Allowing for online processing of the brain signals can drive an automatic control system for unconsciousness. In addition, research in the area of drug interactions is important to assure patient's safety, considering the three parts of anesthesia. The modeling of drug interactions and patients' variability on the hemodynamic responses and brain signals is of great importance when modeling the patient's behavior and developing an efficient control system.

The complexity of the situation at hand and the diversity of all the aspects involved have to be adequately integrated in an identification and control systems problem with high levels of uncertainty. The methods to be used may integrate several advanced techniques of signal processing, fault detection, modeling, and control.

3.3.4 Modeling Results: An Overview

A model for anesthetic drug interactions can prove to be very useful in understanding the full relationship between the concentrations of the two drugs and drug effects. Minto and Schnider [39] modeled the pharmacodynamic interaction for propofol and alfentanil using a response surface model for loss of consciousness, but no other effect was considered. Nieuwenhuijs et al. [40] also used a response surface model for remifentanil and propofol; measuring its effect on cardiorespiratory control and on BIS in volunteers, they found a synergistic interaction on respiration but no effect of remifentanil was found on BIS. These studies use the same modeling technique considering different effects and drugs. However, there is a clear lack of studies in this area during general anesthesia, and the use of different techniques could prove to be very useful.

The goal is to develop a patient model that could describe a typical patient's behavior under general anesthesia. This model should take into consideration the interactions between drugs, variability between patient's, and surgical stimulation. One of the objectives was to inspect if neurofuzzy techniques could be used to model the effect of the two drugs on BIS and if one could attempt to generalize between patients. If so, then this would be a preliminary but important step to understanding the effect of pharmacodynamic drug interactions on the depression of the CNSs as measured by BIS. Another objective was to determine if one could use information acquired during the induction and maintenance of anesthesia to predict recovery of consciousness (ROC), and determine if the pharmacodynamics interactions would have any effect on ROC [41]. One of the concerns of the anesthetist is to provide a comfortable and precise ROC after surgery. If the drug concentration at which ROC occurs can be estimated, then time to ROC can be predicted from the drug's elimination speed. Although these two objectives may not seem related, a model for maintenance of anesthesia would help to control the drugs administration and combination during surgery not only to maintain an adequate and stable BIS value but also allow for a fast and comfortable ROC.

3.3.4.1 *Neurofuzzy Modeling of Pharmacodynamic Drug Interactions on BIS*

The aim of this preliminary study was to try to identify the effect of propofol and remifentanil on BIS, and to determine if a model can be generalized between different patients, during general anesthesia in the presence of surgical stimuli. The data collected during three neurosurgical interventions were used to model a typical patient's reaction to the use of a combination of propofol and remifentanil. Both infusions were manually controlled by the anesthetist according to the patient's needs. Only the maintenance phase of anesthesia was considered in the modelling, which was considered to start just before incision.

A Takagi–Sugeno–Kang (TSK) fuzzy model was used to describe the effect of the interaction between propofol and remifentanil on BIS. An adaptive network-fuzzy inference system (ANFIS)[42] was used to model the parameters of the TSK model. The obtained TSK model was trained with clinical data obtained from two patients (Pat1 and Pat2) and tested with data from a third patient (Pat3). The overall model structure is shown in Figure 3.27. The drugs infusion rates were used as the input to the system. Then a pharmacokinetic model was used to determine the plasma concentration of both the drugs, independently. An effect compartment transformed the plasma concentrations into effect concentrations and which were used as the input to the ANFIS to train the fuzzy TSK model for BIS. The drugs plasma and effect concentrations were modeled independently. Therefore, no pharmacokinetic interactions were considered. This study addressed pharmacodynamic interactions reflected in the drug effects, because these are the most important in clinical practice. The pharmacokinetic models of the two drugs were constructed using a three-compartment model based on the pharmacokinetic/pharmacodynamic models for remifentanil [31,32] and propofol [33,34], respectively. These models are adjusted according to patient's age, gender, and lean body mass (LBM).

A fuzzy inference system (FIS) was used to model BIS into a fuzzy TSK model [43]. The ANFIS identifies a set of parameters through a hybrid learning rule combining the back-propagation gradient descent and the least squares method. This was used to determine the parameters for the fuzzy

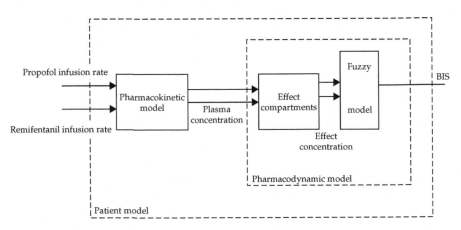

FIGURE 3.27
Block diagram of the BIS model.

TSK models [42]. The TSK model uses the effect concentrations of remifentanil and propofol as inputs and BIS as output. The data were divided into training and testing data sets. The data from patients Pat1 and Pat2 were used as the training set, and the data for patient Pat3 was used as the testing set. The ANFIS system was built through the fuzzy toolbox available for MATLAB. The following properties were found to provide the best results for the models:

- A grid partition or subtractive clustering on the training data to generate the initial FIS structure
- Gaussian input membership functions
- Hybrid optimization method

The FIS used by ANFIS was generated using a grid partition according to the specified number of membership functions. A number of 2, 3, 4, and 5 input membership functions were tested, i.e., TSK models with 4, 9, 16, and 25 rules, respectively. In addition, subtractive clustering was used to determine the input membership functions, the result was 3 input functions and a set of 3 rules. Table 3.2 presents the mean absolute error of the five TSK models on the training and testing data sets. Note that the best model for the training data set had the worst performance on the testing data set, that is, the TSK model with 25 rules was overfitted/overtrained to the training data (i.e., it is a biased model). In addition, 4 rules and 3 rules with subtractive clustering were insufficient to describe the BIS signal.

The results of the training and testing data sets for the TSK models with 9 and 16 rules are presented in Figures 3.28 and 3.29, respectively. The TSK model with 9 rules had a better performance; it reflected the data trend and captured the oscillations in the BIS signal between samples 1 and 50, and between samples 800 and 1200. The TSK model with 16 rules did not capture the initial BIS values; however, it had a better fitness between samples 200 and 800.

In conclusion, subtractive clustering did not improve the performance of the model and a large number of rules led to overtraining and lack of generalization. Overall, the fuzzy TSK model using 9 rules had a better performance

TABLE 3.2

Mean Absolute Error of the TSK Models for BIS on the Training and Testing Data Sets

TSK Models	Training Data Set	Testing Data Set
4 rules	3.62	8.25
9 rules	3.24	5.47
16 rules	3.14	5.68
25 rules	3.08	41.18
3 rules[sc]	3.86	9.54

Note: sc, subtractive clustering.

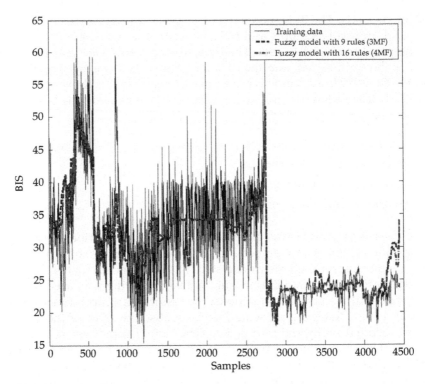

FIGURE 3.28

Results on the training data set (data from patient Pat1 and Pat2) with the TSK fuzzy models using 9 (dashed line) and 16 rules (dashed-dot line). Data points are consecutive samples. Samples from 1 to 2740 are from patient Pat1 and the last 1740 samples are from patient Pat2.

on the testing data set; however, the TSK model with 16 rules also reflected the data trend and was able to capture the synergistic interaction between propofol and remifentanil. The ANFIS was able to optimize the parameters of the fuzzy TSK models, leading to an adequate description of the effects and interaction of the two drugs. The fact that the model was trained with the data of two patients and tested with the data of a third patient is of a special relevance. Different patients have different characteristics and the model results suggest that the BIS values are independent of the patients' variability. However, this model needs to be improved using a larger set of data with more patients. The use of data from more patients can help in a better selection between the TSK models with 9 and 16 rules. In future, the final model will be used to construct a control system algorithm relating to the administration of both drugs.

The study of the interactions between the two drugs helps to determine the ideal combination of infusion rates. This will be a practical guide for the anesthetist, decreasing the amount of drug infused and increasing the comfort of the patient.

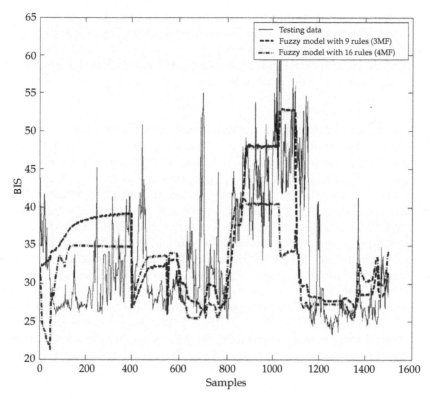

FIGURE 3.29
Results on the testing data set with the TSK fuzzy models using 9 (dashed line) and 16 rules (dashed-dot line). Data points are consecutive samples with a total of 1520 samples.

3.3.4.2 Stochastic, Fuzzy, and Neural Networks Prediction Models for Return of Consciousness

Statistical correlation analysis was performed to determine the clinical variables related to the propofol concentration at ROC. Stochastic regression models were built relating the variables with high correlation and the concentration at ROC. Radial basis functions (RBF) neural networks were built relating different sets of clinical values and the concentration at ROC. TSK fuzzy models were developed with the same objective, also using a combination of clinical variables.

Data collected during 20 surgical interventions were used to model a typical patient's return to consciousness using the effect concentration of the anesthetic drug propofol. All 20 patients were subjected to general anesthesia using the anesthetic drug propofol and the analgesic drug remifentanil. The level of unconsciousness (DOA) was manually controlled by the anesthetist using the patient's vital signs and the BIS monitor as references. The following clinical signs were recorded during the surgery every 5 s: BIS, infusion rate of propofol and remifentanil, and hemodynamic parameters.

The infusion rates were used to calculate the plasma and effect concentration of both drugs. The 20 patients studied were American Society of Anesthesiology status 1/2, 45 ± 17 years, 66 ± 10 kg, 162 ± 8 cm, 13 females, and 7 males. The data from 16 patients were used to develop the prediction models and the remaining 4 patients were used to test the performance of the models.

3.3.4.2.1 Correlation Analysis

A set of clinical variables was analyzed for correlation with the propofol effect concentration at ROC. Table 3.3 shows the correlation coefficients for minimum BIS value at induction, time to reach minimum BIS value at induction, mean BIS value during surgery, propofol mean dose during surgery, remifentanil mean dose during surgery, duration of surgery, patient age, and patient LBM. These clinical variables were chosen for several reasons. The BIS minimum value at induction represents the maximum initial CNS depression (considering that all patients received the same initial target concentration of 5 $\mu g \ ml^{-1}$). The time to reach the minimum BIS value at induction represents the speed of the initial response. These two clinical variables are related to the patient's initial response and were investigated because of the possible relation between the initial response of the patient (loss of consciousness) with his/her recovery characteristics (ROC). The propofol and remifentanil mean dose during surgery represents the dose requirements of each patient (intervariability) to maintain a stable DOA. The duration of surgery was used to evaluate if the elimination of propofol was influenced by the duration of infusion. The patient's age and LBM also represent the patient's individual parameters and may have influence on the drugs clearance and distribution. All analyzed variables are available to the anesthetist before the recovery phase and, therefore, can be used to predict. Analyzing Table 3.3, one can see that only the mean propofol dose and the mean remifentanil dose during surgery and the age of the patient had high correlation coefficients with statistical significance (*). In fact, the patient's age had the strongest correlation.

TABLE 3.3

Correlation Coefficients between Several Clinical Variables and the Propofol Effect Concentration at ROC

Variable	Correlation Coefficient	Statistical p-Value
BIS minimum at induction	0.073	0.788
Time to BIS minimum at induction (min)	0.282	0.29
Mean BIS during surgery	0.012	0.963
Mean propofol dose during surgery (mg kg^{-1} min^{-1})	0.504	0.046*
Mean remifentanil dose during surgery (μg kg^{-1} min^{-1})	0.688	0.003*
Duration of surgery (min)	0.262	0.326
Age	0.691	0.003*
LBM	0.192	0.486

* $p < 0.05$

3.3.4.2.2 Stochastic Models

Regression models were obtained for the three variables with significant correlation with the propofol concentration at ROC using the data from 16 patients (training data set). Equation 3.20 presents the stochastic model for the propofol concentration at ROC (y) using the propofol mean dose during surgery (x); this model shall be referred to as Stochastic Model 1:

$$y = 9.9556x + 0.1451 \tag{3.20}$$

Equation 3.21 presents the stochastic model for the propofol concentration at ROC (y) using the remifentanil mean dose during surgery (x); this model shall be referred to as Stochastic Model 2:

$$y = 7.6479x + 0.4164 \tag{3.21}$$

Equation 3.22 presents the stochastic model for the propofol concentration at ROC (y) using the age of the patient (x); this model shall be referred to as Stochastic Model 3:

$$y = -0.0204x + 2.2999 \tag{3.22}$$

A statistical t-test was used to determine the level of confidence in the slope of each regression model. The slope of Stochastic Model 1 was proved to be significantly positive ($p < 0.025$). Therefore, the propofol mean dose during surgery has a positive relation with the propofol effect concentration at ROC. The slope of Stochastic Model 2 was proved to be significantly positive ($p < 0.01$), meaning that the remifentanil mean dose during surgery has a positive relation with the propofol effect concentration at ROC. The slope of Stochastic Model 3 was proved to be significantly negative ($p < 0.01$), leading to the conclusion that age has a negative relation with propofol effect concentration at ROC (i.e., as age increases, the value of the propofol concentration decreases).

The mean absolute error was determined for the results of the models on the training and testing data set, i.e., the 16 patients used to develop the models and the remaining 4 patients, respectively (Table 3.4). The Stochastic Model 1 has the smallest testing error and the highest training error.

TABLE 3.4

Mean Absolute Error of the Stochastic Models for the Propofol Effect Concentration at ROC on the Training and Testing Data Sets

Model	Training Data Set	Testing Data Set
Stochastic Model 1	0.35	0.25
Stochastic Model 2	0.28	0.27
Stochastic Model 3	0.29	0.32

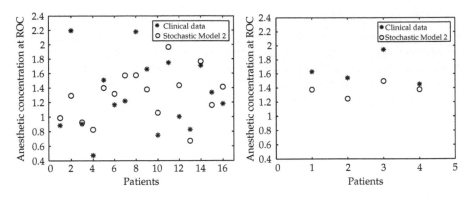

FIGURE 3.30

Results for the Stochastic Model 2 using the remifentanil mean dose during surgery (μg kg^{-1} min^{-1}) to predict the propofol effect concentration (μg ml^{-1}) at ROC on the training and testing data sets.

However, Stochastic Model 2 has a balanced performance in both sets of data. Figure 3.30 shows the results of the training and testing data set for the Stochastic Model 2.

3.3.4.2.3 *Fuzzy Models*

Three different sets of the clinical variables evaluated for statistical correlation were used as inputs to the fuzzy TSK models to estimate the propofol effect concentration at ROC. The variables with significant correlation coefficients were used, because the fuzzy logic and neural networks are more powerful for data fitting and may use extra information in a productive way. In addition, variables may have a nonlinear correlation. The fuzzy TSK models were named according to the input set used: Fuzzy Model 2 uses the propofol and the remifentanil mean dose during surgery as inputs; Fuzzy Model 4 uses the propofol and the remifentanil mean dose during surgery and patient's age as inputs; Fuzzy Model 6 uses the propofol and the remifentanil mean dose, the mean BIS value during surgery, and the duration of surgery as inputs. The data were divided into training and testing data sets. These sets are constituted by the data of 16 and 4 patients, respectively. The best performance was achieved for Fuzzy Model 2 with 25 rules (grid partition), Fuzzy Model 4 with 10 rules (subtractive clustering), and Fuzzy Model 6 with 16 rules (subtractive clustering) [41]. Table 3.5 shows the mean absolute errors for Fuzzy Model 2 to Fuzzy Model 6, considering the training and testing data sets for the structure with the best performance for each model. Figure 3.31 shows the results of Fuzzy Model 2 and Fuzzy Model 6 on the testing data set.

3.3.4.2.4 *Radial Basis Functions Neural Networks*

The RBF uses a set of clinical parameters as inputs and the propofol effect concentration at ROC as output. The RBF network models were named in a

TABLE 3.5

Mean Absolute Error of the Fuzzy Models for the Propofol Effect
Concentration at ROC on the Training and Testing Data Sets

Model	Training Data Set	Testing Data Set
Fuzzy Model 2	$3.01e-04$	0.41
Fuzzy Model 4	$3.16e-05$	0.47
Fuzzy Model 6	$1.92e-05$	0.3

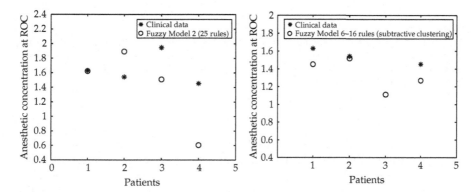

FIGURE 3.31

Results for the Fuzzy Model 2 (using the propofol [mg kg^{-1} min^{-1}] and remifentanil [µg kg^{-1} min^{-1}] mean dose as inputs during surgery) with 25 rules (grid partition) and Fuzzy Model 6 (using the propofol [mg kg^{-1} min^{-1}] and remifentanil [µg kg^{-1} min^{-1}] mean dose as inputs during surgery, the BIS mean value during surgery and duration of surgery [min]) with 16 rules (subtractive clustering) to predict the propofol effect concentration (µg ml^{-1}) at ROC on testing data set.

TABLE 3.6

Mean Absolute Error of the RBF Models for the Propofol Effect
Concentration at ROC on the Training and Testing Data Sets

Model	Training Data Set	Testing Data Set
RBF Model 2	0.09	0.65
RBF Model 4	$5.68e-06$	1.16
RBF Model 6	$1.76e-12$	0.31

similar manner to the fuzzy TSK models, according to the same three data sets used as inputs: RBF Model 2, RBF Model 4, and RBF Model 6. Table 3.6 shows the mean absolute errors for RBF Model 2 to RBF Model 6, considering the training and testing data sets. Figure 3.32 shows the results of the RBF Model 6 on the training and testing data sets, respectively.

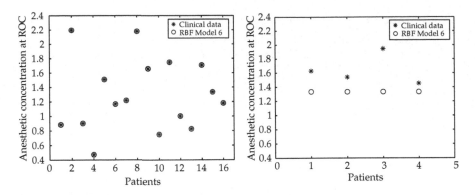

FIGURE 3.32
Results for the RBF Model 6 (using as inputs the propofol mean dose [mg kg^{-1} min^{-1}] during surgery, the remifentanil mean dose during surgery [μg kg^{-1} min^{-1}], the BIS mean value during surgery, and duration of surgery [min]) to predict the propofol effect concentration (μg ml^{-1}) at ROC on the training and testing data sets.

3.3.4.2.5 Discussion

The correlation analysis showed that duration of surgery and patient's LBM are not statistically related with the propofol effect concentration at ROC. This suggests that the clearance of propofol in the body is not related with duration of infusion (i.e., does not accumulate) and with the LBM of the patient. The drugs' mean dose during surgery proved to be important variables in the prediction process. In addition, patient's age had the highest correlation coefficient with the propofol effect concentration at ROC.

The best model structure for the training data set does not have the best performance on the testing data set. This leads to the conclusion that it may have been overfitted/overtrained to the training data (i.e., it is a biased model); this is the case of RBF Model 6. The models perform well on the training data set, reaching almost perfection. The results on the testing data set reflect the effectiveness of the models when generalizing. The clinical variables mean propofol dose, mean remifentanil dose during surgery, and patient's age were the most relevant when analyzing the propofol effect concentration at ROC using the models, in accordance with the correlation analysis. For the RBF models, the best performance in training and testing data sets was achieved with RBF Model 6; however, Figure 3.32 shows that the model does not capture the data. In fact, the RBF Model 6 gives the same output for all four patients. For the TSK models, the best-balanced performance was achieved with Fuzzy Model 6 with 16 rules using subtractive clustering. The use of subtractive clustering improved the results in some cases, since it tries to capture the data unique distribution and relation. Comparing the results of all prediction models, Stochastic Model 2 has the smallest testing error and Fuzzy Model 6 and Fuzzy Model 2 also present a good balance between training and testing data sets. These fuzzy models have good prediction properties and are able to capture efficiently the information in the data, and may be a more adequate technique for such size of data sets.

3.3.5 Multivariable Control: The Future

Different control algorithms and classification techniques are used by researchers. However, these systems are used to control the administration of the anesthetic drug (i.e., the hypnotic), while other adjuvant drugs such as the analgesic are administered by the anesthetist. Most of these systems are advisors except the ones tested in animals.

Webb et al. [44] describes a preliminary study of automatic control of anesthetic drug during surgery, in an attempt to improve the quality of control, to minimize drug usage, and to minimize the recovery time from surgery. The technique combines a fuzzy logic controller and a neural-based processing of the AEP. Mortier et al. [45] used BIS to control propofol administration by a patient individualized adaptive model-based controller incorporating TCI technology combined with a pharmacokinetic–pharmacodynamic model. The closed-loop system was able to sedate patients undergoing surgery under spinal anesthesia. This reduced the clinical workload and the amount of drug infused. Morley et al. [46] also used BIS as the target of a PID controller algorithm for drug administration, to free the anesthetist from the task of adjusting anesthetic depth manually. No clinical utility beyond this was demonstrated. Furthermore, the closed-loop system worked well in clinical practice but it did not perform better than the anesthetist manual control.

Neural networks constitute popular approaches for estimating DOA from AEP features. Nayak and Roy [47] used this approach to control inhalational anesthetic concentration (isoflurane) delivered to a patient, based on a fuzzy logic controller. The system had a clinically acceptable performance in dogs. The AEP index was also used by Kenny [27] in a closed-loop control system to administer propofol in 100 patients breathing spontaneously and also in patients who received a paralyzing drug during surgery. The quality of anesthesia was judged to be satisfactory as assessed by scores of automatic activity, cardiovascular stability, and minimal movement during surgery. No occurrence of awareness was reported. Zhang and Roy [48] used an ANFIS to integrate EEG extracted features for decision making in anesthesia. By eliciting fuzzy IF–THEN rules, the model provides a way to address the DOA estimation problem. This monitoring system was implemented in experiments with dogs under propofol anesthesia.

In the operating theater, an anesthetist still controls the DOA manually based on the clinical signs. There is still a long way to go before closed-loop systems can be implemented in clinical practice. One of the problems is the lack of a reliable patient model that can be used to develop a robust controller, and this model must take into account all the different drugs (effects and interactions), the surgical stimuli, and the patient's individual characteristics.

Majority of the researches in the area are mainly concerned with the automatic control of the anesthetic drug, whereas the analgesic is controlled manually by the anesthetist. However, in this research, the objective is a multivariable control structure for both drugs. A closed-loop control system for anesthesia allows more frequent and accurate adjustments to the DOA (i.e., automated and online measurement of DOA). As a result, better control is possible. The anesthetic drug is titrated according to the individual patient, allowing for interpatient variability of pharmacokinetics and pharmacodynamics, and responding to changes of surgical stimuli. General anesthesia would be safer for the patient as it helps to reduce the incidence of awareness and overdose, and optimizes the recovery times. If automated, the system would help the anesthetist by reducing his workload and supplementing clinical signs (which the anesthetist has access to) with information from within the CNS.

In general, propofol is used for maintenance of anesthesia in combination with an opioid, hence, the anesthetist is confronted with the dilemma of whether to vary propofol or the opioid. Zhang et al. [49] reported on a closed-loop system for total intravenous anesthesia by simultaneously administering propofol and fentanyl. They studied the interaction between propofol and fentanyl for loss of response to surgical stimuli using an unweighted least squares nonlinear regression analysis with human data. A look-up table of optimal and awakening combinations of concentrations was built, and used to determine the fentanyl set point according to the propofol set point. To our knowledge, this is one of the studies in simultaneous control of anesthetic and analgesic drugs.

Multivariable controllers are not easy to design. There are several studies on the design of multivariable controllers; however, very few target biomedical systems [50–52]. Many researches rely on mathematical models. The complexity of biomedical systems renders the task of control design more difficult. The highly nonlinear system behavior and partly unknown dynamics are the areas where fuzzy modeling and control methods can play an important role. The anesthetist's knowledge and experience may be incorporated into the fuzzy control system as a set of linguistic rules. In addition, the interaction between remifentanil and propofol introduces information that could be used to determine the adequate combination of the two drugs.

3.4   Final Remarks

The objective of this research is to solve part of the problems that concern the anesthetist in the operating theater. The study was based on clinical data gathered during surgical procedures with the muscle relaxant atracurium, the anesthetic propofol, and the analgesic remifentanil.

The development of a patient model with an adequate performance reflecting the effect of the drugs and of the surgical stimuli is one of the objectives.

A model describing the drug effect and interactions is very important for the development of a closed-loop system in anesthesia. The aim of a control system for neuromuscular blockade and DOA is to determine the infusion rates of the muscle relaxant, anesthetic and analgesic drugs, helping the anesthetist to decide which drug should be changed in response to different events.

The complexity of DOA, and the unavailability of large data sets, make this area ideal for the application of modeling and multivariable control-based concepts. The objective is to develop a hybrid structure that could provide a closed-loop simulation platform for anesthesia. This system could act as an advisor to anesthetists in the operating theater. In conclusion, a closed-loop system may

- Decrease the amount of drugs infused
- Reduce the anesthetist workload
- Be used as an alarm system
- Process and provide information about neuromuscular blockade
- Process and provide information about the brain signals (i.e., BIS or AEP)
- Detect and use the patient's individual characteristics adequately in the control algorithm
- Detect unexpected changes in steady-state conditions
- Model the interaction between propofol and remifentanil
- Administer the three drugs simultaneously, taking advantage of the existent synergism and adjusting to the presence of stimulus
- Lead to an overall increase in the patient's comfort and safety

A closed-loop simulation system helps to train the anesthetist in different aspects of the three components of anesthesia, and to develop and test different control structures. It also plays an important role in the study of drug interactions and possible side effects. However, the final objective is the implementation of such a system in the operating theater. The advantages of such an advisor system as a source of information and rapidly achieving optimal conditions would improve the quality of general anesthesia.

References

1. Wait C, Goat V, Blogg C. 1987. Feedback control of neuromuscular blockade: a simple system for infusion of atracurium. *Anesthesia*, **42**, 1212–1217.
2. Jaklitsch R, Westenskow D. 1987. A model-based self-adjusting two-phase controller for vecuronium-induced muscle relaxation during anesthesia. *IEEE Trans. Biomed. Eng.*, **34**, 583–594.

3. Lemos J, Mendonça T, Mosca E. 1991. Long-range adaptive control with input constraints. *Int. J. Cont.*, **54**, 289–306.
4. Schwilden H, Olkkola K. 1991. Use of a pharmacokinetic-dynamic model for the automatic feedback control of atracurium. *Eur. J. Clin. Pharmacol.*, **40**, 293–296.
5. Lago P, Mendonça T, Lemos J, Seabra M, Esteves S, Araújo MS. 1994. A β-robust controller for closed loop drug delivery systems: application to the infusion of atracurium in general anaesthesia. *Proceedings of IFAC Modeling and Control in Biomedical Systems* (Galveston, TX), pp. 175–177.
6. Linkens DA. 1994. *Intelligent Control in Biomedicine* (London: Taylor & Francis).
7. Kansanaho M, Olkkola K. 1996. Performance assessment of an adaptive model-based feedback controller: comparison between atracurium, mivacurium, rocuronium and vecuronium. *J. Clin. Monit. Comput.*, **13**, 217–224.
8. Mason D, Edwards N, Linkens D, Reilly C. 1996. Performance assessment of a fuzzy controller for atracurium induced neuromuscular block. *Br. J. Anaesth.*, **76**, 396–400.
9. Lago P, Mendonça T, Gonçalves L. 1998. On-line autocalibration of a PID controller of neuromuscular blockade. *Proceedings of the 1998 IEEE International Conference on Control Applications* (Trieste, Italy), pp. 363–367.
10. Mendonça T, Lago P. 1998. Control strategies for the automatic control of neuromuscular blockade. *Cont. Eng. Pract.*, **6**, 1225–1231.
11. Lendl M, Schwarz U, Romeiser H, Unbehauen R, Georgieff M, Geldner G. 1999. Nonlinear model-based predictive control of non-depolarizing muscle relaxants using neural networks. *J. Clin. Monit. Comput.*, **15**, 271–278.
12. Mason D, Ross J, Edwards N, Linkens D, Reilly C. 1999. Self-learning fuzzy control with temporal knowledge for atracurium-induced neuromuscular block during surgery. *Comput. Biomed. Res.*, **32**, 187–197.
13. Shieh JS, Fan AZ, Chang LW, Liu CC. 2000. Hierarchical rule-based monitoring and fuzzy logic control for neuromuscular block. *J. Clin. Monit. Comput.*, **16**, 583–592.
14. Lago P, Mendonça T, Azevedo H. 2000. Comparison of on-line autocalibration techniques of a controller of neuromuscular blockade. *Proceedings IFAC Modeling and Control on Biomedical Systems* (Karlsburg/Greifswald, Germany), pp. 263–268.
15. Mendonça T, Lago P, Magalhães H, Neves AJ, Rocha P. 2002. On-line multiple model switching control implementation: a case study. *Proceedings of 10th IEEE Conference on Control and Automation* (Lisbon, Portugal).
16. Manuelli C, Mosca E. 2003. A reduced-complexity adaptive switching supervisory control of neuromuscular blockade. *Proceedings of the 16th International Conference on Systems Engineering* (Coventry, U.K.), pp. 463–466.
17. Silva ME, Mendonça T, Silva I, Magalhes H. 2005. Statistical analysis of neuromuscular blockade response: contributions to an automatic controller calibration. *Comput. Stat. Data Anal.*, **49**, 955–968.
18. Stadler K, Leibundgut D, Schumacher P, Bouillon T, Glattfelder A, Zbinden A. 2003. Modeling and closed-loop control of skeletal muscle relaxation during general anaesthesia using mivacurium. *Proceedings of the European Control Conference, ECC* (Cambridge, U.K.).
19. Weatherley B, Williams S, Neill E. 1983. Pharmacokinetics, Pharmacodynamics and dose-response relationships of atracurium administered i.v. *Br. J. Anaesth.*, **55**, 39s–45s.

20. Ward S, Neill A, Weatherley B, Corall M. 1983. Pharmacokinetics of atracurium besylate in healthy patients (after a single i.v. *bolus* dose). *Br. J. Anaesth.*, **55**, 113–116.

21. Rabiner L, Sambur S, Schmidt C. 1975. Applications of a nonlinear smoothing algorithm to speech processing. *IEEE Trans. Acoust. Speech, Signal Process.*, **23**, 552–557.

22. Lemos J, Magalhães H, Dionísio R, Mendonça T. 2003. Control of physiological variables in the presence of interrupted feedback measurements. *Proceedings of 16th International Conference on Systems Engineering, ICSE'2003* (Coventry, U.K.), pp. 433–438.

23. Neves A, Mendonça T, Rocha P. 2000. A switching scheme for neuromuscular blockade control. *Proceedings of 14th International Conference on Systems Engineering, ICSE'2000* (Coventry, U.K.), pp. 651–654.

24. Morse AS. 1996. Supervisory control of families of linear set-point controllers—part 1: exact matching. *IEEE Trans. Autom. Contr.*, **41**, 1413–1431.

25. Narendra KS, Balakrishnan J. 1997. Adaptive control using multiple models. *IEEE Trans. Autom. Contr.*, **42**, 171–187.

26. Mendonça T, Magalhães H, Lago P, Esteves S. 2004. Hipocrates: a robust system for the control of neuromuscular blockade. *J. Clin. Monit. Comput.*, **18**, 265–273.

27. Kenny G. 2000. Closed loop control of intravenous agents. *EuroSiva Vienna 3rd Annual Meeting* (Vienna, Austria).

28. Thornton C, Sharpe R. 1998. Evoked responses in anaesthesia. *Br. J. Anaesth.*, **81**, 771–781.

29. Kelley S. 2004. Monitoring consciousness during anesthesia and sedation, a clinician's guide to the bispectral index (Norwood, Massachusetts: Aspect Medical Systems).

30. Van den Nieuwenhuyzen M, Engbers F, Vuyk J, Burn A. 2000. Target-controlled infusion systems: role in anaesthesia and analgesia. *Clin. Pharmacokinet.*, **38**, 181–190.

31. Minto C, Schnider T, Egan T, Youngs E, Lemmens H, Gambus P, Billard V, Hoke J, Moore K, Hermann D, Muir K, Mandema J, Shafer S. 1997. Influence of age and gender on the pharmacokinetics and pharmacodynamics of remifentanil. I. Model development. *Anesthesiol.*, **86**, 10–23.

32. Minto C, Schnider T, Shafer S. 1997. Pharmacokinetics and pharmacodynamics of remifentanil. II. Model application. *Anesthesiol.*, **86**, 24–33.

33. Schnider T, Minto C, Gambus P, Andersen C, Goodale D, Shafer S, Youngs E. 1998. The influence of method of administration and covariances on the pharmacokinetics of propofol in adult volunteers. *Anesthesiol.*, **88**, 1170–1182; 1999. The influence of age on propofol pharmacodynamics. *Anesthesiol.*, **90**, 1502–1516.

34. Glass P. 1998. Anesthetic drug interactions: an insight into general anesthesia—its mechanism and dosing strategies. *Anesthesiol.*, **88**, 5–6.

35. Olkkola K, Ahonen J. 2001. Drug interactions. *Curr. Opin. Anaesth.*, **14**, 411–416.

36. Vuyk J. 1997. Pharmacokinetic and pharmacodynamic interactions between opioids and propofol. *J. Clin. Anesth.*, **9**, 23S–26S; 2000. The effect of opioids on the pharmacokinetics and pharmacodynamics of propofol. *On the Study and Practice of Intravenous Anaesthesia* (Dordrecht: Kluwer Academic Publishers), pp. 99–111.

37. Olivier P, Sirieix D, Dassier P, D'Attelis N, Baron J. 2000. Continuous infusion of remifentanil and target-controlled infusion of propofol for patients undergoing cardiac surgery: a new approach for scheduler early extubation. *J. Cardiol. Vasc. Anesth.*, **14**, 29–35.

38. Vuyk J. 1998. TCI: supplementation and drug interactions. *Anaesthesia*, **53**, 35–41.
39. Minto C, Schnider T. 2000. Hypnotics and opioids in the elderly. *Curr. Anesthesiol. Rep.*, **2**, 482–489.
40. Nieuwenhuijs D, Olofsen E, Romberg R, Sarton E, Ward D, Engbers F, Vuyk J, Mooren R, Teppema L, Dahan A. 2003. Response surface modeling of remifentanil–propofol interaction on cardiorespiratory control and bispectral index. *Anesthesiol.*, **98**, 312–322.
41. Nunes C, Mendonça T, Amorim P, Ferreira D, Antunes A. 2004. Stochastic and neuro-fuzzy modeling for predicting return of consciousness after general anaesthesia. *Proceedings of the 3rd European Symposium on Intelligent Technologies, Hybrid Systems and their implementation on Smart Adaptive Systems, EUNITE-2004* (Aachen, Germany).
42. Jang J. 1993. ANFIS: adaptive-network-based fuzzy inference system. *IEEE Trans. Sys. Man. Cybern.*, **23**, 665–685.
43. Takagi T, Sugeno M. 1985. Fuzzy identification of systems and its applications to modeling and control. *IEEE Trans. Sys. Man. Cybern.*, **SMC-15**, 116–132.
44. Webb A, Allen R, Smith D. 1996. Closed-loop control of depth of anaesthesia. *Meas. Cont.*, **29**, 211–215.
45. Mortier E, Struys M, DeSmet T, Versichelen L, Rolly G. 1998. Closed loop controlled administration of propofol using bispectral analysis. *Anaesthesia*, **53**, 749–754.
46. Morley A, Derrick J, Mainland P, Lee B, Short T. 2000. Closed loop control of anaesthesia: an assessment of the bispectral index as the target of control. *Anaesthesia*, **55**, 953–959.
47. Nayak A, Roy R, 1998. Anaesthesia control using midlatency auditory evoked potentials. *IEEE Trans. Biom. Eng.*, **45**, 409–421.
48. Zhang X, Roy R. 1999. Depth of anesthesia estimation by adaptive-network-based fuzzy inference system. *Proceedings of the 1st Joint BMES/EMBS Conf., IEEE Eng. in Med. and Bio. 21st Annual Conf. and the Annual Fall Meet. of the Bio. Eng. Society*, Vol. 1(Piscataway, USA), p. 391; 2001. Derived fuzzy knowledge model for estimating the depth of anesthesia. *IEEE Trans. Biom. Eng.*, **48**, 312–323.
49. Zhang X, Roy R, Huang J. 1998. Closed-loop system for total intravenous anesthesia by simultaneously administering two anesthetic drugs. *Proceedings of the 20th Annual Int. Conf. of the IEEE Eng. in Med. and Bio. Society*, Vol. 20 (Piscataway, New Jersey), pp. 3052–3055.
50. King R, Magoulas G, Stathaki A. 1994. Multivariable fuzzy controller design. *IFAC Cont. Eng. Prac.*, **2**, 431–437.
51. Viljamaa P. 2000. Fuzzy gain scheduling and tuning of multivariable fuzzy control—methods of fuzzy computing in control systems. *PhD Thesis* (Tampere University of Technology: Tampere, Finland).
52. Yeh Z. 1999. A systematic method for design of multivariable fuzzy logic control systems. *IEEE Trans. Fuz. Sys.*, **7**, 741–752.

4

Control Methods and Intelligent Systems in the Quality Control of Megavoltage Treatment Machines

John A. Mills, Robert C. Crichton, Phil Sharpe, and William M. Kelly

CONTENTS

4.1 Overview

Quality control (QC) in radiotherapy has become increasingly significant over the past 15 years. It has been recognized that to achieve and maintain high standards of safe and effective dose delivery there is a need for both quality assurance systems and quality control checks. The international committee for radiation protection (ICRP) [1] recognized the need for protection of the patient in radiotherapy and the role that QC played in this. The hazards of radiation to normal tissue were well known and the risk of oncogenesis [2–4] further complicates the use of radiation in cancer treatment. These concerns were reinforced by accidents [5], some of which were due to safety failures, others due to performance failures, and of course some due to human error. It has been clear that the absence or deficiencies of QC had an adverse effect in some of these accidents. Many accidents highlighted the need for standards to be established in both safety and performance.

The acceptable variation in dose delivery has been identified, from clinical experience and radiobiological studies, to be ±5% [6,7]. This gives some perspective on, for example, the Exeter accident in 1988 [8] when a 25% overdose was delivered to many patients between February and July. Systems of quality control checks thus endeavor to ensure that treatment equipment operates both safely and effectively.

Radiation dose is delivered by several techniques in radiotherapy. These techniques take three main physical forms. The first and oldest technique utilizes radioactive sources either introduced into a body cavity or surgically inserted inside the tumor or target volume. Cavities such as the esophagus, vagina, uterus, mouth, bronchus, or rectum are used and such treatments are still commonplace as they provide the radiation dose very locally with little irradiation of normal tissue. To insert radioactive sources into the tumor, needle guide devices are used and lesions in the tongue, rectal wall, and superficial tumors are treated in this manner when very localized dose delivery is required.

A second technique utilizes radioactive substances, which are administered to the patient orally and intravenously or injected into a body cavity such as the knee joint. With this therapy, the isotope itself or the pharmaceutical to which it is attached becomes part of the metabolic process of the organ or tissue to be treated. The result is that although there is a widespread distribution of the isotope throughout the body there is also a concentration and retention at the site of treatment.

The third and most widespread radiotherapy technique, accounting for over 95% of treatment involves the irradiation of the tumor by high-energy photon beams, typically in the range between 5 and 25 MV. This is referred to as teletherapy with the radiation beam being fired through normal tissue to the target. Superposition of several such beams provides a high dose at the target while sparing the normal tissue an unacceptably high dose.

Although there is a variety of equipment associated with all these techniques, the teletherapy equipment is extremely complex and widespread (Figure 4.1).

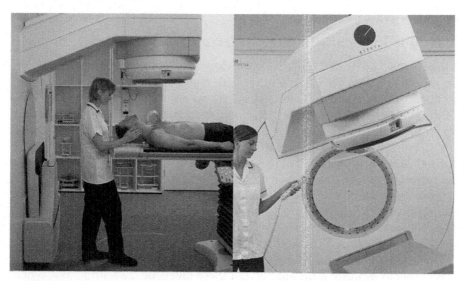

FIGURE 4.1
An Elekta precise medical linear accelerator for megavoltage teletherapy.

These megavoltage teletherapy machines require self-regulatory control systems to accelerate electron beams, produce and control the photon beams, and direct these beams onto the target accurately and reproducibily.

With more than 95% of radiotherapy treatment provided by these types of machines, the extent of QC required to cover all aspects of their complexity and the volume of work to cover so much equipment is considerable.

The international electrotechnical committee (IEC) has published standards to cover the two main aspects required for QC: Safety under IEC60601:2.1 [9] and performance under IEC60601:3.1 [10,11]. Both aspects are essential for safe and effective treatment. This chapter looks at the potential and problems of applying processing techniques to the QC of megavoltage radiotherapy treatment machines.

Radiotherapy, often thought of as a somewhat crude technique for the treatment of cancer, has continued to be used extensively despite the alternative developments in medical and surgical techniques, such as chemotherapy, hormone therapy, and for example, functional magnetic resonance imaging (MRI)-guided neurosurgery. Radiotherapy often provides an adjuvant contribution in a multicomponent treatment. Technical developments, particularly in computing, have also led to significant advances in the way that all forms of radiotherapy are delivered and planned. In teletherapy, this has led to very complex treatment machines, which are required to provide a high level of performance if treatments are to fulfill the anticipated improvements. This complexity of performance arose from the desire to shape and modulate the intensity of the radiation beam. In turn, the machines require more effective QC, which can rapidly determine deterioration in performance within ever-tighter limits. In short, the demand for effective QC has grown.

4.2   Quality Control of Radiotherapy Treatment Machines

Effective QC needs to fulfill two roles. One is to demonstrate the correct performance of a system and the other is to detect deterioration of the system to take corrective action. Demonstration and detection are in fact two sides of the same coin. The boundary between each side is set in accordance with appropriate limits; see Figure 4.2.

QC immediately following a repair of equipment demonstrates that performance has been restored or it detects problems with the repair. Constantly monitoring a parameter demonstrates performance until the point at which it is considered that unacceptable performance has been detected. Hence, the data collected through QC measurements can be used not simply in a binary fashion but to examine trends in performance and if possible use this as predictor for maintaining the performance of equipment.

One routinely used but simple trend technique is when a double tolerance level is set to monitor a parameter. Let these levels be $\pm a\%$ and $\pm b\%$, where $b > a$. Results outside $\pm b\%$ require immediate attention whereas results outside $\pm a\%$ and also within $\pm b\%$ must occur on a number of subsequent occasions for action to be taken.

There are three major categories of QC. Each involves drawing conclusions from measurement data obtained from different methods of monitoring the machine: immediate monitoring, short-term monitoring and long-term monitoring. Each provides different but appropriate means to evaluate and maintain machine performance.

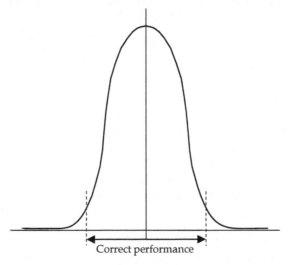

FIGURE 4.2
Illustration of the assumed variation in a performance parameter.

Immediate monitoring. In this category, the immediate effect on the performance of the machine is determined from response to changing a control parameter. Using this approach, the optimum tuning of a circuit or system can be determined. The monitoring of the machine can involve both external and internal output signals. Internal parameters, which alter as a consequence of the changed parameters can provide system response information and externally monitored characteristics of the beam, such as beam symmetry and energy.

Long-term monitoring. Many parameters on a machine alter slowly as components age or their performance deteriorates. To correct and compensate for the deterioration, including replacement of a component, long-term data trends provide an indication as to when action is required. For example, this may be used to indicate the need to adjust a parameter, which requires adjustment continuously over time to maintain the adequate performance, or it might indicate when the end of life of the device can be expected to schedule replacement and avoid treatment delays and interruption due to machine breakdown or lack of component availability. The data might again come from internal parameters of the machine or from measurements made independent of the machine. Inevitably, data is difficult to record, log, and process. In particular for measurements on the radiation beam, the results cannot be gathered in an entirely automatic manner and any system, which can assist with the recording and processing of this data is invaluable.

Short-term monitoring. In contrast to immediate monitoring and long term monitoring, short term endeavors to determine the performance of the machine during a treatment day. Only internal parameters of the machine are recorded during the day for subsequent processing. Analyzing such recorded data makes it possible to identify the deteriorating overall performance of the machine and schedule periods for attending to the machine. Although only internal parameters are monitored, it can be useful to compare the processed results with appropriate daily measurements made of external beam characteristics. This method attempts to exploit the enormous amount of data available from the machine during the treatment period.

This chapter provides three insights; one into each category of monitoring and analysis for QC. A systematic approach to the development of QC in the earlier context is only in its infancy. In radiotherapy there is a need to progress from simply seeing QC as a fault detection system and embrace it as an indicator, which can direct effective intervention to maintain machine performance and contribute to the improvement of radiotherapy treatment apace with other technological advances. Although none of these insights can be totally comprehensive, they endeavor to demonstrate in real terms some of the problems and potential for the future QC of radiotherapy treatment machines.

4.3 Gun Current Tuning: Immediate Monitoring Quality Control

Medical linear accelerators are radiotherapy treatment machines, which produce high-energy beams of electrons or photons [12]. In electron mode they are used to treat superficial lesions and in photon mode to treat deep-seated tumors while delivering a low dose to the skin. The photon beam is generated by high-energy electrons striking a tungsten target; thus both modes require an accelerated electron beam. The electrons are generated by thermionic emission from a cathode formed from a tungsten filament. On an Elekta SL linear accelerator (Elekta Limited, U.K.), the filament wire is 0.31 mm in diameter and the total diameter across the filament spiral is 6.0 ± 0.25 mm. A normal operating current of approximately 7.5 A is applied to the wire to raise its temperature and release electrons that are injected into a traveling waveguide. In the waveguide, the electrons are accelerated with microwaves that are generated by a high power magnetron. As the electron beam travels down the waveguide it is focused and steered by sets of magnetic coils around the waveguide. At the end of the guide, the accelerated electron beam passes through a thin copper window in electron mode and in photon mode it strikes a tungsten target, producing a high-energy photon beam (x-rays); see Figure 4.3.

Linear accelerators are continuously subject to changes in operating conditions. These include variation in ambient temperature, local geomagnetic fields, mechanical deflection, and the electrical circuitry, which may not be perfectly stable. All these changes have a significant impact on the radiation beam produced and its control mechanism. To ensure that the quality of

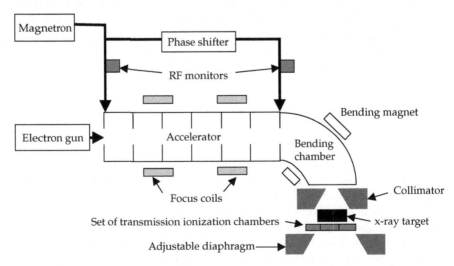

FIGURE 4.3
Schematic diagram representing the major components of a medical linear accelerator.

the radiation beam delivered to the patient is acceptable, control parameters have to be adjusted to compensate for changing conditions.

One of the most important factors that ensures the accuracy and quality of treatment is the need for a consistent radiation beam. The filament wire, which forms the electron gun of the linear accelerator is gradually thinned due to the loss of material. As the wire thins, the filament resistance increases raising the filament temperature for a fixed filament current. This in turn increases the number of electrons released into the accelerating waveguide. Because the microwave power that is injected into the guide is constant, the final energy per electron will decrease. To maintain a constant electron beam current energy, the gun current has to be reduced as the filament ages.

During normal operation the electron beam current is controlled by a servo system. For the correct operation of the servo, the aim level, which is the point around which the servo operates, is manually adjusted on a regular basis as the gun filament ages. The influence of the adjustment of the aim level for the gun is critical for optimal operation of the servo system. Provided that the gun aim is within ±0.15 A from optimal, the servo will operate correctly. However, outside this range, the gun control mechanism fails to adequately control the beam. This gives rise to poor performance and an inconsistent beam for treatment.

To illustrate one of the effects that suboptimal machine running parameters have, radiation beam symmetry has been shown to be sensitive to changes in machine performance. This is shown by the time required for the radiation beam to achieve a steady symmetrical profile. This is shown in the following graphs of radiation beam symmetry. The symmetry assessment was performed using a custom built, full field radiation monitoring device. Measurements were taken at the centre of the field and at positions ±9 cm and ±19 cm from the centre towards the machine gantry and away from the machine gantry. This line of measurement is routinely used to check the performance of the beam and is referred to as the GT axis. Changes were made to the gun aim level to give optimal and suboptimal set-up conditions. Figure 4.4 shows the response for an optimal gun aim setting whereas Figure 4.5 shows the response for a suboptimal gun aim setting. The corresponding real-time variation in microwave (Mag Tune), gun current (GunI), and Doserate (Plate Sum) machine parameters were taken directly from the accelerator and are shown in Figures 4.6 and 4.7, respectively.

For optimal gun aim, beam symmetry is achieved between 1 and 5 s of the radiation start. It can be seen that for a change of ±0.05 A from optimal gun aim level, the required beam symmetry is achieved between 5 and 10 s. If the aim level is changed by +0.1 A, beam symmetry is not achieved within 15 s. Changes in the aim level can produce greater or less variations in beam profiles orientations other than that measured.

The foregoing results demonstrate that beam symmetry in the first 15 s is sensitive to small changes in one of the machine's major control systems of the linear accelerator. With the optimal gun aim (Figure 4.6), both Gun I and Plate Sum values reach their steady state within approximately 5 s of the gun

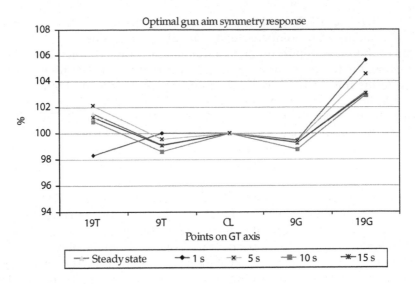

FIGURE 4.4
Start-up field symmetry for optimal gun aim setting.

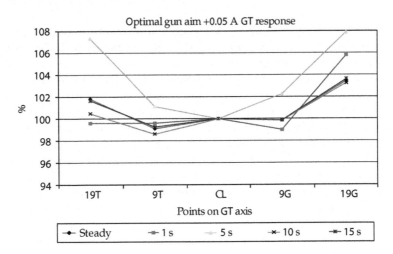

FIGURE 4.5
Start-up field symmetry for a change of +0.05 A from optimal gun aim setting.

being switched on. However with a suboptimal gun aim (Figure 4.7), both Gun I and Plate Sum outputs take much longer time to reach their steady state.

A noticeable difference in the Plate Sum output of the linac when it operates with suboptimal gun aim level can be seen. Applying a systems modeling approach to the data obtained directly from the machine, a model has been found to exist between the gun current and the output of the Plate Sum.

A systems modeling approach, using a single-input-single-output system was used to evaluate if a model could be used to determine whether the linear accelerators gun current was operating at an optimal value.

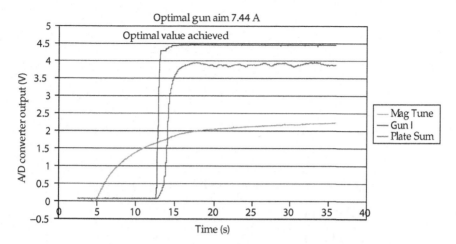

FIGURE 4.6
Real-time machine running parameters for optimal gun aim.

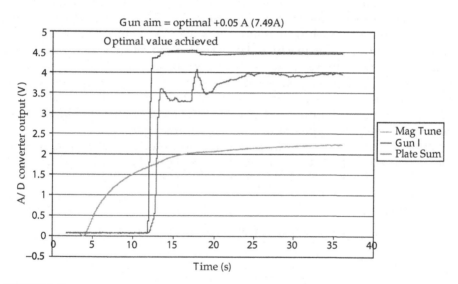

FIGURE 4.7
Real-time machine running parameters for suboptimal gun aim.

A model having an ARMAX structure is adopted as follows:

$$A(q^{-1})y(t) = q^{-k} B(q^{-1})u(t) + \varepsilon(t)$$

where the polynomials $A(q^{-1})$, $B(q^{-1})$ are defined by

$$A(q^{-1}) = a_0 + a_1q^{-1} + a_2q^{-2} + \cdots + a_{n_a}q^{-n_a} \tag{4.1}$$

and

$$B(q^{-1}) = b_0 + b_1 q^{-1} + b_2 q^{-2} + \cdots + b_{n_b} q^{-n_b} \tag{4.2}$$

with

$$a_0 = 1, \quad b_0 \neq 0$$

where

n_a and n_b are, respectively, the order of $A(q^{-1})$ and $B(q^{-1})$, with $n_b \leq n_a$

$k = 1$

q^{-1} is the backward shift operator defined by $q^{-i} y(t) \equiv y(t - i)$

$u(t) = $ sampled input

$y(t) = $ sampled output

$\varepsilon(t) = $ noise signal

System parameters were estimated using the CALCE software package [13] applied to the sampled input and output data obtained from linear accelerator measurements. These measurements were made for optimal and suboptimal operation of the machine. The ARMAX model structure was varied and the best estimate was found to be a third-order model with $n_a = 3$ and $n_b = 2$. The resulting mean coefficients from ten measurements at each gun aim current setting are given in Table 4.1. This indicates the variation for changes in gun aim within ± 100 mA. Table 4.2 indicates the standard deviation for the coefficients determined for the optimal gun aim.

Figure 4.8 shows the variation of coefficients a_1, a_2, and b_0 graphically for comparison. This clearly shows that taking into account the standard

TABLE 4.1

Model Parameters (Mean) for Optimal and Suboptimal Gun Aim

Gun Aim Change	a_1	a_2	a_3	b_0	b_1
0.1	−2.1217	1.5051	−0.4452	−0.0419	0.0592
0.05	−2.0623	1.0417	−0.1843	0.0268	0.0226
0	−1.8148	0.7278	−0.0338	−0.0199	0.0505
−0.05	−1.5172	0.5648	0.0686	0.0363	−0.0246
−0.1	−1.2154	0.0196	0.3187	0.016	−0.0124

TABLE 4.2

Standard Deviation of Coefficients for Optimal Gun Aim

a_1	a_2	a_3	b_0	b_1
0.1204	0.6425	0.2458	0.0387	0.0478

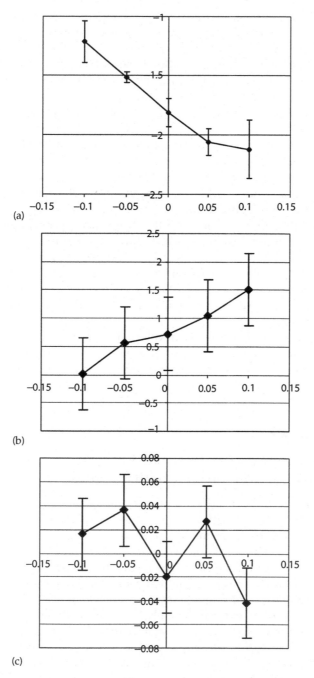

FIGURE 4.8
ARMAX coefficients with change from optimal gun current (a) a_1, (b) a_2, and (c) b_0.

deviation of the coefficients over the measurements, which were made, the a_1 coefficient can be used as an indicator that optimal gun aim is not set on the linear accelerator from -100 to $+50$ mA.

When the gun aim is below or above the optimum, a_2 and the absolute value of a_1 are below or above the value obtained for the average model with an optimal gun aim. Although the other parameters are also found to vary, it is the consistent nature of the variation in a_1, which is considered to be a good indicator.

These models can be used to determine automatically whether the gun aim needs to be adjusted for the linear accelerator to operate optimally and whether it should be increased or decreased. Further work includes the refinement of the model and a fuzzy-based approach to assess the effect of parameter variation during the day.

Improvements and refinements to this approach may include the use of additional machine parameters, which have a critical effect on machine performance. A system to automatically detect nonoptimal machine performance could be developed. This system could be used to indicate when the machine is performing poorly or when a servo system is working hard to achieve optimal conditions. Such a system would serve as a very useful tool for the service engineer.

This approach could be further developed with a system that would not only monitor machine parameters to indicate its working condition but also make adjustments to restore its performance when required. In effect a self-tuning linear accelerator.

4.4   Beam Steering, Logging, and Analysis: Short-Term Monitoring Quality Control

By recording machine control parameters during the daily operation of the machine in routine treatment, it may be possible to predict future faults and performance deterioration; thus, action must be taken before they become a major problem such as breakdown.

The short-term monitoring technique forms the basis for providing the data to be studied in such a project. The realization of the work described in the following section was made possible by a facility developed at Coventry Radiotherapy and Oncology Center to routinely record the main beam control parameters of a clinical radiotherapy accelerator during daily treatments. Short-term monitoring of the main beam control parameters was implemented on one of the 6-MV Philips SL75-5 machines using an industrial standard logger and in-house signal conditioning electronics. Modern computer-controlled accelerators can be easily configured to accumulate large amounts of such data, but neither the use and analysis of that data has been studied nor its potential realized; and linear accelerator manufacturers are now aware of this potential.

4.4.1 Beam Control Parameters Investigated

The main beam control parameters that were studied in this work are as follows:

Gun current. An increase or decrease in the gun filament current increases and decreases the number of electrons emitted into the accelerator guide. The radio-frequency (RF) energy fed into the guide remains constant and so the final energy of the accelerated electrons is decreased or increased depending on the number of electrons emitted. Owing to the 90° bending section in these machines, a change in the trajectory onto the x-ray target that affects the symmetry of the beam takes place.

2R current. Changes in the 2R beam centering or steering coil currents create changes in the beam symmetry due to beam alignment in the GT direction (Figure 4.9).

2T current. Changes in the 2T beam centering or steering coil currents create changes in the beam symmetry due to beam alignment in the AB direction (Figure 4.9).

Gantry angle. Movement of the gantry affects the waveguide, which is housed inside the gantry arm. The change of the waveguide in space means that the electron beam is acted upon by changing forces dependent on the gantry angle, which can affect beam energy and symmetry. Gravity and mechanical movements between sections of the guide structure can affect the beam. For example, the magnetron filament moves when the gantry rotates, which changes the impedance of the magnetron and increases or decreases the energy transferred into the guide.

It is known that these parameters affect machine performance and may interact with each other in many ways. These parameters were monitored throughout the clinical operation of the treatment machine using an industrial standard data logger attached to the linear accelerator on a daily basis.

The data from this study was obtained for over 3 years and each treatment day during that time was monitored and recorded. On average, 21,500 readings were taken for each treatment day.

The data logger (Figure 4.10) was a Smartreader 7 Process Signal Logger manufactured by ACR Systems. The device itself is approximately 11 cm × 6 cm × 2 cm in size, which is built using solid state components including an 8-bit analog to digital converter, an accurate clock crystal, and a thermistor for ambient temperature measurement. It is powered by a 10-year life span lithium battery. It has seven input channels, of which five record voltage and two record currents.

Each of the four beam parameters were processed and calibrated into usable values (i.e., current for Gun, 2R and 2T, and angle for gantry).

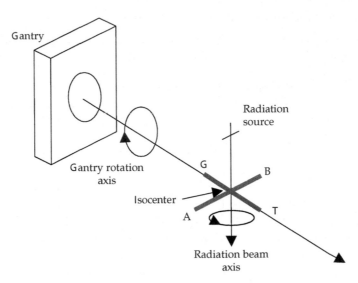

FIGURE 4.9
AB and GT directions on a linear accelerator.

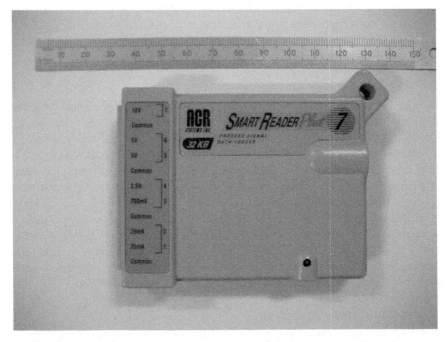

FIGURE 4.10
Smart reader data logger.

The definition of the normal operation of a linear accelerator was an important issue to resolve during this study. Defining the "normal" operating times of the machine dictates the abnormal operating times and creates a benchmark of comparison for the data obtained.

4.4.2 Satisfactory and Unsatisfactory Periods of Machine Performance

It was necessary to build up a log or diary of the accelerator's running incidents over the 3 years it was being monitored to use as an indication of the normal and abnormal periods of operation. These incidents included adjustments, faults, breakdowns, and component replacement (Table 4.3).

The next stage was to pick out major breakdowns and changes in the list, which might have had a greater influence on the beam control parameters being logged. It was believed that a change in the main structure or main components of the beam control section might have the greatest effect on the beam due to the possible physical changes in this area.

4.4.2.1 Possible Major Machine Replacements

- *Magnetron replacement.* Changing the magnetron would change the optimum frequency for the RF field created to transport the electrons along the guide and could result in new running parameters for the magnetron.
- *Gun filament replacement.* Changing the filament would change the current produced by the gun and affect the energy of the electrons

TABLE 4.3

Record of Breakdowns and Component Replacements on the SL75-5

Date	Problem Occurred	Action Taken
01/08/95	Ion chamber failed	Ion chamber replaced
05/01/96	Gun current running low	Gun filament replaced
14/02/96	Reverse diode overload, noisy magnetron	Magnetron replaced
14/05/96	Bending section	Bending section replaced
13/08/96	Reverse diode overload and noisy magnetron	Magnetron replaced
11/10/96	Magnetron change required	Magnetron replaced
18/04/97	Fluctuating gun current	Gun filament replaced
21/05/97	Low fluctuating dose rate	Magnetron replaced
21/06/97	Magnetron change required	Magnetron replaced
05/08/97	Gun current running low	Gun filament replaced
25/08/97	Ion chambered failed	Ion chamber replaced
12/06/98	Unstable 2R current	Magnetron replaced
17/09/98	Gun current running low toward predicted level	Gun filament replaced
16/10/98	Unstable ion chamber readings	Ion chamber replaced
15/07/99	Flatness faults, 2R instability, and low gun current	Gun filament replaced
31/07/99	Flatness faults, 2R instability	Magnetron replaced

coming from it. It might also affect the collection of electrons into the waveguide and their trajectory on entering it.

- *Ion chamber replacement.* Monitoring the beam that comes from the radiation head. Therefore, a replacement can have different working tolerances and running parameters.
- *Bending section replacement.* The bending section is where the radiation beam is bent through 98° to strike the target and produce the x-ray beam through the radiation head. Therefore, a new bending section may have structural differences and different magnetic tolerances.

During each section of result analysis on the data logger, it was noticed that 2R was a significant factor in the results. The 2R parameter was chosen because of its known variation, sensitivity, and interaction with the gun current and gantry angle. As the gun current slowly decays through use and age, 2R must alter to match its change because it controls the radiation beam in the GT direction. The 2R current is also one of the most common beam control parameters being adjusted on the linear accelerator on a regular basis. By defining these periods of frequent adjustment as unsatisfactory periods of operation, it was possible to define stable periods of operation as satisfactory.

Some rules of monitoring were created that could be easily performed each night after the cessation of a day's treatment and adjustments made the following morning before that day's treatment.

The following sections discuss the procedures undertaken to determine whether it was possible to predict faults from the daily recording of the data logger or not.

4.4.3 Studies to Predict Faults from Short-Term Data Logging

To evaluate if the logged data could be used for future predictions, a number of data analysis were undertaken. One was an analysis of the daily variance of a particular machine parameter. The others examined how the mean of this daily parameter correlated with itself, under different machine conditions, and with a routinely measured parameter. The following section describes the data analysis, which was applied, and the indicators provided by this.

4.4.3.1 Variation of Parameters with Gantry Angle

For the satisfactory period, the mean values of currents at all major gantry angles: 0°, 90°, 180°, and 270° were correlated. For this period the correlation coefficient was found to be 0.65. However, during what was judged to be an unsatisfactory period, the correlation coefficient was 0.12. This indicated that the correlation between the steering currents for different gantry angles was sensitive to the satisfactory performance of the machine.

4.4.3.2 Determination of 2R Standard Deviation and Action Level Cut-Off Point

Because the machine was used extensively at gantry angle 0°, most data occurred at this gantry position and so data from this gantry angle was studied for most of the analysis.

The values of 2R were recorded for the day's treatment at the same time as the other three parameters involved in the project and downloaded from the data logger every night. The 2R current was isolated and matched up with the different gantry angles the machine had been set up on that day. The mean of each 2R value at gantry zero was then calculated with a corresponding standard deviation to show how the 2R current varied. The range of values of 2R at particular gantry angles shows how much the current has altered throughout the day; and this is indicated by the standard deviation.

Visual examination of the standard deviation of the 2R steering current values, with the gantry at 0° during the satisfactory and unsatisfactory periods, suggested that there was a difference. This was borne out in a statistical assessment of the incidence of the standard deviation during these periods as shown in Figure 4.11. The range of standard deviation was greater during the unsatisfactory period and the frequency of occurrence of these higher variances was greater.

A cut-off point for the standard deviation was determined as a rule of adjustment, which could be adhered to for making a decision on whether to adjust or not. To come up with a relevant cut-off point, the satisfactory period of the machine's performance was examined and the two peaks identified (Figure 4.12). Peak 1 was identified as representing the minimal variation of the machine. The 2R current mean value and its corresponding standard deviation were determined to set a reasonable cut-off level.

The range of standard deviations was then split evenly and the first half was used to determine a mean standard deviation (Table 4.4).

The calculated limit was based on ±2 standard deviations of the mean standard deviation derived from the histogram of Peak 1. Therefore, when deciding whether to adjust the machine or not, the standard deviation calculated for a particular day was checked with the cut-off limit of 0.025; and this was used to make the final decision.

4.4.3.2.1 Daily Monitoring

Once the adjustment level had been set, the machine was checked every night for its mean 2R value and its corresponding standard deviation at gantry zero to make an adjustment decision.

This adjustment procedure was carried out continuously for approximately 3 weeks until a sufficient amount of data had been obtained. The machine was then left alone for another 3 weeks and the data logger information was taken but not acted upon using the 2R standard deviation rule. This was done so that a comparison between the presence and absence of daily monitoring could be done to see any observable differences (Table 4.5).

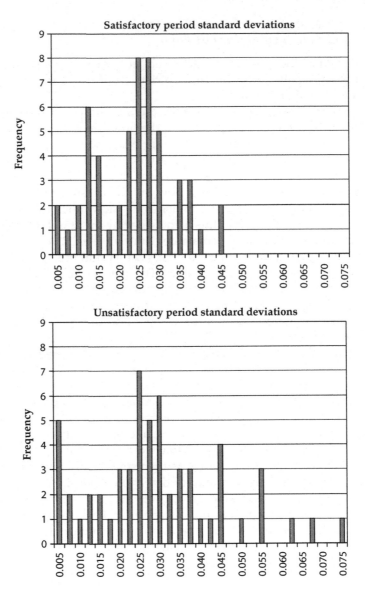

FIGURE 4.11
Frequency distributions of 2R standard deviations during satisfactory and unsatisfactory periods of operation.

There was a higher percentage of readings taken than would have required adjustment in the "no monitoring" section. When the machine was being monitored the readings above and below were evenly split. The machine was actually adjusted twice during the no-adjustment period due to significant 2R variations during normal quality assurance checks. The monthly QC was performed during this time and a symmetry and flatness check was done,

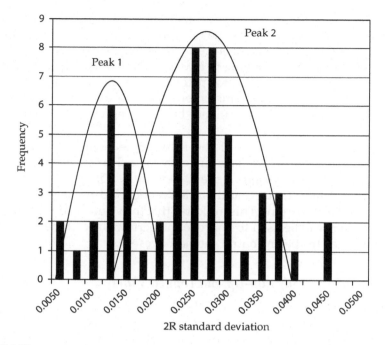

FIGURE 4.12
Identifiable peaks within the satisfactory period.

TABLE 4.4

Standard Deviation of Frequency Distributions to Calculate Cut-Off Point

Overall mean 2R standard deviation	0.023
Mean 2R standard deviation for Peak 1	0.014
Standard deviation of mean 2R standard deviations for Peak 1	0.006
Set limit for 2R standard deviation (+2 s.d.)	0.025

TABLE 4.5

Comparison of 2R Standard Deviation during Adjustment and No-Adjustment Periods

	Period of Monitoring and Adjustment	Period with Monitoring and No Adjustment
Number of days with standard deviations above the tolerance level	13	16
Number of days with standard deviations below the tolerance level	13	6
Percentage of days above the tolerance level (%)	50	73
Percentage of days below the tolerance level (%)	50	27

which revealed the need for the gun aim to be rebalanced. Later, 2R was running out of tolerance during the morning run-up, so a symmetry check was performed and 2R was rebalanced. These two dates coincided with the 2R standard deviations dropping to lower levels. This could not be avoided as the machine was being used for patient treatment and could not be left to run out of tolerance. The machine monitoring showed that it was possible to improve the machine's performance by adjusting when the 2R standard deviation was out of tolerance.

The frequency distributions of the resulting standard deviations were compared using a Chi-squared contingency table. This showed a significant difference between the periods at the 5% level ($p < .05$).

This result demonstrates that this intervention did reduce the standard deviation in the 2R current. This indicates that the standard deviation for 2R steering current can be used to improve the performance of the machine by indicating when the beam control should be reset.

4.4.3.3 Ratio of Daily Run-Up and Logged Value

The mean 2R steering value during the day for the gantry at $0°$ was examined with respect to the value of the 2R steering current when the machine was run-up before the eight-hour treatment day. This indicated a good agreement of 1.05 ± 0.08 during the satisfactory period. However, during the period considered to be unsatisfactory, this ratio was 1.15 ± 0.14. These mean ratios were determined from 55 measurements in each period and a student t-test indicated that this difference was significant at $p < 0.001$. This ratio therefore indicates that satisfactory and unsatisfactory operation of the machine can be distinguished.

4.4.4 Conclusion

Data can be logged from a medical linear accelerator to provide an indicator about the machine's performance on a daily basis. Variation in the beam control parameters reflect a variation in the radiation field and so maintenance of consistent radiotherapy treatment requires the standard deviation in these control parameters to be maintained within a limit that can be related to the radiation field. Daily monitoring of the standard deviation of a beam control parameter can indicate when there should be intervention to make adjustments and reset the beam control. This can improve the stability of the machine's performance.

During the entire four-year period, a loose gun filament was not encountered. Such a fault may have exhibited a gun current variation with gantry angle and made it possible to study another parameter.

Although this has been demonstrated for only the 2R steering current on a Philips SL75-5 accelerator, the principle could be extended to other machine components, parameters, or systems to improve the stability of the performance of these and other treatment machines.

4.5 A Cyber Assistant: Long-Term Monitoring Quality Control

The Cyber Assistant was developed as a software tool for use in routine QC for radiotherapy. QC in radiotherapy consists of a standard set of tests, which aims to ensure that the treatment machine is performing within acceptable tolerances. The results of the tests may be recorded to indicate if there are any trends in results that indicate further action is required. The Cyber Assistant system needs to be user independent and portable to allow work to be undertaken efficiently with "hands-free" operation of the equipment. It integrates voice recognition and text to voice technology into a structured package to provide guidance and ensures consistency in undertaking checks and to ease the recording of data and its analyses as the checks were being done. The system presents the user with instructions for completing the test and records the results of the test. In presenting the instructions for the test, the system allows the experienced user to operate with minimal instructions, whereas the less experienced user may select a more comprehensive set of instructions. The users must be able to "script" their own tests and establish the parameters to be recorded. The results of tests are recorded and checked for compliance to the tolerance acceptable for the test. The system will also allow diagrams or photographs to be shown to illustrate the setup for the QC test and besides voice control, it allows keyboard and mouse operation.

4.5.1 System Description

The Cyber Assistant contains a database of QC tests and results. The tests are broken down into a series of verbal instructions to aid the setting up and measurement of the equipment to be tested and the tools used for the QC. These instructions are "read" to the operator completing the testing. The operative controls the QC session by issuing commands to the Cyber Assistant by voice or mouse and keyboard. The aim of using speech recognition technology is to promote hands-free operation and the ability to move around the room or use tools, while still using the system. The Cyber Assistant is installed on a laptop computer. The laptop has a close talk wireless microphone and is equipped with a speech recognition engine and a text to speech engine.

4.5.2 Methodology

The Cyber Assistant was developed to run on Microsoft Windows. However, the early Windows 98 platform operating system was not capable of supporting the application running for extended periods due to a memory leak problem. The current version runs on Microsoft Windows 2000 and has excellent stability. The user interface has been developed using Visual Basic and the database server has been developed using Alaska Xbase++. The voice recognition engine used is developed by Microsoft [14–16]. The user has a choice of voices for the Microsoft text to speech engine. Also in the early development of the system, the database was included within the main application. However, this degraded the performance of the system, and in the current version the database is a separate application.

4.5.3   System Design

To ensure portability, the application has been designed to be operated on a laptop computer. Figure 4.13a illustrates the selection of a "Field Size" test with Figure 4.13b showing the illustration accompanying the audio information.

A design implication of the laptop platform is the requirement for the voice recognition engine and the text to speech engine to operate in half duplex mode. This means that each engine operates the sound card of the laptop exclusively, therefore the system is either "speaking" or "listening"; once it has began to speak it will complete its sentence before listening for a response, similarly the system will listen until it hears a "speechlike" sound and then it will try to understand the sound that it heard.

A major challenge of any system that uses voice recognition as a part of the user interface is ensuring the accuracy of the system to understand correctly what the user said [17]. Speech recognition systems are often disappointing in use and unless the local environment is suitable, are usually inaccurate. Most systems will require training to get the best from the system, but in practice the system still has a high failure rate. The challenge of producing a system that will be acceptable, particularly in the noisy areas where QC is performed, requires an approach that would limit the work that the recognizer has to perform.

The accuracy of speech recognition can often be disappointing. The early testing, with the Cyber Assistant using the first grammar, realized an accuracy of recognition of 74%. In noisy conditions this would drop to 62%. This performance was below an acceptable level. Speech recognition engines use an algorithm called the Markov method, which produces the best results when polyphonetic words are used. Therefore, a change in the grammar from simple words to more complex words improved the recognition rate to 91% with a similar level of 88% in the noisy environment. It was possible to use other sets of grammar and find different rates of recognition. This has resulted in the sample grammars that are in the system.

Hence the Cyber Assistant uses a customized command and control speech engine. There is a limited set of commands that are used within the system. The system is also designed to be training free, thus increasing the accessibility of the system. Using a phonetic approach to the structure of the commands enables distinct phonetic combinations to the words used during program operation. By increasing the number of connected phonemes, the likelihood of pattern matching and accurate recognition rose. Despite this approach, the early recognition results were still lower than required. To overcome this handicap, the system allows for "custom grammar" sets to

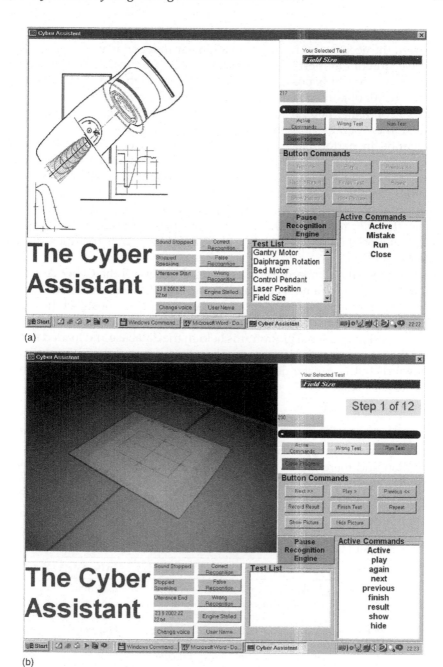

FIGURE 4.13
Cyber Assistant running on a laptop. (a) Cyber Assistant running a Field Size check. (b) Illustration of Cyber Assistant Field Size check.

be designed by the user to operate the program. This approach has greatly improved the accuracy of recognition and user independence.

By selection of the "best" set of words for user pronunciation, from any of the grammar sets, it is possible to achieve a very high and acceptable rate. The best result obtained has been 98% in normal conditions and 96% in noisy conditions. Furthermore, if the noise is mainly mechanical or electrical in nature, rather than the spoken word or music with lyrics, there is very little degradation in performance. It is better to use a "close talk" microphone rather than either handheld or desk mounted. The system is designed to give a response to any utterance detected. Although this gives the user sufficient feedback to confirm operation of the system, it can also give a "false detection" in the noisy environment. The use of a microphone with a switch enables the user to prevent the system from picking up extraneous noise. The use of close talk equipment also facilitates hands-free operation during the test execution.

Table 4.6 lists the standard commands used to control the Cyber Assistant during the main program execution after a specific QC test has been selected.

The users may select any of the standard grammars or construct their own using any words from any of the alternative grammars or a completely new set of words. This personal grammar is held in the database and loaded for the users when they log on to the system. Therefore, the users have the ability to select words with a greater degree of accuracy to suit their pronunciation.

4.5.4   Program Flow

The database module must be started before the Cyber Assistant application. Provided that the speech recognition engine has been installed with the application and there are no errors initializing the engine, program execution will proceed. A program flow diagram is shown in Figure 4.14.

TABLE 4.6

Standard Grammars within the Cyber Assistant

Command	Standard	Alternate 1	Alternate 2	Alternate 3
Mistake	Mistake	Error	Blunder	Inaccurate
Close	Close	Discontinue	Shut	Culminate
Run	Run	Execute	Progress	Process
Active	Active	Operational	Working	Functional
Clock	Time	Chronometer	Now	Watch
Play	Play	Engage	Execute	Perform
Again	Again	Repeat	Once more	Another time
Next	Next	Subsequent	Step	After
Previous	Previous	Preceding	Before	Former
Finish	Finish	Terminate	Cease	Conclude
Result	Result	Findings	Record	Input
Show	Show	Illustrate	Demonstrate	Display
Hide	Hide	Conceal	Disappear	Secrete

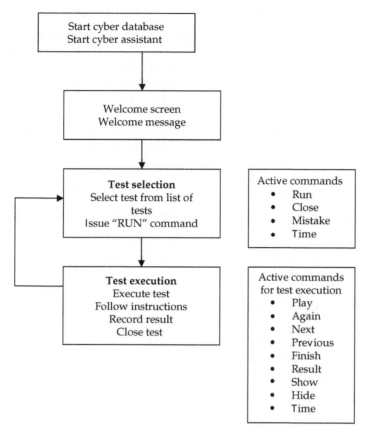

FIGURE 4.14
Cyber Assistant program flow.

This indicates that the application is functionally split into two sections:

- Test selection
- Test execution

4.5.4.1 Test Selection

In this section, the acceptable inputs are reduced to five commands and test name selection:

- *Run*—will execute the selected test
- *Close*—will close the application
- *Mistake*—will close the selected test and allow another test to be selected

- *Time*—will give the current time
- *Active*—will repeat the active commands for the selected grammar

In this section of the program it is also possible to change the grammar used, log on to the system as a user, and change the voice with which the system speaks to the user. The user may also record the accuracy of the recognition to a log file.

It is important that the test names are not only descriptive, but are also multisyllabic. This will ensure a unique match to the desired test.

4.5.4.2   Test Execution

In this section of the program there are ten commands to control the system:

- *Active*—will repeat the active commands for the selected grammar.
- *Time*—will give the current time.
- *Play*—the test instruction will be "read" by the Cyber Assistant to the user.
- *Again*—the test instruction will be "read" by the Cyber Assistant to the user. It repeats the current instruction.
- *Next*—will select the next instruction and "read" it to the user.
- *Previous*—will "read" the previous instruction to the user.
- *Finish*—will close the test and return the user to the test selection section.
- *Result*—will record the result of a test if it has been setup in the database. It will switch control to the results engine.
- *Show*—will provide an on screen diagram or photograph to assist with the test execution. These have to be set in the database.
- *Hide*—this will remove the diagram or photograph from the display.

The results engine is a subsection of the test execution where the system will listen for a numeric input to record to the database. If the database has been setup for tolerance values for the result, some guidance will be fed back to the user.

4.5.5   Program Modules

The Cyber Assistant has two software components: a user interface module and a database module. In addition to the user interface and database, the speech recognition engine and text to speech engine are freely available from Microsoft. The relationship between them is shown in Figure 4.15.

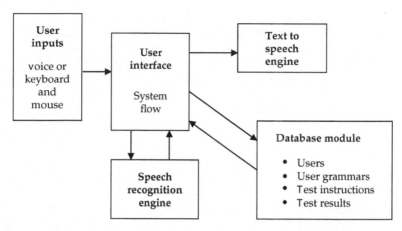

FIGURE 4.15
Cyber Assistant modules.

4.5.5.1 The User Interface

The user interface is written in Visual Basic and provides a front end for the Cyber Assistant. It provides the input and output for the system and is enabled for voice operation and mouse or keyboard use. During the execution of QC tests, voice control is expected to be the main user interface. The user interface is also linked to the voice recognition engine and the text to speech engine.

The user interface contains a set of internal commands and procedures, which may be triggered by voice command, keyboard, or mouse control. The set of commands is very basic and similar in concept to the operation of a video player. The user is able to select the test to run and then navigate through its operation by issuing simple commands such as play, next, previous, repeat, and stop. The Cyber Assistant can display digital images, which may assist the operator during test execution. These images may be photographs or diagrams to illustrate the QC test and are listed in the Cyber Assistant database. The user interface also communicates with the database module.

4.5.5.2 The Database Module

The database module is a separate application that contains all of the test scripts, illustrations, results of the tests, and user grammars. The database module receives request messages from the user interface. The database module responds to the request by returning the required information.

4.5.6 System Operation

When the cyber assistant is switched on, it displays a welcome screen with a "welcome message." The Cyber Assistant will initially load the default grammar set for system operation and then the users may log on and choose their grammar set. The first-time users are recommended to try the "standard" grammars on the system, see Table 4.6. The users will quickly be able to establish which grammar works best for them or design their own. It is best to use multisyllable words with discrete phonetic sounds. The voice recognition engine listens for phonemes, if the combination of phonemes is unique and distinct from the other words in the grammar they are easy to match. The amount of words in the grammar has also been very limited; this aids recognition by limiting the search for matches and provides a quicker response time. During program execution the number of options that the users have available to them varies; the grammar changes dynamically to match these options. The system operates in a series of loops until the operator completes the required testing procedure. The results obtained are added to the database and may be used for further analysis. If the test has been scripted to allow for results analysis, the system will indicate to the user-required actions for marginal or out of tolerance results.

In radiotherapy there is a standard set of tests to perform. The tests have to be translated into a set of verbal instructions, which can be any information that is required. This was done on the basis of how the test process would be explained to a person who has not seen it before. Each process was broken down into small discrete steps, which explain the test procedure. It was possible to augment each of the verbal instructions with a diagram or photograph.

However this resulted in a simple test having many steps to go through, which was frustrating and time-consuming for the experienced user. The system permits some steps to be classified, as "experienced" user and the system will only present the minimal set of instructions rather than the complete set. The following is an example of the test instructions for a simple test that will check the accuracy of the field delineation:

Place template on couch surface

Set distance to 100 cm

Set field X to 8 cm

Set field Y to 12 cm

Set diaphragm rotation to zero

Set crosswires to intersect the center of the template

Ensure field margins fit to the template

Ensure that the field margins are within the indicated tolerance

All test procedures were broken down in this manner. To enable recording of the result of a test, an identifier is placed in the database, which indicates that the result should be recorded. The result is recorded by using the appropriate grammar for the "result" command. It is possible to record both logical (yes or no; true or false) or numeric results, and the database can be set up to the appropriate units and scale of the result value.

4.5.7 Conclusion

The Cyber Assistant is a simple system, which satisfies the aim of reducing the effort to record data from QC measurements and help them to be undertaken and analyzed more efficiently. The user interface for both voice and keyboard control is stable and accurate. Further development of the results engine and utilities to script the tests is required. The Cyber Assistant could be used in areas other than radiotherapy, which require a quality assurance testing approach and hands-free operation.

4.6 The Future Potential

The manufacturers of treatment machines provide help desk support for the assistance in diagnosis of problems, which can be either breakdown- or performance related. Already, with the availability of a great wealth of machine data recorded from computer-controlled machines, these manufacturers are looking to provide diagnostic assistance based on accumulated data collected from many machines. This data may well be compared with information collected on-line from a specific machine or even externally from measurement of beam characteristics. Following the installation of a new machine, there is inevitably a learning period during which familiarity is gained concerning the control and operation of the machine as well as its particular idiosyncrasies. The diagnosis systems might need to be even smarter than simply drawing comparison with known data variations and be able to intelligently deal with specific rogue machines, outlying in the distribution.

Nevertheless the potential for aiding human intervention in the future through the analysis of long- and short-term trends looks very promising.

Perhaps the immediate monitoring trends hold even more exciting promise. By fitting and understanding the mathematical models that can relate one or several parameters to another, it may be possible to develop methods by which a machine may be able to tune itself. This may ultimately be through adaptive algorithms enabling it to store its own individual characteristics as it does so. Any of the categories of monitoring: long, short, or immediate might provide the trigger for the machine to inform the engineer of the machine status and then provide an option for self-tuning either with or without human intervention.

References

1. ICRP Publication 44. *Protection of the Patient in Radiation Therapy*. Pergamon Press. Oxford.
2. Travis L.B., Hill D., Dores G.M., Gospodarowicz M., van Leeuwen F.E., Holowaty E., Glimelius B., Andersson M., Pukkala E., Lynch C.F., Pee D., Smith S.A., Van't Veer M.B., Joensuu T., Storm H., Stovall M., Boice J.D. Jr., Gilbert E., and Gail M.H., Cumulative absolute breast cancer risk for young women treated for Hodgkin lymphoma. *J. Natl. Cancer Inst.* 97(19):1394–1395, 2005.
3. Lorigan P., Radford J., Howell A., and Thatcher N., Lung cancer after treatment for Hodgkin's lymphoma: a systematic review. *Lancet Oncol.* 6(10):773–779, 2005.
4. Amemiya K., Shibuya H., Yoshimura R., and Okada N., The risk of radiation-induced cancer in patients with squamous cell carcinoma of the head and neck and its results of treatment. *Br. J. Radiol.* 78:1028–1033, 2005.
5. Lessons learned from accidental exposures in radiotherapy. IAEA Safety Report Series No 17. 2000.
6. Dutreix A., When and how can we improve precision in radiotherapy? *Radiother. Oncol.* 2:275–292, 1984.
7. Herring D.F. and Compton D.M.J., The degree of precision required in the radiation dose delivered in cancer radiotherapy. *Br. J. Radiol.* (Suppl. 5): 51–58, 1971.
8. The Report of the Committee of Enquiry into the overdoses administered in the Department of Radiotherapy in the period February to July 1988. The Exeter District Health Authority, Exeter, U.K., 28th November 1988.
9. Medical electrical equipment. General requirements for safety. Collateral standard. Safety requirements for medical electrical systems. BSEN 6061-1-1, British Standards Institute. 2002.
10. Medical electrical equipment—medical electron accelerators—functional performance characteristics. BSEN 60976, British Standards Institute. 2001.
11. Medical electrical equipment—Part 3: Particular requirements for performance—Section 3.1 methods of declaring functional performance characteristics of medical electron accelerators in the range 1 MeV to 50MeV—Supplement 1: Guide to functional performance values. BS 5724-3.1, Supplement 1, British Standards Institute. 1990.
12. Greene D. and Williams P.C., *Linear Accelerators for Radiation Therapy*, 2nd Edition, Medical Science Series, Institute of Physics Publishing, 1997.
13. Gonzalez-Quiros C., Haas O.C.L., and Burnham K.J., *Bilinear extension to estimation and control software package, Computer Based Experiments, Learning and Teaching*, Wroclaw University of Technology, pp. 131–139, 1999.
14. Microsoft research, http://research.microsoft.com/, accessed on May 16, 2007.
15. HTK3, Speech recognition engines, http://htk.eng.cam.ac.uk/links/speech_comp.shtml, accessed on May 16, 2007.
16. The International Phonetic Association, http://www2.arts.gla.ac.uk/IPA/ipa.html, accessed on May 16, 2007.
17. HCI Bibliography: Human-Computer Interaction Resources, http://www.hcibib.org/, accessed on May 16, 2007.

Part II

Artificial Intelligence Applied to Medicine

5

Neural, Fuzzy, and Neurofuzzy Systems for Medical Applications

António Dourado, Jorge Henriques, and Paulo de Carvalho

CONTENTS

5.1 Introduction

An immense quantity of information is available in all sectors of human activity, especially in the healthcare and medical sector. The processing of that information is a challenge to the human user—the medical doctor. The challenge is to develop tools (systems, procedures, and methods), to support clinicians, which are more exact, cost-effective, and friendly to use. Information of several kinds is available.

- Numeric analytic from known cause–effect relations that can be formalized through a mathematical (in the classical sense) relation.
- Numeric empirical, issued from experimentation and practical work but to which there is no known cause–effect relation.
- Linguistic (qualitative), expressed in an approximate way by the user, with several levels of granularity (detail). Granularity may be connected to the words of common language (big, small, strong, weak, etc.) or to intervals whose limits are not clearly defined.

In most cases the available information is empirical or linguistic, issued from complex and imprecisely known relations in complex systems, such as those with which the clinicians work. To process all these types of information, several approaches have been developed, each one more appropriate for a certain context:

- The integral–differential approach, purely numeric, aiming at the determination of a set of mathematical equations building a model of the process (in the classical sense). This model assumes usually the form of integral or differential equations, and is seldom applicable in a medical context.
- The empirical-data approach, using some basic tools for nonlinear function approximations. The aim is to obtain a compact tool able to predict behaviors of systems (in the more general sense), after the tool has been trained with past known data. Because most real systems (such as the human biologic systems) are nonlinear, nowadays the most-used tool to synthesize relations between sets of data is the artificial neural network (ANN) formalism, giving space to the so-called neurocomputing body of knowledge.
- The linguistic approach where tools enabling computers to compute with words are used. These tools enable computers to process the language of the clinician and to make inferences and deductions. Fuzzy logic is the most-used framework for that purpose. Because of the type of information, computing with fuzzy logic is sometimes called granular computing.
- Finally, because in real situations the available information is a mix of these three types, combinations of these tools are used to build flexible systems capable of working in a diversity of situations with an acceptable degree of efficiency.

Many situations require searching for the "best" solution or, at least, a "good" solution. To search for better solutions with a computer, one must have some mathematical way of expressing what is good or not good, and an analytical formalism to search iteratively from a solution to another, improving its quality.

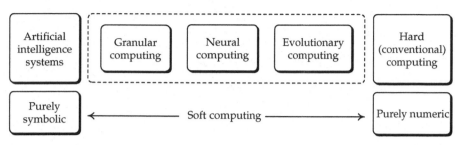

FIGURE 5.1
The place of soft computing.

Unfortunately, the body of knowledge of traditional mathematical optimization requires formal constraints and the possibility to write a mathematical criterion for comparison between solutions. In real life this classical framework is of limited applicability, because most of the real problems are ill-conditioned in the sense that they cannot be mathematically formalized in a proper way.

Soft computing techniques are a family of tools to "exploit the tolerance for imprecision, uncertainty and partial truth to achieve tractability, robustness and low solution cost" [1]. Besides neurocomputing and fuzzy computing, it also includes genetic (evolutionary) computing and other techniques able to deal with incomplete knowledge. Figure 5.1 represents the soft computing family of techniques among the computing discipline at present times (adapted from Ref. 208 and according to Ref. 207). Numeric computation deals with numbers, whereas symbolic computation deals with symbols (for example, letters) and its mathematical manipulation [2].

A brief introduction to these techniques will be presented in the following section. Section 5.3 discusses some medical applications involving these techniques, covering the following domains: modeling and biosignal processing and interpretation, biological system control and prognosis, image-processing and decision-support systems.

5.2   Soft Computing Techniques

5.2.1   The Data Paradigm: Artificial Neural Networks

This representational/computational tool derives its name from its similarities with the natural neuron (Figure 5.2). The basic element of an ANN is a single neuron, shown in Figure 5.3. It is inspired by the natural neuron and its first use in the academic community was in 1944, more or less at the same time as the birth of the digital computer. For a brief history of ANNs, see Ref. 3.

The natural neuron receives electrical impulses from its neighbors through dendrites, these impulses being combined in the cell body that attains a certain

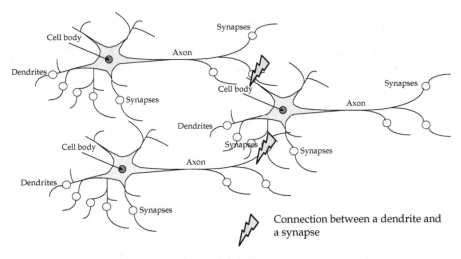

FIGURE 5.2
The biological neuron and its connections. One neuron is composed of dendrites, cell body, and synapses. A synapse transmits a signal to another neuron by contacting one dendrite.

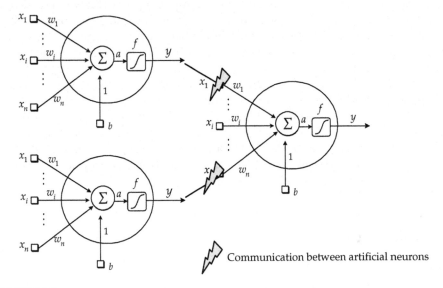

FIGURE 5.3
Artificial neuron and its connections.

activation degree and an electrical impulse is then transmitted through the axon to a synapse of the following neuron. In the artificial neuron, signals represented by numeric values are presented to the input, weighted by the artificial synapses (called weights for this reason), are combined (usually summed), and the resulting signal is the argument of a certain function—the activation function—that produces a transformed signal as output [4].

Figure 5.3 shows the usual representation of an artificial neuron. A single artificial neuron is a simple and powerful computational tool. The weighted sum a of several inputs is passed through an activation function f to produce the output y of the neuron; Equation 5.1.

$$a = w_1 x_1 + w_2 x_2 + \cdots + w_n x_n + b \Leftrightarrow a = w^T x + b$$
$$y = f(w^T x + b)$$

$$(5.1)$$

An alternative architecture is the radial basis neuron measuring a radial distance between the input presented to it and an interior center (the radbas function is illustrated in Figure 5.4d).

$$y = \text{radbas}(a) = \text{radbas}(\|w - x\|b)$$

$$(5.2)$$

$$y = \text{radbas}\left(\sqrt{(w_1 - x_1)^2 + (w_2 - x_2)^2 + \cdots + (w_n - x_n)^2}\right)$$

The special input of constant value 1 is called the bias of the neuron, allowing a nonzero output for a zero input. One neuron has the following degrees of freedom: the number of inputs, the value of the weights, the type and parameters of the activation function f, and the value of the bias weight b. It is possible to use a multiplicity of activation functions. Figure 5.4 shows some of them [5].

A neuron can be combined (networked) in an arbitrary way in series and in parallel, giving place to structures that can model any nonlinear relations between a set of inputs and a set of outputs, with or without feedback. One of the most used structures is shown in Figure 5.5, the multilayer feedforward neural network (MLFNN), also known as perceptron [6].

Other well-known structures are radial basis function neural networks (RBFNN), having only a radial hidden layer, and recurrent neural networks (RNN), which involve dynamic elements and have feedback connections. Each structure has its own potentialities and is more adequate for certain types of applications. In general terms, ANN is used to find relations between two sets of data, the input set and the desired output set (the target set). An input set is presented and the network is trained to reproduce at its output the target set, or to classify the input set among a finite number of target classes. Training means to use its degrees of freedom to find the configuration that best fits the situation (best according to some criteria). Figure 5.6 illustrates how this is done in the case of modeling biological systems. The same input set is given to the system

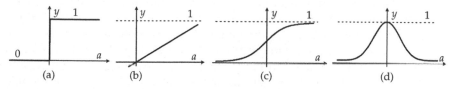

FIGURE 5.4

Some types of activation functions. (a) Binary; (b) linear; (c) logistic; and (d) radial based.

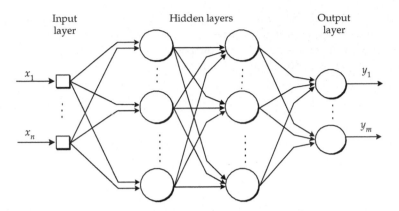

FIGURE 5.5
Multilayer feedforward neural network (in this case with two hidden layers).

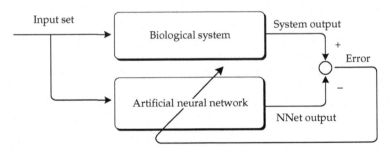

FIGURE 5.6
Training the artificial neural network to model biological systems.

to be modeled and to the network. The output of the network is compared with the system output (the output set) and an error information signal is fed back to the network, where a training algorithm changes its degrees of freedom (usually the weights) until the error is minimized.

The MLFNN and RBFNN are appropriate for function approximation such as in modeling biological systems. There is one weight for each input for each neuron, resulting in a high number of weights even for small networks. For a particular problem, the training of the network is just the procedure to find the best set of values of these weights such that the network is able to mimic the response to a certain history of inputs. After the training phase, the network is able to predict the future behavior, or give answers to new inputs (data generalization). Of course the generalization capability of the network depends on many factors, namely the quantity and quality of inputs and the particular architectures. Similarly, because of the complexity and variety of the human body, doctors make decisions that are not based on a single symptom [4]; a doctor with more experience is more likely to make correct decisions than a newcomer because of his learnings from past mistakes and

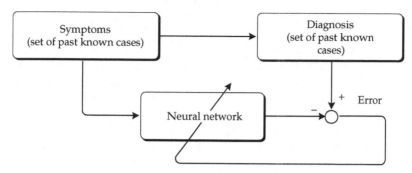

FIGURE 5.7
Training the artificial neural network for diagnosis tasks.

successes (he has more training data). To have a good set of examples is decisive for the "experience" of the neural network.

For pattern recognition, some kinds of single-layer neural networks, see for example Hebb, Widrow–Hoff [6], can also be adequate. In medical decision problems, the medical symptoms such as "stabbing pain in head," are presented to the inputs (after being codified by a numerical value) as training examples. For a set of past data, containing symptoms for which the correct diagnosis is known, the weights of the network are varied in a systematic way such that the output of the network produces the correct diagnosis (also codified by a numeric value)—"brain tumor," "stress," etc. The hidden layers are just used for the computation of the mapping between the inputs and the outputs (Figure 5.7).

Therefore, the network is trained just like a doctor, being presented with a large number of known cases (training set) with known outcome. The most used algorithm for training the multilayer network, that is, the adjustment of its weights, is the backpropagation algorithm [7], using the backward propagation of the information. Basically it is a computational procedure that varies the weights to progressively reduce the distance between the correct answer and the network answer. The information is sent backward because only at the end of the network (its output) this distance can be computed and the changes of the weights in the beginning of the network depend on that information. If properly trained, the network can give answers to new unknown cases with some reliability. If the training is made constantly, as new cases happen, the network becomes adaptive and with improved capabilities. For a good review see Ref. 8.

The simple structure of a neural network has a high number of degrees of freedom: type of activation function, number of layers, and number of neurons per layer. Moreover there are two fundamental learning approaches: supervised learning and unsupervised learning. In supervised training a desired behavior of the network output is previously specified, as in Figure 5.6. In unsupervised learning no such behavior is provided; unsupervised learning can be used for analyzing (clustering, for example) the input data and to find features embedded in the data expressing properties of the system of interest. Self-organizing maps (SOM), introduced by Kohonen [9] in 1982, are originated from learning vector quantization (LVQ) that was also introduced by Kohonen [10]. If systems continue to remain adaptive, care must be taken such that previously learned information is not lost. Grossberg [11] has coined the term stability–plasticity dilemma for this problem introducing adaptive resonance theory (ART). Regarding the flow of information, basically, neural networks can be classified as static (feedforward) and dynamic (recurrent). RNN were first introduced by Hopfield [12], and then developed by some other authors. Unlike the neurons in MLFNN, the Hopfield network consists of only one layer whose neurons are fully connected with one another. Owing to their intrinsic abilities to incorporate time, they have some advantages with respect to static neural networks (feedforward multilayer perceptrons), particularly for modeling dynamic processes.

Additionally, other properties justify their application in the medical domain, [13–17]: noise is quite well-managed by neural networks, their prediction capabilities are well-suited for regressive models, the online learning capabilities allow to face the possibilities of automatic analysis and diagnosis with updated knowledge. Online learning means that data are processed iteratively as it is obtained, as opposed to off-line learning where complete data are firstly obtained and then the learning process is launched with all data simultaneously considered.

The total number (on June 20, 2006) of references in Pub Med under "artificial AND neural AND networks" was 2617 ([18], in all fields). For the past 10 years, the number is indicated in Table 5.1. Under "neural AND networks," the total number was 11,293 (most of these are related with artificial networks) and for the past 10 years the number of publications is shown in Table 5.2.

The main drawback of ANNs is that they are "black boxes," that is, they do not give any understandable explanation for the relation between its inputs and outputs. They just give numbers that cannot be interpreted in terms of a natural language. If the model would become transparent, explaining the reasons for the diagnosis, then it would increase its importance and usability.

TABLE 5.1

Publications in Pub Med with "Artificial AND Neural AND Networks"

Year	Number of Papers
1996	168
1997	209
1998	233
1999	229
2000	252
2001	298
2002	311
2003	321
2004	498
2005	428

TABLE 5.2

Publications in Pub Med with "Neural AND Networks"

Year	Number of Papers
1996	632
1997	641
1998	728
1999	809
2000	895
2001	998
2002	996
2003	1185
2004	1542
2005	1548

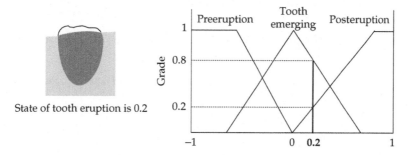

FIGURE 5.8
Membership functions of teeth development phases.

Fuzzy systems allow making this evolution of the machine because fuzzy logic is a way to compute with words.

5.2.2 The Linguistic Paradigm: Fuzzy Logic and Fuzzy Systems

Fuzzy logic was born in 1965 by the pioneering work of Lofti Zadeh [1], at MIT, United States, as a mathematical tool for dealing with uncertainty. In fuzzy logic, statements are not "true" or "false" (as in the Aristotelian bivalued logic), but they may have several degrees of truth and several degrees of false. Fuzzy sets do not have a well-defined frontier, but an imprecise (fuzzy) one. It is not only black and white but it has many levels of gray in between. Consider, for example, the classification of teeth development in preeruption, emerging, and posteruption. Is there a well-defined frontier between these phases? If a tooth has a state of eruption 0.2, what is its state? It is still emerging, but has it already emerged! How can we represent this in the classical binary Aristotelic logic (true or false, 0 or 1, black or white)? Fuzzy sets are very convenient to represent the situation (see Figure 5.8) [19].

The membership functions (*mf*), representing the membership degree may have several shapes (see Figure 5.9).

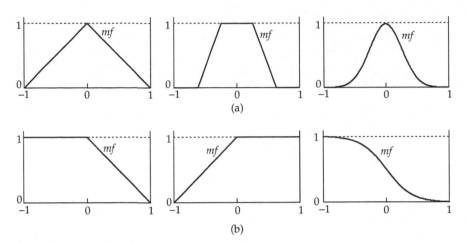

FIGURE 5.9
Membership functions: (a) symmetrical (triangular, trapezoidal, Gaussian) and (b) nonsymmetrical.

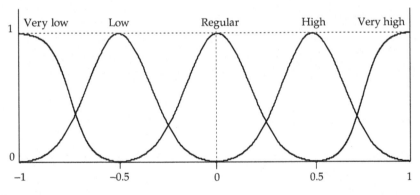

FIGURE 5.10
A universe of discourse (one variable) divided into five fuzzy sets. The first and last ones are nonsymmetrical.

For any variable (for example, temperature) its universe of discourse is divided into several labels, each one corresponding to a fuzzy set (e.g., "very low," "low," "regular," "high," and "very high").

The fuzzy sets must overlap and they must cover completely the universe of discourse (all the intervals of possible temperatures, in the example, as illustrated in Figure 5.10). Usually they must overlap in such a way that the sum of the membership degrees for any point is 1, and at most two sets are valid for that point. Fuzzy logic is the logic of fuzzy sets. In fuzzy logic there are many levels of truth and of false, in the real interval [0,1]. A value in the universe of discourse belongs simultaneously to several fuzzy sets, and usually with different membership values.

There are some characteristics of our perception systems that can be seen as fuzzy sets. For example, according to Sir Thomas Young's (1802) theory

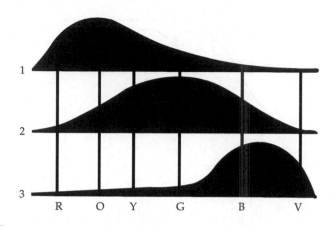

FIGURE 5.11
Illustration of the theory of Young. The curves show the amount of activity of each of the three visual receptors types by each wavelength (of the several colors). Maximum excitation in each is produced by one wavelength, and adjacent wavelengths produce progressively less activity. (From Erikson, R., Chelaru, M., and Buhusi, C., *Fuzzy and Neurofuzzy Systems in Medicine*, CRC Press, Boca Raton, FL, 1999. With permission.)

[20] for color perception, there are three principal colors—red, green, and blue—and three types of visual receptors.

The way these visual receptors vibrate with the colors' wavelengths is illustrated in Figure 5.11. Maximum excitation in each is produced by one wavelength; adjacent wavelengths produce progressively less activity in the particular receptor.

This seems to be the case of all sensorial systems: although we have many neurons, there are not enough such that each neuron has a specific function (e.g., to encode red apples) distinct and disjoint from that of every other neuron. There is too much information to be encoded. The sensitivity functions of all individual neurons in all sensory systems are bell-shaped at a first approximation and have been referred to as neural response functions (NRF) [18]. There are few neural resources to represent many stimuli. So the few neurons available must have fuzzy sets (NRF) that can cover, as broadly as possible, all stimuli (Figure 5.12).

Fuzzy systems work in a similar way. Using fuzzy sets and fuzzy logic, fuzzy inference systems may be built enabling to compute a decision based on a set of rules. Fuzzy rule–based systems perform a sequence of fuzzy logical operations: fuzzification, conjunction, inference, and defuzzification [21].

A fuzzy system consists of three stages: the fuzzification, the deffuzification, and the inference procedure (Figure 5.13). The fuzzification stage determines the membership degrees of the input values in the antecedent fuzzy sets, converting numerical values of patient data (symptoms that mainly define the patient's state of health) into linguistic variables. The inference mechanism combines input information with the knowledge stored in the fuzzy rules and determines the output of the rule-based system. The knowledge base contains the formalization of the existing knowledge about the case, by expressing

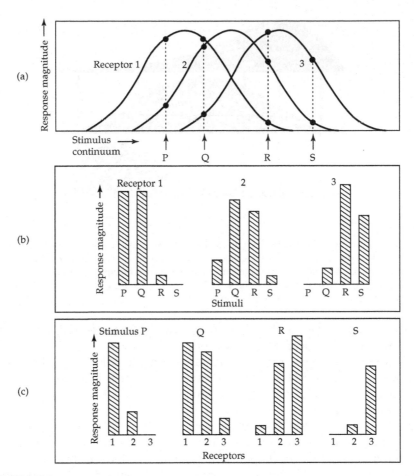

FIGURE 5.12
Fuzzy sets and neural codes according to Young's theory. (a) Three idealized receptor types (1, 2, 3) and four stimulus (P, Q, R, S); (b) the magnitude of the response of each receptor to each stimulus; (c) the neural codes for P, Q, R, S. The brain interprets these codes. (From Erikson, R., Chelaru, M., and Buhusi, C., *Fuzzy and Neurofuzzy Systems in Medicine*, CRC Press, Boca Raton, FL, 2000. With permission.)

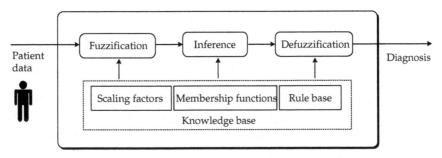

FIGURE 5.13
Different modules of a fuzzy system.

associations between symptoms and diseases by means of fuzzy rules. The diagnosis is obtained by the defuzzification part, which chooses among rules that have been fired simultaneously.

Let us consider a fuzzy system with four rules in the form depicted in Figure 5.14.

Rule 1: If Symptom1 is LOW   and Symptom2 is LOW   then Diagnosis is LOW.
Rule 2: If Symptom1 is LOW   and Symptom2 is HIGH then Diagnosis is HIGH.
Rule 3: If Symptom1 is HIGH and Symptom2 is LOW   then Diagnosis is LOW.
Rule 4: If Symptom1 is HIGH and Symptom2 is HIGH then Diagnosis is HIGH.

Fuzzification is the operation of transforming a numeric value, issued from a measurement, into a membership degree of a fuzzy set. Figure 5.15 shows 2 measurements: Symptom1 = −0.443 and Symptom2 = 0.193. All four rules have some degree of truth and some degree of false. All rules must be fired. To compute the firing intensity of one rule, one may consider the weakest case in the antecedents, corresponding to the application of the minimum

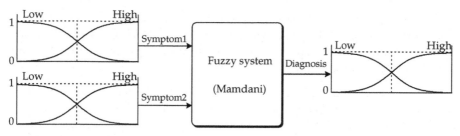

FIGURE 5.14
The fuzzy system and membership functions.

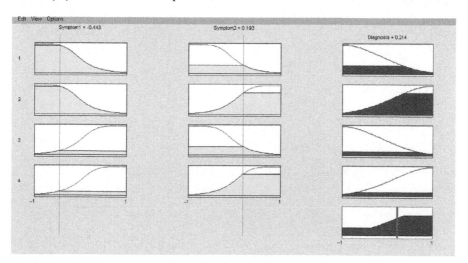

FIGURE 5.15
Firing the rule base: fuzzification, conjunction, inference, defuzzification (obtained with the Fuzzy Logic Toolbox, The MathWorks).

operator. Now we transport these values to the consequents. This is done by cutting the fuzzy set of the consequent at the height equal to the firing intensity of the antecedent. The graph shows that (on the right-hand side) rule 3 is quite truth, rule 1 is about 0.3 truth, rules 2 and 4 are about 0.2 truth.

The final decision is the result of the balanced contribution of the four rules. Defuzzification is this balancing to obtain a numerical value to be assigned to the decision. If the four figures (of the consequents) are superposed, in geometrical terms the point of equilibrium is the center of mass. This is the most used defuzzification method and it is applied in the example. The graphical construction is quite intuitive. Formally, there are some properties and operations of fuzzy logic supporting it. For example, the cutting of the membership function on the consequent is made by a minimum operator:

$$DiagnosisOut = \min(sympotom1, symptom2)$$

The operator maximum performs the aggregation of outputs:

$$OutputFinal = \max(diagnosis1, diagnosis2, diagnosis3, diagnosis4)$$

Takagi–Sugeno–Kang (TSK) fuzzy systems are based on rules that have a nonfuzzy consequent. For a zero order TSK system, each consequent is simply a constant: they are in the form (for a similar example, and with constants 0 and 1 in the consequents).

Rule 1: IF Symptom1 is LOW   and Symptom2 is LOW   then Diagnosis is 0.
Rule 2: IF Symptom1 is LOW   and Symptom2 is HIGH   then Diagnosis is 1.
Rule 3: IF Symptom1 is HIGH   and Symptom2 is LOW   then Diagnosis is 1.

Figure 5.16 illustrates how it works. Now for the same measurement values, it proceeds as follows:

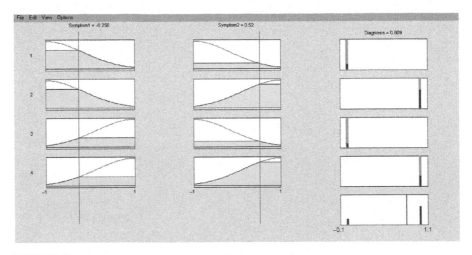

FIGURE 5.16
Firing the rule base in TSK model: fuzzification, conjunction, inference, defuzzification (obtained with the Fuzzy Logic Toolbox, The MathWorks).

Rule 1: firing strength: 0.15   output1 = 0
Rule 2: firing strength: 0.75   output2 = 1
Rule 3: firing strength: 0.15   output3 = 0
Rule 4: firing strength: 0.4   output4 = 1

The overall output is the sum of the individual outputs weighted by their degrees of truth, that is, the firing strength of the respective rule, giving 0.8, shown on the right-hand side of Figure 5.16.

TSK fuzzy systems are simpler to compute than the previous types called Mamdani fuzzy systems. They are particularly important in neurofuzzy systems. A fuzzy system is a set of fuzzy rules describing what is known about some problem. The development of the fuzzy system is basically the writing of the rules. However, how are these rules obtained? Several approaches may be applied:

- Expert interviews (actual medical knowledge)
- Simulation using models of the processes (seldom possible)
- Rule extraction from data (data mining)

The latter is becoming the most important approach, where the machine (computer) learns from data. Two aspects must be analyzed: the determination of an initial set of rules (the initial structure of the system) and the update and optimization of the rules as new data and knowledge become available. For the determination of the initial set of rules, the most important technique is clustering. The second operation, the optimization of the fuzzy structure (i.e., the number of rules, the parameters of the membership function, etc.) is actually carried out in the context of neurofuzzy systems. The first application of fuzzy logic to the medical field dates back to 1969, when Zadeh published a paper on the possibility of applying fuzzy sets in biology [22].

The medical field has several sources of inaccuracy [23]: information about the patient consists of a number of categories, all of which have uncertainties; medical history of patients is most of the times subjective and may include nonunderstandable symptoms (supplied by the patient); and lack of knowledge of previous diseases that usually leads to doubts about the patients' medical history. Additionally, although results of laboratory tests are objective data, they are however dependent on the accuracy of the measurements and on the possible inadequate behavior of the patient during the examination. Fundamentally, fuzzy systems allow transparency in knowledge representation and in the formulation of decision rules that mimic human thinking, justifying its medical application in the representation of narratives and clinical guidelines in decision-support systems [24,25].

The number of papers published on fuzzy logic in medicine during the past 10 years are given in Table 5.3. (Pub Med [18] under "fuzzy" on June 20, 2006, all fields).

TABLE 5.3

Publications in Pub Med with "Fuzzy" (All Fields)

Year	Number of Papers
1996	98
1997	128
1998	135
1999	135
2000	152
2001	218
2002	191
2003	201
2004	304
2005	278

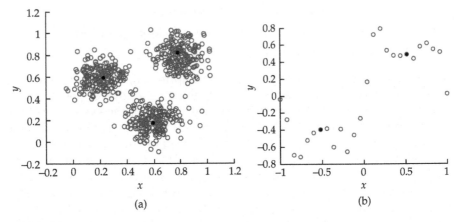

(a)   (b)

FIGURE 5.17

Clustering is a classification of multidimensional points into classes. The black points represent the centers of (a) fuzzy C-means (FCM) used; (b) the classes obtained by subtractive clustering (obtained with the Clustering Interface, The MathWorks).

The first applications were related to assessing of symptoms and the modeling of medical reasoning. For a more detailed historical perspective of the early development stage, see Ref. 26. For the similarity between fuzzy reasoning and the physiology of the nervous system, see Ref. 20, where the dynamic model of sensory systems by the neural response functions, for example, for taste neurons, is made by a TSK fuzzy system.

5.2.3 Clustering and Neurofuzzy Systems

Clustering is basically the detection of similar points in an input–output space data set. Figure 5.17 illustrates the case of a bidimensional data space for a system with two variables (e.g., input x and output y).

A region where there is a concentration of data forms a cluster. A cluster then represents a repetition of facts resulting from some property such as input–ouput relation. Computational clustering techniques form a large body of knowledge. In soft computing, the most used are the FCM, the mountain, and the subtractive clustering [27–29].

Clustering is an operation that requires high-computational resources and currently, the main research direction is to obtain recursively implementable clustering techniques such that the clusters are continuously updated with new data. This is particularly important for systems that operate in changing environments needing permanent adaptation and learning.

A cluster identifies a working region, so it defines a relation between the variables; this relation may be translated into a fuzzy rule. One of the main important applications of clustering is precisely in the development of rules from data, leading to the neurofuzzy systems. Once the clusters have been identified, fuzzy rules can be built based on the identified centers $c_i = (x_i, y_i)$, of the form:

IF Input is X_i THEN Output is Y_i

where X_i and Y_i are the fuzzy sets centered on x_i and y_i, respectively.

Fuzzy systems are designed to work with knowledge in the form of linguistic rules; neural networks to deal with data. The optimization of rules with respect to a set of data or to new data needs an efficient computational tool able to process a nonlinear mapping (a rule is in general a nonlinear mapping). Neural networks enter here in a natural manner. A hybrid technique is defined as any effective combination of different techniques that performs superior or, in a competitive way, over simple standard techniques [15,30]. Neurofuzzy systems are possibly the most promising hybrid soft computing technique, combining the capabilities of neural networks with fuzzy systems, that is, enabling to acquire knowledge (fuzzy rules) from experimental data. Because of the accuracy and the interpretability that they may allow to achieve, neurofuzzy systems have shown a high potential of success when applied in complex domains of application such as in the medical field [23].

The artificial neural fuzzy inference system (ANFIS) [31,32], depicted in Figure 5.18 is probably used the most. It is based on TSK fuzzy rules and has the structure of an MLFNN neural network with five layers.

Each layer computes a fuzzy operation:

- Layer 1—the fuzzification layer (A_i, B_i): each numerical input is presented to each neuron. The neuron output is the membership value of the input. For each input variable there are as many neurons as fuzzy sets in its space.

- Layer 2—the conjunction layer (T): each neuron computes the conjunction of the antecedents of each rule. Usually the conjunction operator is the algebraic product of the antecedent membership values. The output of the neuron is the absolute firing strength of the rule. There are as many neurons as rules.

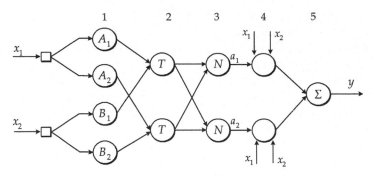

FIGURE 5.18
The ANFIS neurofuzzy system is composed of five layers.

- Layer 3—the normalization layer (N): each neuron computes the relative firing strength of the rule with respect to the sum of all strengths of all fired rules.

- Layer 4—the inference layer: computes the consequent value for each rule weighting the consequent function by the relative firing strength of the rule.

- Layer 5—the defuzzification layer: computes the overall output of all the rules by summing the individual consequents from the previous layer.

The operations of the network are the same as we saw in the graphical inference method (Figure 5.16). The advantage is that we now have a neural network that can adjust its weights to a set of data in a way such that the output of the network approaches optimally the experimental output. We have here simultaneously the advantages of fuzzy logic and the advantages of the neural networks. Neurofuzzy systems have this nice property.

The research of new architectures for neurofuzzy systems is very active. Several developments from the ANFIS architecture can be found: dynamic evolving neural-fuzzy inference system (DENFIS) [33], generalized network based fuzzy inferencing systems (GeNFIS) [34], and others [35].

Under "neurofuzzy," all fields, the number of publications in Pub Med on June 20, 2006, was as indicated in Table 5.4.

5.2.4 Fuzzy Medical Image Processing

Medical images convey uncertainty due to the intrinsic nature of modalities that originate noise, blurring, background variations, partial volume effects (this effect is induced by low-resolution sensors, which induce borders strictly not defined between tissues), low contrast, and certain modality-specific effects. This uncertainty is not always due to randomness but due to ambiguity and vagueness and may propagate to the entire image-processing chain, that is, from the low- to the high-level image-processing stages (see Figure 5.19). According to Refs. 37 and 38, besides randomness three other sources of imperfection can

TABLE 5.4

Publications in Pub Med with "Neurofuzzy"

Year	Number of Papers
1996	1
1997	1
1998	0
1999	6
2000	4
2001	10
2002	10
2003	9
2004	10
2005	12

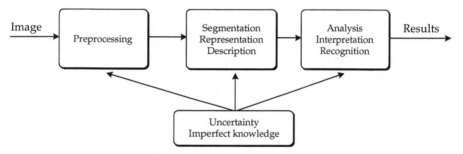

FIGURE 5.19
Imperfect knowledge in image processing. (Adapted from Tizhoosh, H., *Fuzzy Image Processing: Introduction in Theory and Practice,* Springer-Verlag, Heidelberg, 1997.)

be distinguished in images in general: (1) grayness ambiguity, (2) geometrical fuzziness, and (3) vague (complex/ill-defined) knowledge.

These uncertainties are difficult to overcome using the traditional image-processing approaches such as probabilistic and physics-based image inter-pretations. Under these circumstances, expert knowledge can provide a valuable source of information to deal with uncertainty.

Following Tizhoosh's [36,37] definition, fuzzy image processing comprises the collection of all approaches that understand, represent, and process the images, their segments, and features as fuzzy sets. The representation and processing depend on the selected fuzzy technique and on the problem to be solved. From this definition it becomes clear that to integrate the fuzzy framework into image processing, a new image definition has to be applied, that is, images and their components have to be fuzzified, whereas relationships between image parts have to be extended into fuzzy sets. During the processing stage, appropriate fuzzy techniques modify the membership values. These can be fuzzy cluster-ing, fuzzy rule–based approaches, fuzzy-integration approaches, or others. As one would expect, a defuzzification stage has to be performed to obtain crisp results. This general procedure is illustrated in Figure 5.20.

Typical fuzzifiers depend on the task at hand. For instance, to perform global image-processing tasks, that is, point operations, each pixel should be

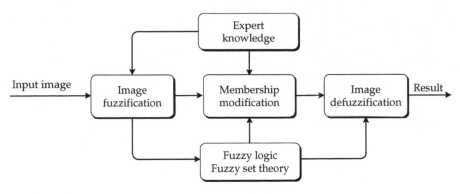

FIGURE 5.20
The general structure of fuzzy image processing.

assigned one or more membership values. This is known as histogram-based gray-level fuzzification that can easily be extended for color image point processing. For typical neighborhood-based pixel operations, such as in filtering (e.g., noise filtering and edge detection) and local contrast enhancement, image fuzzification usually takes into account the same neighborhood applied during the processing step. For intermediate- and high-level image-processing tasks, fuzzification of the extracted image features is required (e.g., shape descriptors, corners, curvature, texture, and motion). Again the fuzzifier is application dependent and should be setup based on expert knowledge, and eventually combined with a learning strategy [36–38].

Fuzzy processing is performed by modifying the membership values of pixels or features by means of a suitable fuzzy approach. The most common modification principles are [37] (1) aggregation using, for instance, fuzzy integrals, (2) membership value transformation (this is usually the case for contrast enhancement), (3) classification by means of fuzzy classifiers such as fuzzy clustering or syntactic approaches, and (4) inference by means of if–then rules. Low- and intermediate-level image-processing operations usually require crisp outputs; these may be computed during a defuzzification stage. For image processing two general groups of defuzzifiers exist: (1) conventional defuzzifiers such as center of area and mean of maximum and (2) inverse mapping for point-based operations.

For further reading on fuzzy image-processing principles and theory refer to Refs. 36,37,39,40.

5.3 A Brief Review of Applications in the Medical Field

Regarding medical domain applications, handled with soft computing schemes, numerous approaches have been presented in the literature. Significant medical applications that make use of neural networks, fuzzy systems, and both involve the following among others:

Medical Application	References
Bacteriology	41
Cardiology	39,40,42–44
Dentistry	45
Drug and anesthesia delivery	24,46
Gastroenterology	47
Genetics	48,49
Intensive care	50,51
Neurology	52,53
Nuclear medicine	54
Obstetrics and gynecology	55
Oncology	56–59
Ophthalmology	60,61
Otology-rhinology-laryngology	62
Pathology	63,64
Radiology	65–68
Sleep research	69
Urology	70,71

The review presented in this work is structured on medical application domains, covering the following areas: modeling and biosignal processing and interpretation, biological system control and prognosis, and image-processing and decision-supporting system [42].

5.3.1   Modeling and Biosignal Processing and Interpretation

Our understanding of biological systems is incomplete. There are features and information hidden in the physiological signals that are not clear, and effects between the different subsystems that are not evident. Moreover, biological signals are characterized by significant variability, caused by impulsive internal mechanisms or external stimulus and, most of the times, are corrupted by noise.

There are two main, recognized advantages of using neural networks for modeling and biosignal processing [72]: one is their capacity to perform any nonlinear mapping between input and output patterns (providing an adequate number, type, and association of neurons). This capacity offers a universal approximation property of unknown systems based on sparse sets of noisy data, such as biological systems. Another advantage is the adaptive learning capacity of neural networks, enabling them to adapt to new input patterns. Unfortunately, it is almost impossible to come to a reasonable and humanly understandable (transparent) interpretation of the overall structure of these networks. Furthermore, the existence of previous knowledge, for instance, the explanation of clinical rules, is not easily incorporated into the neural model. In the context of modeling and biosignal processing, fuzzy systems provide tolerance and partial correctness; thus a suitable way to represent qualitative linguist information. Independently or combined, neural network and fuzzy systems can assist the modeling of their relationships

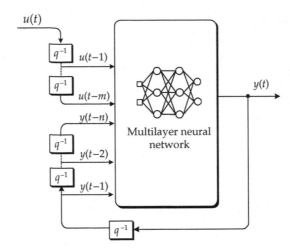

FIGURE 5.21
Multilayer neural network with external recurrence.

containing uncertainty and nonlinearity characteristics, extract parameters
and features, identifying and removing biosignal artifacts.

Multilayer perceptrons with external recurrence have been extensively
applied in the biological systems domain. Using this structure, the nonlinear
mapping between output and past information is implemented by a neural
network, the output $y(t)$ being a function of the m past inputs, $u(t - 1)$, ...,
$u(t - m)$ and the n past outputs, $y(t - 1)$, ..., $y(t - n)$, as described in Equation 5.3
and depicted in Figure 5.21 (q^{-1} represents the unitary delay operator).

$$y(t) = NN\{y(t - 1), y(t - 2), ..., y(t - n), u(t - 1), u(t - 2), ..., u(t - m)\} \quad (5.3)$$

In the field of biosignal processing (mainly for cardiology), soft computing
techniques have been widely used in clinical practice for automatic electro-
cardiographic (ECG) analysis. There have been several attempts to use neural
networks to improve the ECG diagnostic accuracy and achieve more fault-
less operation, even in the presence of complicating factors. In this context,
Lee et al. [73] have studied and compared multilayer RNN with conventional
algorithms for recognizing fetal heart rate abnormality, revealing the excep-
tional performance of neural networks. Multilayer neural networks were
also used to model heart rate regulation [74,75], although Ortiz et al. [76]
have applied them to examine heart failure. Assessment of long-term ECG
recordings (Holter-monitor) is a time-consuming and exhausting procedure
(nearly 90,000 ECG-complexes a day). Neural networks have shown capabilities to recognize disorder events automatically, which occur infrequently with
up to 99.99% sensitivity [77]. For long-duration ECG recordings, Papaloukas
et al. [78] have presented a method that employs neural networks for the
automated detection of ischemic episodes.

Silipo and Marchesi [79] have demonstrated the capabilities of neural networks to deal with the ambiguous nature of ECG signals. In their work they have used static and RNN architectures and explored ECG analysis for arrhythmia detection, myocardial ischemia recognition, and chronic alterations. Janet et al. [44] discuss a neural network that has been trained to detect acute myocardial infarction. They have used ECG measurements from more than 1,000 patients who had suffered a heart attack, and more than 10,000 healthy persons, with no history of heart attack. They have concluded that neural networks were 15.5% more sensitive than an interpretation program and 10.5% more sensitive than experienced cardiologists in diagnosing any abnormalities. However, the cardiologist was slightly better at recognizing ECGs with very clear-cut acute myocardial infarction changes.

Waltrous and Towell [80] reported the use of a neural network, synthesized from a rule-based classifier, applied to an ECG patient monitoring task. Serum enzyme–level analysis forms the basis of acute myocardial infarction diagnostics. The neural network has been trained based on the analysis of these heart enzyme levels, showing a diagnostic accuracy of 100% with an 8% false-positive rate. The neural beat classifier was integrated into a four-stage procedure for the diagnosis of ischemic episodes.

When conditions are such that an RBFNN can act as a classifier [81], an advantage of the local nature of radial basis function networks, compared with multilayer neural networks, is that a new set of input values that falls outside all the localized receptor fields could be recognized as not belonging to any of the existing classes. Employing an RBFNN, Bezerianos et al. [82] have approximated the nonlinearity of heart dynamics, using the local reconstruction of the dynamics in the space spanned by each basis function. Fraser et al. [83] have investigated the effectiveness of radial basis function networks for diagnosis of myocardial infarction. Their method achieved a sensitivity of 85.7%. However, as studied by Tarrassenko [84], an RBFNN may not perform as well as a multilayer network. For example, in an electroencephalogram (EEG) application an RBFNN has shown a shortly increased misclassification (11.6%) when compared to a multilayer neural network.

Lagerholm et al. [85] employed self-organizing neural networks in conjunction with Hermite basis function, for the purpose of beat clustering to identify and classify ECG complexes in arrhythmia. As claimed by the authors, self-organizing networks benefit in interpreting ECG data, allowing to extract the most relevant information from it, outperforming other supervised learning methods.

Hu et al. [86] have studied the feasibility of neural networks applied to a patient-adaptable ECG beat classification algorithm. Their approach consists of an SOM/LVQ–based scheme, easily adapted to other existing automated patient monitoring algorithms. Their analysis reveals that the performance of the patient-adapted network was improved due to their ability to adjust the boundaries between classes, although the distributions of beats were distinct for each patient.

Several neurological disorders are routinely examined by EEG analysis and the differentiation between physiological and pathological alterations requires the flexibility and excellent capability and recognition of various EEG-complexes. In this context, Schetinin [87] has developed an algorithm to classify artifacts and normal segments in clinical EEGs. This method involves evolving cascade neural networks, ensuring a nearly minimal number of input and hidden neurons as well as connections. The algorithm was successfully applied, classifying correctly 96.69% of the testing segments. Singh [88] has developed a polygon feature selection method for the classification of temporal data from two or more sources, with emphasis on the analysis of EEG data. A feature classification, using a modified fuzzy nearest-neighbor method, was used and a recognition rate varying from 90–99% was achieved. Millan et al. [89] have proposed a local neural classifier for the recognition of mental tasks from online spontaneous EEG signals, allowing to recognize three mental tasks. Leichter et al. [90] have developed and applied a classification of EEG data based on independent component analysis (ICA) as a feature extraction technique, and on evolving fuzzy neural networks as a classification modeling technique.

5.3.2 Biological System Control and Prognosis

Imprecisely defined processes for which clinical model–based control techniques are impractical but can be satisfactorily controlled by physicians, fuzzy logic is of particular interest. Fuzzy control can be described as a "control with sentences rather than equations" providing natural-to-use sentences or rules as the control strategy written in terms of if–then clauses.

A fuzzy controller system is usually used in feedback configuration (Figure 5.22). The fuzzy controller establishes a relationship, expressed using the if–then formalism, between inputs (the desired output or set point and the actual output) and the output, the control action.

The field of anesthesia is one of the most relevant concerning applications of fuzzy control in the clinical domain [24]. It involves monitoring the vital parameters of the patient and controlling the drug infusion to maintain the anesthetic-level constant. It includes depth of anesthesia [91], muscle relaxation [46,92], hypertension during anesthesia [93], arterial pressure

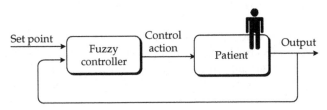

FIGURE 5.22
Fuzzy controller.

control [94], mechanical ventilation during anesthesia [95], and postoperative control of blood pressure [96].

Another example of application of fuzzy control is to develop a computer-based system for control of oxygen. Sun et al. [97] have applied a fuzzy control system delivery to ventilated infants. A successful example is VentPlan, a ventilator management advisor that interprets patients' physiological data to predict the effect of proposed ventilator changes [98] and NéoGanesh, a program for automated control of assisted ventilation in intensive care units [99].

An open-loop system for treatment of diabetic outpatients was developed for calculating the insulin dose [100]. Advisory expert systems can also be considered as an open-loop controller for advising on drug administration in general anesthesia [101]. Carollo et al. [102] have proposed a fuzzy pain control and Ying et al. [103] a fuzzy blood pressure control.

Most of the fuzzy logic control applications in the field of artificial organs are concerned with artificial hearts. In this context, a fuzzy controller has been implemented for adaptation of the heart pump rate to body perfusion demand by pump chamber filling detection [104]. A more advanced system, based on neural and fuzzy controller for an artificial heart, was developed by Lee et al. [105].

In Ref. 25, a combination of fuzzy logic and neural networks is used to develop an adaptive control system for arterial blood pressure using the drug nitroprusside. Another hybrid intelligent system based on a neuro-fuzzy approach can be found in Ref. 106. The system consists of an adaptive fuzzy controller and a network-based predictor for controlling the mean arterial blood pressure of seriously ill patients. The system has the ability to learn the control rules from an off-line training process as well as to adjust the parameters during the control process.

Neural networks are able to provide prognostic information based on retrospective parameter analysis. Given the ability of neural networks to identify patterns or trends in data, they are well suited for prediction or forecasting. In medical applications, neural networks can help clinicians, for example, to investigate the impact of parameters after certain conditions or treatments; they supply clinicians with information about the risk or incoming circumstances.

Patients who are hospitalized for having high-risk diseases require special monitoring. Neural networks have been used as a tool for patient diagnosis and prognosis to determine patients' survival. In this context, Bottaci and Drew [107] have investigated multilayer neural capabilities to predict survival and death of patients with colorectal cancer. Pofahl et al. [108] have implemented a neural network scheme for predicting the length of stay (more than 7 days) for acute pancreatitis patients, having achieved the highest sensitivity (75%). Ohlsson et al. [109] have presented a study for the diagnosis of acute myocardial infarction. In their study a multilayer neural network has been applied to predict whether the patient suffered from acute myocardial infarction or not.

Neural networks have also been successfully applied to other clinical problems [110]. Abidi and Goh [111] proposed a multilayer neural network as a forecaster for bacteria–antibiotic interactions of infectious diseases. Their results have shown that the 1-month forecaster produces a correct output (within occurrences of sensitivity) although predictions for the 2 and 3 months are less accurate. Prank et al. [112] have also used neural networks for predicting the time course of blood glucose levels from the complex interaction of glucose counterregulatory hormones and insulin. Benesova et al. [113] have developed a neural network scheme to predict the teratogenity of perinatal administrated drugs. Lapeer et al. [55] applied neural networks for similar predictive tasks, attempting to pick out perinatal parameters influencing birthweight.

5.3.3   Image Processing

Medical imaging has revolutionized medical practice by providing new, noninvasive, and probably, the most effective tools for diagnosis. Today any medical expert may rely on multiple imaging modalities such as ultrasound (US), projection x-ray, computer tomography (CT), magnetic resonance imaging (MRI), single photon emission computed tomography (SPECT) and positron emission tomography (PET) to obtain detailed morphological (structural), functional, and pathological insight on several aspects of the human body. Besides their diagnosis function, these systems are of great help for other medical tasks such as treatment and surgery planning. To be helpful for healthcare, medical images have to be interpreted either qualitatively or quantitatively in a timely and accurate manner. In this context, medical image processing is increasingly an important tool to aid the medical professional in managing and extracting valuable information from these data sets. Typical useful processing operations on these images are as follows [36]:

Image compression. Most medical images are high-resolution images (see Table 5.5). Hence, image compression is an imperative operation in

TABLE 5.5

Some Typical Characteristics of Medical Images

Modality	Image Matrix	Bytes/Pixel	Megabytes/Study
Digital radiography (DR)	2048 × 2580	2	20
CT	512 × 512	2	30
MR	256 × 256	2	25
US	512 × 512	3	10
Mammography	4096 × 6144	2	192
Angiography	1024 × 1024	2	30
Fluoroscopy	1024 × 1024	1	10
PET/SPECT	256 × 256	2	2

many medical contexts to ensure fast interactivity during browsing of large sets of images (e.g., volumetric data sets, time sequences of images, and image data bases), their efficient storage management in picture archiving and communication systems (around 3.5 TB of data per year may be collected for a medium size hospital) and their application in teleradiology over low or moderate bandwidth networks such as integrated services digital network (ISDN) and satellite networks. For medical image applications, special care has to be devoted to lossy compression schemes to avoid permanent loss of their diagnostic value.

Image preprocessing. The three most frequent image preprocessing operations under the medical context are image restoration in general, image reconstruction, and contrast enhancement. Distortion is an intrinsic property of most medical imaging modalities. In medical images, distortions may be both due to the electronics and the characteristics of the human body. In images where the distinction between normal and abnormal tissues is subtle, accurate interpretation may become difficult in the presence of distortions. Under these circumstances, image enhancement is usually applied to obtain clearer images for medical observation as well as for most automated or semiautomated diagnosis systems. Another common image preprocessing operation for medical applications is image reconstruction. The output from some modalities is not directly observable. For instance, the output from CT scanners is sinograms (collection of projections for different angles) that have to be backprojected to reconstruct the image. Owing to randomness, special algorithms have to be designed to avoid the cost of important details during reconstruction.

Image registration. Registration of images from different modalities is essential in several applications where the correspondence between the images conveys the desired medical information. These images may convey different information such as structural (e.g., CT) and functional (e.g., SPECT) information obtained from the same body part at different instances. Registration algorithms have to account for different types of geometrical and modality-specific distortions as well as distortions due to soft tissue elasticity to properly align the data sets for medical observation.

Image segmentation. Image segmentation is one of the most important processing steps in the analysis of patient image data. The main goal of segmentation algorithms is to divide the image into sets of pixels with strong correlation to significant structures such as organs, lesions, and different tissues that are a part of the image. These sets of segmented regions may be used to aid the medical professional in identifying important features in the image or to extract the necessary features for their automatic classification and disease diagnosis.

5.3.4 Neural and Fuzzy Applications in Medical Image Processing

5.3.4.1 Medical Image Compression

Although there is a considerable research effort concerning medical image compression, most compression approaches reported in the literature do not rely on fuzzy methods. In this context, neural networks (see, e.g., Refs. 114–123) are much more common than fuzzy techniques.

Application of neural network (NN) for data compression always relies on the principle of space reduction. According to Egmont-Petersen et al. [124], two different types of image-compression approaches can be identified using neural networks: direct pixel-based encoding/decoding by one ANN [114,119] and pixel-based encoding/decoding based on a modular approach [124]. Concerning architecture and principle, the major types of NNs that have been adapted for image compression are feedforward networks [116,119,120,122,123], SOMs [125,126], a learning vector quantifier [123], and a radial basis function network [126]. For a more extensive overview see Refs. 111 and 124.

A few attempts to combine fuzzy techniques for image compression have been reported. For example, Karras et al. [127] achieve higher lossy compression thresholds for wavelet coefficients in each discrete wavelet transform (DWT) band in terms of how they are clustered according to their absolute value. Kaya [128] introduces a fuzzy Hopfield neural network for the same purpose as that described by Karras et al. [127]. Fuzzy vector quantization for image compression is performed by Karayiannis et al. [129].

5.3.4.2 Image Enhancement

The majority of applications of ANNs in medical image preprocessing are for image restoration [121–134] and enhancement of specific image features [135]. The goal in image restoration is to compensate for the image distortion introduced by the physical measurement device. Besides noise, the major distortions introduced by the acquisition system are motion blur, out-of-focus blur, and distortion caused by low resolution (e.g., in SPECT). Image restoration is an intrinsically ill-posed problem, since conflicting criteria need to be accomplished, that is, resolution versus smoothness.

Lee and Degyvez [133] introduced color image restoration based on cellular NN (CNN). The generalized adaptive neural filter (GANF) is reported in Ref. 136, which has been applied for noise suppression. A GANF is build up on a set of neural operators, based on a stack of filters. Hopfield networks are a common use of NN for deblurring and diminishing out-of-focus effects (see, e.g., Ref. 137). This problem is usually addressed using maximum *a posteriori* probability (MAP) and regularization. These objective functions can be mapped onto the energy function of Hopfield networks. Usually it is observed that some architecture modifications are required to enable the mapping operation.

Regarding image feature enhancement, most NN applications reported in literature are for edge enhancement. Few exist for other tasks. Usually,

NN approaches for medical image enhancement rely on regression ANNs and classifiers [138,139]. In the latter, typically binary image outputs are obtained. For instance, Shih et al. [139] report an ART network for binary image enhancement. Regarding edge enhancement two approaches can be distinguished: (1) filter approximation [138] and (2) edge classification [140].

Other applications of NNs in this context are the implementation of morphological operators with modified feedforward networks [135] and the use of Grossberg's center-surround shunting feedforward network for contrast enhancement.

Fuzzy techniques have mainly been introduced for noise suppression [141–143], edge, and contrast enhancement [141,144]. Noise reduction in medical images is not a trivial task. The filter should be able to distinguish between unwanted information (noise) and image details that have to be preserved and ideally, be enhanced. From this contradictory objective, it is seen that nonlinear filters based on expert knowledge tend to outperform conventional methods. This is the main reason why fuzzy reasoning is one of the main supporting tools for fuzzy applications to noise reduction—the fuzzy inference ruled by else-action (FIRE) class of filters [145]. Another class of fuzzy filters is the fuzzy weighted filters [146,147]. This approach applies one or more fuzzy systems to evaluate weights of a weighted linear filter. These weights may be associated with the inputs (fuzzy weighted mean filters) [146,147] or with the outputs of different operators (fuzzy selection filters) [148,149]. Other fuzzy filtering approaches rely on the generalization of classical filtering methods such as median and order statistics filters [150–152]. Although a majority of these filters have not been specifically developed for medical image processing, they are very well suited for the task, given their ability to incorporate expert information. An example of their application in a medical context can be found in Zeng et al. [153]. For further information and recent reviews on fuzzy filters, see Ref. 141.

Fuzzy contrast enhancement for medical image processing has been attempted using global transformations, that is, histogram transformations and local adaptive transformations. Global contrast transformations have been reported in Refs. 154–157, whereas local contrast transformations are introduced in Refs. 158 and 159. In Ref. 144, the possibility distribution is applied together with four hard if-then-else rules to stretch the histogram of the input image. A similar approach using an intensification operator over fuzzified image pixel values is presented in Ref. 153. In Ref. 157, histogram hyperbolization is extended for fuzzy image coding. Fuzzy inference is applied in Ref. 144 to globally and locally enhance the image contrast. These techniques are local adaptations, using a small neighborhood of the global algorithms previously mentioned. A completely different approach is presented by Krell et al. [159], who combine histogram hyperbolization with a modified associative memory network to implement a local contrast enhancement algorithm for feature matching in radiotherapy.

5.3.4.3 *Image Registration*

The only known method for image registration using fuzzy techniques is the one reported by Maintz et al. [160]. Their algorithm is a surface-based method for registration of SPECT and MR images. In particular they propose to use the "surfaceness" computed from morphological operators as a fuzzy surface measure that is able to retain more information than concurrent algorithms based on binary segmentation. The registration is performed by optimizing the cross-correlation between the registered "surfaceness" spaces. Although several NNs applications for image registration exist in literature, few are for medical imaging purposes. A rare example is the algorithm reported by Rangarajan and Chui [161]. These authors formulate the registration problem as a feature-based matching approach with correspondences as a mixed variable objective function. Optimization is performed based on a neural-network approach.

5.3.4.4 *Image Segmentation*

Applications of the major fuzzy theoretical principles for image segmentation have been reported. From these, fuzzy clustering is the most straightforward and probably the most applied fuzzy technique for image segmentation in medical contexts. Typical application of this clustering principle is to divide the image into clusters and interpreting the class membership as a correlation or similarity with an ideal anatomical structure or its property. Although several variations on fuzzy clustering exist, the most applied principles for medical image segmentation are FCM [162,163] and the maximum entropy principle–based fuzzy clustering (MEP-FC) [164]. Other fuzzy algorithms applied in this context are possibilistic neuro-fuzzy C-means (PNFCM) [162] and fuzzy hidden Markov chains (FHMC) [165]. For an introduction to these algorithms see Ref. 162.

Fuzzy clustering has been extensively applied for medical image segmentation using two main strategies: (1) as the main segmentation algorithm and (2) as a preprocessing for nonfuzzy segmentation strategies or directly combined with them. In the first class of algorithms, clustering is usually performed directly on the intensity data, although other features may be applied (see, e.g., Ref. 166). A comparative performance analysis for the multimodal image segmentation problem using this approach is presented in Ref. 162. In Ref. 167, FCM is applied to extract the ventricular region in angiography images, whereas Ref. 168 uses a modified FCM to segment brain images obtained from noisy CT scans and one-channel MRI scans. Automatic identification of brain tumors using FCM is discussed in Ref. 169 (for a survey paper on fuzzy applications in brain-related topics, namely on its segmentation, see Ref. 170). Other fuzzy clustering applications to CT and MR image segmentation can be found in Refs. 171 and 172. Ghafar et al. [173] apply FCM for Pap smear image segmentation, whereas tracking of vessels in retinal images using FCM is reported in Ref. 174. Several medical domain applications of fuzzy clustering for unsupervised and supervised image segmentation are reviewed in Ref. 42.

In the second class of segmentation algorithms, clustering is combined with nonfuzzy approaches. For instance, a neurofuzzy segmentation technique for radiographic images is proposed in Ref. 175 based on the clustering of a feature space obtained from a wavelet decomposition of the image. Zhang et al. [176] report a multiresolution approach for cluster identification. In their work intra- and interscale properties are formulated as fuzzy functions, being significant clusters obtained from the minimization of their combined effect. A combined multiresolution FCM algorithm for breast calcifications was recently introduced by Sentelle and Sutton [177]. Fuzzy clustering is applied by Schüpp et al. [178] to initialize seed regions for active contours. Karayiannis and Pai [179] describe a hybrid algorithm for MR image segmentation based on fuzzy algorithms for LVQ, whereas Derrode et al. [165] combine fuzzy and hidden Markov chains to segment ill-defined images.

Other fuzzy image segmentation principles that can be found in literature are methods based on fuzzy integrals (e.g., applied for fuzzy feature weightening), fuzzy geometry (e.g., compactness and connectness) [178], and fuzzy entropy and divergence. However, these principles are less common for medical image segmentation. For a review on these techniques, see Ref. 37.

Algorithms for medical image segmentation using NN can be broadly divided into two classes [124]: (1) pixel-based algorithms [180–188], and (2) feature-based algorithms [189,190]. Regarding the underlying NN, most existing types have been applied: feedforward NNs [185,189], SOMs [168,183–185,188,189], Hopfield networks [186], and constraint satisfaction networks. For medical purpose, most NN-based algorithms have been trained to operate on texture [168,189] and a combination of texture and shape [187].

Regarding the application area, most of these segmentation algorithms have been developed for MR image segmentation [184,185] (a comparison between neural and fuzzy techniques for MR image segmentation is presented in Ref. 182), digital radiology [189], and multimodal images [186].

5.4 Decision Support Systems

Applications in clinical areas often involve analysis and classification of the outcome of an experiment. Clinical diagnosis systems aim at offering suggestions and help in arriving at a diagnosis based on patient data. However, biosignal processing and interpretation in medicine involve a complex analyses of signals, image processing and interpretation, graphic representations, and pattern classification. Consequently, even experienced physicians could misinterpret the available data [191,192].

Diagnosis of diseases is an important and difficult task in medicine. In fact, detecting a disease from several factors or symptoms is a many-layered problem that also may lead to false assumptions with often unpredictable effects. Therefore, the attempt of using the knowledge and experience of many specialists collected in databases to support the diagnosis process seems reasonable. Fuzzy systems are well suited to tasks that heavily rely on human experience and intuition, which is the case of clinical diagnosis systems. Unfortunately in many cases, experts may not know, or may not be able to formulate, what knowledge they actually use in solving their problems. Given a set of clinical cases that act as examples, learning in soft computing can be achieved, for example, with a neurofuzzy methodology.

One of the most widely known applications of neural networks in medicine is the Papnet system [193]: a commercial neural network-based computer program for assisted screening of Pap (cervical) smears. If detected early, cervical cancer has an almost 100% chance of cure. With this system, a Pap smear test examines automatically cells taken from the uterine cervix for signs of precancerous and cancerous changes, thus enabling to detect very early precancerous changes.

Another diagnostic system is presented by Blekas et al. [194], employing a fuzzy neural-network approach, for the discrimination of benign from malignant gastric lesions. The input to the system consists of images of routine-processed gastric smears, stained by Papanicolaou technique. The analysis of the images provides a data set of cell features, being the fuzzy min–max classification network based on hyberbox fuzzy sets that can be incrementally trained. The application of the fuzzy min–max neural network has shown high rates of correct classification (both at cell- and patient level). Alonge et al. [195] presented a neurofuzzy scheme, able to perform focal lesions classification in MR images of brain tissues affected by multiple sclerosis disease. Images are first segmented using a fuzzy technique; then each cluster is processed to classify and label nonpathologic tissues and to locate all possible candidates to be sclerosis lesions. Finally, the neural classification step is implemented using a multilayer neural network, providing an estimate of the position and the shape for each lesion.

Lee et al. [196] have proposed the combination of a multimodule contextual neural network and spatial fuzzy rules and fuzzy descriptors for automatically identifying abdominal organs from a series of CT image slices. With this approach, the difficulties arising from partial volume effects, gray-level similarities of adjacent organs, and contrast media effect can be highly reduced. Basically the multimodule contextual neural network segments each image slice through a divide-and-conquer concept, and the variations in organ position and shape are addressed with spatial fuzzy rules and fuzzy descriptors, along with a contour modification scheme implementing consecutive organ region overlap constraints.

A three-dimensional (3-D) visualization fuzzy-based decision support system to timely detect glaucoma in older patients, as well as to optimize the monitoring process, allowing measuring the progress of the disease is presented in Ref. 197. The practical application of the system at the Department of Ophthalmology and the Eye Hospital of the University of Saarland in Homburg has proven that the optimized support, enhanced by fuzzy methods, for an accurate decision making in disease monitoring can offer direct benefits for the level of medical care and the interactive 3-D visualization might substantially enhance the doctor's involvement in the treatment of patients threatened by glaucoma.

The objective of the work presented by Cherrak et al. [198] was to test the performances of a computer system that was designed to analyze and quantify lesions on two-dimensional renal arteriograms. The system is based on a fuzzy automaton and performs a syntactic analysis of the arterial segment providing automatic and reproducible quantification of lesions. When compared to individual radiologists, the computer system gave a more precise estimation of percent stenosis and did not over- or underestimate the severity of the lesion.

Dutch et al. [199] have studied several systems for extraction of logical rules from data, and applied to the analysis of the melanoma skin cancer data. These systems include, among others, neural networks, enabling a very simple and accurate classification for the four types of melanoma. Clark et al. [200] have presented a knowledge-based paradigm that combines fuzzy techniques, multispectral analysis, and image-processing algorithms, to produce an unsupervised system capable of automatically segmenting and labeling complete glioblastoma-multiforme tumor volumes from transaxial MR images over a period of time during which the tumor is treated.

Zhang and Berardi [201] have investigated the potential of ANNs in diagnosing thyroid diseases. The robustness of neural networks with regard to sampling variations were examined using a cross-validation method. They have demonstrated that for medical diagnosis problems, where the data are often highly unbalanced, neural networks can be a promising classification method for practical use.

Pesonen et al. [202] have presented a neural network–based decision support system for the diagnosis of acute abdominal pain. Namely, two neural network algorithms, backpropagation and LVQ were studied, and the k-nearest neighbors in deciding the correct class for the LVQ network was used. The evaluation of the network with different databases as well as the comparison to statistical analyses has shown the effectiveness of the neural network scheme. Smith and Arabshahi [203] report the development of a fuzzy decision system to semiautomate ultrasonic prenatal examinations. The main goal is to reduce costs and minimize exposure time of the fetus to ultrasonic radiation. Varachiu et al. [204] have proposed the use of a knowledge discovery process to develop a fuzzy logic inference system for diagnosis and prediction of glaucoma.

Aphasia is a disturbance in the communicative use of language (disability to use or comprehend words), which can occur in different forms and results from brain damage. Jantzen et al. [205] have explored the capabilities of neurofuzzy system to classify several types of aphasia, showing their effectiveness for aphasia diagnosis.

Sordo et al. [206] have implemented a knowledge-based neural network for classification of phosphorus (31P) magnetic resonance spectra from normal and cancerous breast tissues. *A priori* knowledge of metabolic features of normal and cancerous breast tissues was incorporated into the structure of the neural network to overcome the scarcity of available data. The knowledge-based neural network proposed has outperformed conventional neural networks revealing that the combination of symbolic and neural techniques is more robust than a neural technique alone.

5.5   Conclusion

Computational intelligence theories have undergone important developments during the past years. They provide techniques and tools that may support, in a very useful way, human decisions in complex contexts. Complexity here means high number of factors, changing conditions, imprecise knowledge, vagueness, lack of data, etc. The medical and healthcare domains are probably those with the highest potential for these techniques. An intense research has been and is going on worldwide concerning neural networks, fuzzy systems, and their combinations for applications covering practically all activities in these areas. This chapter provides a brief overview for these techniques and applications. The use of these techniques in the daily life of clinicians is in progress and it is expected that, with demonstrable confidence, massive utilization will result in real benefits for the patients and for the clinicians.

References

1. Zadeh, L. (1965). Fuzzy sets, *Information and Control*, 8, 338–353.
2. Noor, A. (1996). A hard look at soft computing. *Aerospace America*, September.
3. Hagan, M. (1997). Neural networks design, MIT Press, Cambridge, MA.
4. Brause, R. (2001). Medical analysis and diagnosis by neural networks. *Proceedings of the 2nd International Symposium on Medical Data Analysis*, Vol. 2199, Lecture Notes in Computer Science, Springer Verlag, London, 1–13.
5. MATLAB (2002). *Users Manual of Neural Networks Toolbox*, The MathWorks, Natick, MA.
6. Hassoun, M. (1995). Fundamentals of artificial neural networks, MIT Press, Cambridge, MA.
7. Rumelhart, D., Hinton, G. and Williams, R. (1986). *Learning internal representations by error propagation*, PDP Research Group, Parallel Distributed Processing, MIT Cambridge, MA.
8. Lisboa, P. (2002). A review of evidence of health benefit from artificial neural networks in medical intervention. *Neural Networks*, 15(1), 11–39.
9. Kohonen, T. (1982). Self-organized formation of topologically correct feature maps. *Biol. Cybern.*, 43, 59–69.

10. Kohonen, T. (1978), *Associative Memory: A System-Theoretical Approach*, Springer-Verlag, Heidelberg.
11. Grossberg, S. (1976). Adaptive pattern classification and universal recoding: I. parallel development and coding of neural feature detectors. *Biol. Cybern.*, 23, 121–134.
12. Hopfield, J. (1982). Neural networks and physical systems with emergent collective computational abilities. *Proc. Natl. Acad. Sci. USA*, 9(2554), 2554–2558.
13. Partridge, D., Abidi, S. and Goh, A. (1996). Neural network applications in medicine. *Proceedings of National Conference on Research and Development in Computer Science and Its Applications* (REDECS'96), Universiti Pertanian Malaysia, Kuala Lumpur, 20–23.
14. Papik, K., Molnar, B., Schaefer, R., Dombovari, Z., Tulassay, Z. and Feher, Z. (1998). Application of neural networks in medicine, a review. *Med. Sci. Monit.*, 4(3), 538–546.
15. Douglas, R. (2001). Mathematical modeling and simulation of biological systems in renal transplantation and renal disease, www.uninet.edu/cin2001/html/conf/mclean/node3.html.
16. Ishak, W. (2001). The potential of neural networks in medical applications. Faculty of Information Technology, Universiti Utara Malaysia, 06010 Sintok, Kedah, Malaysia, http://www.generation5.org/content/2004/NNAppMed.asp.
17. Dounias, G. (2003). Hybrid computational intelligence in medicine. *Eunite—Intelligent and Adaptive Systems in Medicine*, Prague, March 31–April 1.
18. PubMed public site, http://www.ncbi.nlm.nih.gov/entrez.
19. Ozaki, M. (1999). A fuzzy system for dental development age evaluation, in *Fuzzy and NeuroFuzzy Systems in Medicine*, H.-N. L. Teodorescu, A. Kandel and L. C. Jain (Eds), CRC Press, Boca Raton, FL, pp. 195–209.
20. Erikson, R., Chelaru, M. and Buhusi, C. (1999). The brain as a fuzzy machine, in *Fuzzy and Neuro-Fuzzy Systems in Medicine*, H.-N. L. Teodorescu, A. Kandel and L. C. Jain (Eds), CRC Press, Boca Raton, FL, pp. 17–56.
21. Ross, T. (1995). *Fuzzy Logic with Engineering Applications*, McGraw Hill, New York.
22. Kulkarni, J. and Karwowski, W. (1986). Research guide to applications of fuzzy set theory in human factors, in *Application of Fuzzy Set Theory in Human Factors*, W. Karwowski, and A. Mital (Eds), Elsevier, Amsterdam, pp. 395–446.
23. Warren, J., Beliakov, G. and Zwaag, B. (2000). Fuzzy logic in clinical practice decision support systems. *Proceedings of the 33rd Hawaii International Conference on System Sciences*, Maui, Hawaii, January 4–7.
24. Linkens, D., Abbod, M. and Mahfouf, M. (1999). An initial survey of fuzzy logic monitoring and control utilisation in medicine. *European Symposium on Intelligent Techniques*, Orthodox Academy of Crete, Greece, June 3–4.
25. Teodorescu, H. and Kandel, J. (1998). *Fuzzy and Neuro-Fuzzy Systems in Medicine*, CRC Press, Boca Raton, FL.
26. Theodorescu, H., Kandel, A. and Jain, L. (1999). Fuzzy logic and neuro-fuzzy systems in medicine and bio-medical engineering: A historical perspective, in *Fuzzy and Neuro-Fuzzy Systems in Medicine*, H.-N. L. Teodorescu, A. Kandel and L. C. Jain (Eds), CRC Press, Boca Raton, FL, pp. 17–56.
27. Pedrycz, W. and Gomide, F. (1998). *An Introduction to Fuzzy Sets Analysis and Design*, MIT Press, Cambridge, MA.
28. Hoppner, F., Klawonn, F., Kruse, R. and Runkler, T. (1999). *Fuzzy Cluster Analysis, Methods for Classification, Data Analysis and Image Recognition*, Wiley, New York.
29. Babuska, R. (1998). *Fuzzy Modeling for Control*, Kluwer Academic, Dordrecht.

30. Nauck, D., Klawonn, F. and Kruse, R. (1997). *Foundations of Neuro-Fuzzy Systems*, Wiley, Chichester.
31. Chiu, S. (1994). Fuzzy model identification based on clustering estimation. *J. Intelligent Fuzzy Syst.*, 2, 267–278.
32. Deng, D. and Kasabov, N. (2000). Evolving self-organizing maps for online learning, data analysis and modeling. *Proc. IJCNN'2000*, VI, 3–8.
33. Kasabov, N. and Song, Q. (2002). DENFIS: Dynamic evolving neural-fuzzy Inference system and its application for time-series prediction. *IEEE Trans. Fuzzy Syst.*, 10(2), 144–154.
34. Azeem, M., Hanmandlu, M. and Ahmad, N. (2000). Generalization of adaptive neuro-inference system. *IEEE Trans. Neural Networks*, 11(6), 132–134.
35. Paiva, R. and Dourado, A. (2000). Structure and parameter learning of neuro-fuzzy systems: A methodology and a comparative study. *J. Intelligent Fuzzy Syst.*, 11, 147–161.
36. Haussecker, H. and Tizhoosh, H. (1999). Fuzzy image processing, *Handbook of Computer Vision and Applications*, Vol. 2, B. Jähne, H. Haussecker, P. Geissler (Eds), Academic Press, New York, pp. 683–727.
37. Tizhoosh, H. (1997). *Fuzzy Image Processing: Introduction in Theory and Practice*, Springer-Verlag, Heidelberg.
38. Ishibuchi, H., Nazaki, K. and Yamamoto, N. (1993). Selecting fuzzy rules by genetic algorithms for classification. *IEEE Int. Conf. Fuzzy Syst.*, 2, 1119–1124.
39. Lapuerta, P., Azen, S. and LaBree, L. (1995). Use of neural networks in predicting the risk of coronary artery disease. *Comput. Biomed. Res.*, 28(1), 38–52.
40. Kerre, E. and Nachtegael, M. (2000). *Fuzzy Techniques in Image Processing, Studies in Fuzziness and Soft Computing*, Vol. 52, Springer-Verlag, Heidelberg.
41. Goodacre, R., Neal, M. and Kell, D. (1996). Quantitative analysis of multivariate data using artificial neural networks: A tutorial review and applications to the deconvolution of pyrolysis mass spectra. *Zentralblatt Fur Bakteriologie*, 284(4), 516–539.
42. Sutton, M., Bezdek, J. and Cahoon, T. (2000). Image segmentation by fuzzy clustering: methods and issues, in *Handbook of Medical Imaging: Processing and Analysis*, I. Bankman (Ed.), Academic Press, New York, pp. 87–106.
43. Itchhaporia, D., Snow, P., Almassy, R. and Oetgen, W. (1996). Artificial neural networks: Current status in cardiovascular medicine. *J. Am. Coll. Cardiol.*, 28(2), 515–521.
44. Janet, F. (1997). Artificial neural networks improve diagnosis of acute myocardial infarction. *Lancet*, 350(9082), 935.
45. Brickley, M., Shepherd, J. and Armstrong, R. (1998). Neural networks: A new technique for development of decision support systems in dentistry. *J. Dentistry*, 26(4), 305–309.
46. Linkens, D. and Mahfouf, M. (1988). Fuzzy logic knowledge based control of muscle relaxation. *IFAC Proceedings on Modeling and Control of Systems*, Venice, Italy, 185–190.
47. Hamamoto, I., Okada, S., Hashimoto, T., Wakabayashi, H., Maeba, T. and Maeta, H. (1995). Prediction of the early prognosis of the hepatectomized patient with hepatocellular carcinoma with a neural network. *Comput. Biol. Med.*, 25(1), 49–59.
48. Burstein, Z. (1995). A network model of developmental gene hierarchy. *J. Theor. Biol.*, 174(1), 1–11.
49. Eklund, P. and Forsstrom, J. (1995). Computational intelligence for laboratory information systems. *Scand. J. Clin. Lab. Invest. Suppl.*, 222, 21–30.

50. Chittajallu, S. and Wong, D. (1994). Connectionist networks in auditory system modeling. *Comput. Biol. Med.*, 124(6), 431–439.
51. Burtis, C. (1995). Technological trends in clinical laboratory science. *Clin. Biochem.*, 28(3), 213–219.
52. DeFigueiredo, R., Shankle, W., Maccato, A., Dick, M., Mundkur, P., Mena, I. and Cotman, C. (1995). Neural-network based classification of cognitively normal, demented, Alzheimer disease and vascular dementia from single photon emission with computed tomography image data from brain. *Proc. Natl. Acad. Sci. USA*, 92(12), 5530–5534.
53. Guigon, E., Dorizzi, B., Burnoda, Y. and Schultz, W. (1995). Neural correlates of learning in the prefrontal cortex of the monkey: a predictive model. *Cereb. Cortex*, 5(2), 135–147.
54. Clarke, L. P. and Wei, Q. (1998). Fuzzy-logic adaptive neural networks for nuclear medicine image restorations. *Proceedings of the 20th Annual International Conference of the IEEE Engineering in Medicine and Biology Society*, Vol. 20, Biomedical Engineering Towards the Year 2000 and Beyond, IEEE Piscataway, NJ, 3384–3390.
55. Lapeer, R., Dalton, K., Prager, R., Forsstrom, J., Selbmann, H. and Derom, R. (1995). Application of neural networks to the ranking of perinatal variables influencing birthweight. *Scand. J. Clin. Lab. Invest. Suppl.*, 222, 83–93.
56. Naguib, R. and Sherbet, G. (1997). Artificial neural networks in cancer research. *Pathobiology*, 65(3), 129–139.
57. Baker, J., Kornguth, P., Williford, M. and Floyd, C. (1995). Breast cancer: prediction with artificial neural network based on BI-RADS standardized lexicon. *Radiology*, 196(3), 817–822.
58. Fogel, D. and Wasson, E. (1995). Boughton EM: Evolving neural networks for detecting breast cancer. *Cancer Lett.*, 96(1), 49–53.
59. Moul, J., Snow, P., Fernandez, E., Maher, P. and Sesterhenn, I. (1995). Neural network analysis of quantitative histological factors to predict pathological stage in clinical stage I nonseminomatous testicular cancer. *J. Urol.*, 153(5), 1674–1677.
60. Accornero, N. and Capozza, M. (1995). OPTONET: Neural network for visual field diagnosis. *Med. Biol. Eng. Comput.*, 33(2), 223–226.
61. Maeda, N., Klyce, S. and Smolek, M. (1995). Neural network classification of corneal topography. Preliminary demonstration. *Invest. Ophthalmol. Vis. Sci.*, 36(7), 1327–1335.
62. Leisenberg, M. and Downes, M. (1995). CINSTIM: The Southampton cochlear implant-neural network simulation and stimulation framework: Implementation advances of a new, neural-net-based speech-processing concept. *Ann. Otol. Rhinol. Laryngol. Suppl.*, 166, 375–377.
63. Kolles, H., von-Wangenheim, A., Vince, G., Niedermayer, I. and Feiden, W. (1995). Automated grading of astrocytomas based on histomorphometric analysis of Ki-67 and Feulgen stained paraffin sections. Classification results of neuronal networks and discriminant analysis. *Anal. Cell Pathol.*, 8(2), 101–116.
64. Nazeran, H., Rice, F., Moran, W. and Skinner, J. (1995). Biomedical image processing in pathology: A review. *Australas. Phys. Eng. Sci. Med.*, 18(1), 26–38.
65. Doyle, H., Parmanto, B., Munro, P., Marino, I., Aldrighetti, L., McMichael, J. and Fung, J. (1995). Building clinical classifiers using incomplete observations: A neural network ensemble for hepatoma detection in patients with cirrhosis. *Methods Inf. Med.*, 34(3), 253–258.

66. Gurney, J. and Swensen, S. (1995). Solitary pulmonary nodules: determining the likelihood of malignancy with neural network analysis. *Radiology*, 196(3), 823–829.

67. Tourassi, G., Floyd, C., Sostman, H. and Coleman, R. (1995). Artificial neural network for diagnosis of acute pulmonary embolism: effect of case and observer selection. *Radiology*, 194(3), 889–893.

68. Kahn, C. (1994). Artificial intelligence in radiology: Decision support systems. *RadioGraphics*, 14, 849–861.

69. Robert, C., Guilpin, C. and Limoge, A. (1998). Review of neural network applications in sleep research. *J. Neurosci. Methods*, 79(2), 187–193.

70. Wei, J., Zhang, Z., Barnhill, S., Madyastha, K., Zhang, H. and Oesterling, J. (1998). Understanding artificial neural networks and exploring their potential applications for the practicing urologist. *Urology*, 52(2), 161–172.

71. Tewari, A. (1997). Artificial intelligence and neural networks: Concept, applications and future in urology. *Br. J. Urol.*, 80(3), 53–58.

72. Suzuki, Y., Itakura, K., Saga, S. and Maeda, J. (2001). Signal processing and pattern recognition with soft computing. *IEEE Proceedings of the IEEE*, 89(9), 1297–1317.

73. Lee, A., Ulbricht, C. and Dorffner, G. (1999). Application of artificial neural networks for detection of abnormal fetal heart rate pattern: A comparison with conventional algorithms. *J. Obstet. Gynaecol.*, 19(5), 482–485.

74. Farruggia, S. and Nickolls, Y. (1993). Implantable cardiverter defibrillator electrogram recognition with a multilayer perceptron. *PACE Pacing Clin. Electrophysiol.*, 161(1Pt2), 228–234.

75. Hoyer, D., Schmidt, K. and Zwiener, U. (1995). Principles and experiences for modeling chaotic attractors of heart rate fluctuations with artificial neural networks. *Biomed. Tech. Berl.*, 40(7–8), 190–194.

76. Ortiz, J., Sabbatini, E., Ghefter, C. and Silva, C. (1195). Use of artificial neural networks in survival evaluation in heart failure. *Arq. Bras. Cardiol.*, 64(1), 87–90.

77. Macfarlane, P. (1993). Recent developments in computer analysis of ECGs. *Clin. Physiol.*, 12(3), 313–317.

78. Papaloukas, C., Dimitrios, F., Likasb, A. and Michalisc, L. (2002). An ischemia detection method based on artificial neural networks. *Artif. Intelligence Med.*, 24, 167–178.

79. Silipo, R. and Marchesi, C. (1998). Artificial neural networks for automatic ECG analysis. *IEEE Trans. Signal Proc.*, 46(5), 1417–1425.

80. Waltrous, R. and Towell, G. (1995). A patient-adaptive neural network ECG patient monitoring algorithm, computer in cardiology, Proceedings of Computers in Cardiology, Vienna, Austria, September 10–13, 229–232.

81. Nabney, I. (1999). Efficient training of RBF networks for classification. Technical report NCRG/99/002, Neural Computing Research Group, Aston University.

82. Bezerianos, A., Papadimitriou, S. and Alexopoulos, D. (1999). Radial basis function neural networks for the characterization of heart rate variability dynamics. *Artif. Intelligence Med.*, 15(3), 215–234.

83. Fraser, H., Pugh, R., Kennedy, R., Ross, P. and Harrison, R. (1994). A comparison of backpropagation and radial basis functions, in the diagnosis of myocardial infarction. In *International Conference on Neural Networks and Expert Systems in Medicine and Healthcare*, E. Ifeachor and K. Rosen (Eds), University of Plymouth, Plymouth, England, pp. 76–84.

84. Tarrassenko, L. (1998). *A Guide to Neural Computing Applications*, Arnold, London.

85. Lagerholm, M., Peterson, C., Braccini, G., Edenbrandt, L. and Sörnmo, L. (2000). Clustering ECG complexes using Hermite functions and self-organizing maps. *IEEE Trans. Biomed. Eng.*, 47, 838–848.

86. Hu, Y. Palreddy, S. and Tompkins, W. (1997). A patient-adaptable ECG beat classifier using a mixture of experts approach. *IEEE Trans. Biomed. Eng.*, 44(9), 891–900.

87. Schetinin, V. (2003). A learning algorithm for evolving cascade neural networks, *Neural Process. Lett.* 17(21), 21–31.

88. Singh, S. (2000). EEG data classification with localised structural information, *15th International Conference on Pattern Recognition*, Barcelona, IEEE Press, 2, 271–274, September 3–8.

89. Millan, J., Mouriño, J., Franzé, M., Cincotti, F. and Varsta, M. (2002). A local neural classifier for the recognition of EEG patterns associated to mental tasks, *IEEE Trans. Neural Networks*, 13(3), 678–686.

90. Leichter, C., Cichocki, A. and Kasabov, N. (2001). Independent component analysis and evolving fuzzy neural networks for the classification of single trial EEG data, ANNES01.

91. Abbod, M. and Linkens, D. (1998). Anaesthesia monitoring and control using fuzzy logic fusion. *J. Biomed. Eng.—Applications, Basis & Communications, Special Issues on Control Methods in Anaesthesia*, Taiwan, August.

92. Westenskow, D. (1997). Fundamentals of feedback control: PID, fuzzy logic, and neural networks. *J. Clin. Anesth.*, 9(6 Suppl), 33S–35S.

93. Oshita, S., Nakakimura, K., Kaieda, R. and Hiraoka, T. (1993). Application of the concept of fuzzy logistic controller for treatment of hypertension during anaesthesia. *Masui*, 42(2), 185–189.

94. Zbinden, A., Feigenwinter, P., Petersen-Felix, S. and Hacisalihzade, S. (1995). Arterial pressure control with isoflurane using fuzzy logic. *Br. J. Anaesth.*, 74(1), 66–72.

95. Schaublin, J., Derighetti, M., Feigenwinter, P., Petersen-Felix, S. and Zbinden, A. (1996). Fuzzy logic control of mechanical ventilation during anaesthesia. *Br. J. Anaesth.*, 77(5), 636–641.

96. Ying, H. and L. Sheppard (1994). Regulating mean arterial pressure in postsurgical cardiac patients. A fuzzy logic system to control administration of sodium nitroprusside. *IEEE Engineering in Medicine and Biology Magazine*, 13(5), 671–677.

97. Sun, Y., Kohane, I. and Stark, A. (1994). Fuzzy logic assisted control of inspired oxygen in ventilated new-born infants. *Proceedings of the 18th Annual Symposium on Computer Applications in Medical Care*, Hanley & Belfus, Washington, D.C., Philadelphia, November 5–9, 756–761.

98. Rutledge, G. and Thomsen, G. (1991). Ventplan: A ventilator-management advisor, *15th Annual Symposium on Computer Applications in Medical Care*, McGraw-Hill, New York, 869–871.

99. Dojat, M., Pachet, F., Guessum, Z., Touchard, D., Harf, A. and Brochard, L. (1997). NéoGanesh: A working system for the automated control of assisted ventilation in ICUs. *Artif. Intelligence Med.*, 11, 97–117.

100. Kokutei, K., Yamashita, H., Horie, K. and Funatsu, H. (1993). Diabetic retinopathy score and fuzzy logic in therapy. *Nippon Ganka Gakkai Zasshi*, 97(5), 632–638.

101. Greenhow, S., Linkens, D. and Asbury, A. (1992). Development of an expert system advisor for anaesthetic control. *Comput. Methods Programs Biomed.*, 37(3), 215–229.

102. Carollo, A., Tobar, A. and Hernandez, C. (1993). A rule-based postoperative pain controller: Simulation results. *Int. J. Bio-Med. Comput.*, 33, 267–276.

103. Ying, H., McEachern, M., Eddleman, D. and Sheppard, L. (1992). Fuzzy control of mean arterial pressure in postsurgical patients with sodium nitroprusside infusion. *IEEE Trans. Biomed. Eng.*, 39, 1060–1069.

104. Kaufmann, R., Becker, K., Nix, C., Reul, H. and Rau, G. (1995). Fuzzy control concept for a total artificial heart. *Artif. Organs*, 19(4), 355–361.

105. Lee, M., Ahn, J., Min, B., Lee, S. and Park, C. (1996). Total artificial heart using neural and fuzzy controller. *Artif. Organs*, 20(11), 1220–1226.

106. Xu, Z. and Packer, J. (1995). A hybrid fuzzy-neural control system for management of mean arterial pressure of seriously ill patients, *1995 IEEE International Conference on Neural Networks Proceedings*, IEEE, New York.

107. Bottaci, L. and Drew, P. (1997). Artificial neural networks applied to outcome prediction for colorectal cancer patients in separate institutions. *Lancet*, 350(9076), 469–473.

108. Pofahl, W., Walczak, S., Rhone, E. and Izenberg, S. (1998). Use of an artificial neural network to predict length of stay in acute pancreatitis, *66th Annual Scientific Meeting and Postgraduate Course Program*, Louisville, KY, January 31–February 4.

109. Ohlsson, M., Holst, H. and Edenbrandt, L. (1999). Acute Myocardial Infarction: Analysis of the ECG Using Artificial Neural Networks. LU TP 99-34 (submitted to ANNIMAB-1; Artificial Neural Networks in Medicine and Biology).

110. Nayak, R, Jain, L. and Ting, B. (2001). Artificial neural networks in biomedical engineering: A review. *Proceedings of the 1st Asian-Pacific Congress on Computational Mechanics*, S. Valliappan, N. Khalili (Eds). Sydney, pp. 887–892.

111. Abidi, S. and Goh, A. (1998). Neural network based forecasting of bacteria-antibiotic interactions for infectious disease control. *In 9th World Congress on Medical Informatics (MedInfo'98)*, Seoul, August 18–22.

112. Prank, K., Jurgens, C., Muhlen, A. and Brabant, G. (1998). Predictive neural networks for learning the time course of blood glucose levels from the complex interaction of counterregulatory hormones, *Neural Comput.*, 10(4), 941–954.

113. Benesova, O., Tejkalova, H., Kristofikova, A. and Dostal, M. (1995). Perinatal pharmacotherapy and the risk of functional teratogenic defects. *Cesk. Fysiol.*, 44(1), 11–14.

114. Panagiotidis, N., Kalogeras, G. and Kollias, S. (1996). Neural network-assisted effective lossy compression of medical images, *Proceedings of the IEEE*, 84(10), 1474–1487.

115. Amerijckx, C., Verleysen, M. and Thissen, P. (1998). Image compression by self-organized Kohonen map. *IEEE Trans. Neural Networks*, 9(3), 503–507.

116. Brause, R., Rippl, M. (1998). Noise suppressing sensor encoding and neural signal orthonormalization. *IEEE Trans. Neural Networks*, 9(4), 613–628.

117. Dony, R. and Haykin, S. (1995). Neural network approaches to image compression. *Proc. IEEE*, 83(2), 288–303.

118. Hauske, G. (1997). A self organizing map approach to image quality. *Biosystems*, 40(1–2), 93–102.

119. Skarbek, W. and Cichocki, A. (1996). Robust image association by recurrent neural subnetworks. *Neural Process. Lett.*, 3, 131–138.

120. Dony, R. and Haykin, S. (1995). Optimally adaptive transform coding. *IEEE Trans. Image Process*, 4(10), 1358–1370.

121. Mitra, S. and Yang, S. (1999). High fidelity adaptive vector quantization at very low bit rates for progressive transmission of radiographic images. *J. Electron. Imaging*, 8(1), 23–35.

122. Tzovaras, D. and Strintzis, M. (1998). Use of nonlinear principal component analysis and vector quantization for image coding. *IEEE Trans. Image Process,* 7(8), 1218–1223.
123. Weingessel, A., Bischof, H., Hornik, K. and Leisch, F. (1997). Adaptive combination of PCA and VQ networks. *IEEE Trans. Neural Networks,* 8(5), 1208–1211.
124. Egmont-Petersen, M., Ridder, D. and Handels, H. (2002). Image processing with neural networks—a review. *Pattern Recognition,* 35(10), 2279–2301.
125. Amerijckx, C., Verleysen, M. and Thissen, P. (1998). Image compression by self-organized Kohonen map. *IEEE Trans. Neural Networks,* 9(3), 503–507.
126. Rizvi, S. A., Wang, L. C., Nasrabadi, N. M. (1997), Nonlinear, vector prediction using feed-forward neural networks. *IEEE Trans. Image Process,* 6(10), 1431–1436.
127. Karras, D. A., Karkanis, S. A. and Maroulis, D. E. (2000). Efficient image compression of medical images using the wavelet transform and fuzzy c-means clustering on regions of interest. *Proceedings of The 26th EUROMICRO Conference (EUROMICRO'00)-Volume 2,* The Netherlands.
128. Kaya, M. (2003). A new image clustering and compression method based on fuzzy hopfield neural network. *Proc. Int. Conf. Signal Process.,* 1(2), 1304–2386.
129. Karayiannis, N., Pai, P. and Zervos, N. (1998). Image compression based on fuzzy algorithms for learning vector quantization and wavelet image decomposition. *IEEE Trans. Image Process.,* 7(8), 1223–1230.
130. de Ridder, D., Duin, R. P. W., Verbeek, P. W. and van Vliet, L. J. (1999). The applicability of neural networks to non-linear image processing. *Pattern Anal. Appl.,* 2(2), 111–128.
131. Guan, L., Anderson, J. and Sutton, J. (1997). A network of networks processing model for image regularization. *IEEE Trans. Neural Networks,* 8(1), 169–174.
132. Hanek, H. and Ansari, N. (1996). Speeding up the generalized adaptive neural filters. *IEEE Trans. Image Process,* 5(5), 705–712.
133. Lee, C. and Degyvez, J. (1996). Color image processing in a cellular neural-network environment. *IEEE Trans. Neural Networks,* 7(5), 1086–1098.
134. Russo, F. (2000). Image filtering using evolutionary neural fuzzy systems, in *Soft Computing for Image Processing,* S. K. Pal, A. Ghosh and M. K. Kundu (Eds), Physica-Verlag, Heidelberg, pp. 23–43.
135. Moh, J. and Shih, F. (1995). A general purpose model for image operations based on multilayer perceptrons. *Pattern Recognition,* 28(7), 1083–1090.
136. Zhang, Z. and Ansari, N. (1996). Structure and properties of generalized adaptive neural filters for signal enhancement. *IEEE Trans. Neural Networks,* 7(4), 857–868.
137. Sun, Y. and Yu, S. (1995). Improvement on performance of modified Hopfield neural network for image restoration. *IEEE Trans. Image Process.,* 4(5), 683–692.
138. Pugmire, R., Hodgson, R. and Chaplin, R. (1998). The properties and training of a neural network based universal window filter developed for image processing tasks, in *Brain-like Computing and Intelligent Information Systems,* S. Amari and N. Kasabov (Eds), Springer-Verlag, Singapore, pp. 49–77.
139. Shih, F., Moh, J. and Chang, F. (1992). A new ART-based neural architecture for pattern classification and image enhancement without prior knowledge. *Pattern Recognition,* 25(5), 533–542.
140. Chandrasekaran, V., Palaniswami, M. and Caelli, T. (1996). Range image segmentation by dynamic neural network architecture. *Pattern Recognition,* 29(2), 315–329.

141. Nachtegael, M., Van der Weken, D., Van der Ville, D., Kerre, E., Philips, W. and Lemahieu, I. (2001). An overview of fuzzy filters for noise reduction, *IEEE International Conference on Fuzzy Systems*, Melbourne, Australia, December 2–5, 7–10.

142. Russo, F. and Ramponi, G. (1996). A fuzzy filter for images corrupted by impulse noise. *IEEE Signal Process. Lett.*, 3(6), 168–170.

143. Ville, D., Nachtegael, M., Van der Weken, D., Kerre, E., Philips, W. and Lemahieu, I. (2003). Noise reduction by fuzzy image filtering. *IEEE Trans. Fuzzy Syst.*, 11(4), 429–436.

144. Hassanien, A. and Badr, A. (2003). A comparative study on digital mamography enhancement algorithms based on fuzzy theory. *Stud. Inf. Control*, 12(1), 21–31.

145. Russo, F. (1996). Fuzzy systems in instrumentation: Fuzzy signal processing. *IEEE Trans. Instrum. Measurement*, 45, 683–689.

146. Muneyasu, M., Wada, Y. and Hinamoto, T. (1996). Edge-preserving smoothing by adaptive nonlinear filters based on fuzzy control laws, in *Proc. 1996 IEEE Int. Conf. Image Processing*, ICIP'96, Lausanne, Switzerland, September, pp. 16–19; 785–788.

147. Muneyasu, M., Wada, Y. and Hinamoto, T. (1998). A new edge-preserving smoothing filter based on fuzzy control laws and local features, in *Proc. IEEE Int. Symp. Circuits Systems*, ISCAS'98, Monterey, CA.

148. Choi, Y. and Krishnapuram, R. (1997). A robust approach to image enhancement based on fuzzy logic. *IEEE Trans. Image Process.*, 6, 808–825, June.

149. Keller, J., Krishnapuram, R., Gader, P. and Choi, Y. (1996). Fuzzy rule-based models in computer vision, in *Fuzzy Modeling: Paradigms and Practice*, W. Pedrycz (Ed.) Kluwer, Norwell, MA, pp. 353–374.

150. Plataniotis, K., Androutsos, D. and Venetsanopoulos, A. (1997). Multichannel filters for image processing. *Signal Processing: Image Communication*, 9, 143–158.

151. Taguchi, A. and Izawa, N. (1996). Fuzzy center weighted median filters, in *Proc. VIII Eur. Signal Processing Conf.*, EUSIPCO'96, Trieste, Italy, September 10–13, 1721–1724.

152. Yu, P. and Chen, R. (1996). Fuzzy stack filters—their definitions, fundamental properties and application in image processing. *IEEE Trans. Image Process.*, 5, 838–854.

153. Zeng, N., Taniguchi, K., Watanabe, S., Nakano, Y. and Nakamoto, H. (2000). Fuzzy computation for detecting edges of low-contrast substances in urinary sediment images. *Med. Imaging Technol.*, 18(4), 217–229.

154. Zadeh, L. (1972). A fuzzy-set-theoretic interpretation of linguistic hedges. *J. Cybern.* 2, 4–34.

155. Pal, S. and King, R. (1981). Image enhancement using smoothing with fuzzy sets. *IEEE Trans. on Systems, Man Cybern.*, SMC-11(7), 494–501.

156. Banks, S. (1990). *Signal Processing, Image and Pattern Recognition*, Prentice Hall, New York.

157. Tizhoosh, H. and Fochem, M. (1995). Image enhancement with fuzzy histogram hyperbolization. *Proc. EUFIT'95*, 3, 1695–1698.

158. Gonzalez, R. and Woods, R. (1992). Digital Image Processing, Addison-Wesley, Reading, MA.

159. Krell, G., Tizhoosh, H., Lilienbum, T., Moore, C. and Michaelis, B. (1997). Fuzzy enhancement and associative feature matching for radiotherapy, *Int. Conf. on Neural Networks*, Houston, TX, June 9–12.

160. Maintz, J., Elsen, P. and Viergever, M. (1996). Registration of SPECT and MR images using fuzzy surfaces, in *Medical Imaging: Image Processing*, Vol. 2710, M. H. Loew and K. M. Hanson (Eds), SPIE, Bellingham, WA, pp. 821–829.

161. Rangarajan, A. and Chui, H. (2000). *Applications of Optimizing Neural Networks in Medical Image Registration, Artificial Neural Networks in Medicine and Biology*, Springer Verlag Series: Perspectives in Neural Computing, Springer Verlag, London.

162. Masulli, F., Schenone, A. and Massone, A. (2000). Fuzzy clustering methods for the segmentation of multimodal medical images, in *Fuzzy Systems in Medicine*, P. S. Szczepaniak, P. J. G. Lisboa and S. Tsumoto (Eds), Series Studies in Fuzziness and Soft Computing, J. Kacprzyk (Ed.), Springer-Verlag, Heidelberg (Germany), pp. 335–350.

163. Bezdek, J., Hall, L. and Clarke, L. (1993). Review of MR image segmentation techniques using pattern recognition. *Med. Phys.*, 20, 1033–1048.

164. Krishnapuram, R. and Keller, J. (1993). A possibilistic approach to clustering. *IEEE Trans. Fuzzy Syst.*, 1, 98–110.

165. Derrode, S., Carincotte, C. and Bourenanne, S. (2004). Unsupervised image segmentation based on high-order hidden Markov chains *ICASSP'04*, Montreal, Canada.

166. Yazdan-Shahmorad, A., Soltanian-Zadeh, H., Zoroofi, R. and Zamani, P. (2003). MRSI brain lesion characterization using wavelet transform and fuzzy clustering, *World Congress on Medical Physics and Biomedical Engineering*, Sydney, Australia, August 24–29.

167. Rubén, M., Mireille, G., Diego, J., Carlos, C. and Javier, T. (1999). Segmentation of ventricular angiographic images using fuzzy clustering, http://www.ing.ula.ve/~dmiranda/page20.html.

168. Ahmed, M., Yamany, S., Mohamed, N., Farag, A. and Moriarty, T. (2002). A modified fuzzy c-means algorithm for bias field estimation and segmentation of MRI data. *IEEE Trans. Med. Imaging*, 3(21), 193–199.

169. Fletcher-Heath, L., Hall, L., Goldgof, D. and Reed Murtagh, F. (2001). Automatic segmentation of non-enhancing brain tumors in magnetic resonance images. *Artif. Intelligence Med.*, 21, 43–63.

170. Bay, O. and Usakli, A. (2003). Survey of fuzzy logic applications in brain-related researches. *J. Med. Syst.*, 27(2), 215–223.

171. Cosic, D. and Loncari, S. (1997). Rule-Based labeling of CT head image, in *Proceedings of the 6th European Conf. of AI in Medicine Europe (AIME97)*, Grenoble.

172. Jiang, L. and Yang, W. (2003). A modified fuzzy c-means algorithm for segmentation of magnetic resonance images, *Proceedings of the VIIth Digital Image Computing: Techniques and Applications*, C. Sun, H. Talbot, S. Ourselin and T. Adriaansen (Eds), CSIRO Publishing, Macquarie University, Sydney, Australia, pp. 10–12.

173. Ghafar, R., Mat, N., Ngah, U., Mashor, M. and Othman, N. (2003). Segmentation of stretched pap smear cytology images using clustering algorithm, *World Congress on Medical Physics and Biomedical Engineering*, Sydney, Australia, August 24–29.

174. Tolias, Y. and Panas, S. (1998). A fuzzy vessel tracking algorithm for retinal images based on fuzzy clustering. *IEEE Trans Med. Imaging*, 17(2), 263–273.

175. Pemmaraju, S., Mitra, S., Shieh, Y. and Roberson, G. (1995). Multiresolution wavelet decomposition and neuro-fuzzy clustering for segmentation of radiographic images, *Proceedings of the 8th Annual IEEE Symposium on Computer-Based Medical Systems*, IEEE Computer Society, Washington, DC, pp. 142–149.

176. Zhang, H., Bian, Z., Yuan, Z., Yea, M. and Ji, F. (2003). A novel multiresolution fuzzy segmentation method on MR image. *J. Comput. Sci. Technol. archive* 18(5), 659–666.

177. Sentelle, S. C. and Sutton, M. (2002). Multiresolution-based segmentation of calcifications for the early detection of breast cancer. *Real-Time Imaging,* 8(3), 237–252.

178. Schüpp, S., Elmoataz, A., Fadili, J., Herlin, P. and Bloyet, D. (2000). Image segmentation via multiple active contour models and fuzzy clustering with biomedical applications, *International Conference on Pattern Recognition (ICPR'00)-Volume 1,* Barcelona, Spain, September 3–8.

179. Karayiannis, N. and Pai, P. (1999). Segmentation of magnetic resonance images using fuzzy algorithms for learning vector quantization. *IEEE Trans. Med. Imaging,* 18(2), 172–180.

180. Ahmed, M. and Farag, A. (1997). Two-stage neural network for volume segmentation of medical images. *Pattern Recognition Lett.* 18(11–13), 1143–1151.

181. Chiou, G. and Hwang, J. (1995). A neural network based stochastic active contour model (NNS-SNAKE) for contour finding of distinct features. *IEEE Trans. Image Process.,* 4(10), 1407–1416.

182. Hall, L., Bensaid, A., Velthuizen, R., Clarke, L., Silbiger, M. and Bezdek, J. (1992). A comparison of neural network and fuzzy clustering techniques in segmenting magnetic resonance images of the brain. *IEEE Trans. Neural Networks,* 3(5), 672–682.

183. Handels, H., Busch, C., Encarnação, J., Hahn, C. and Kühn, V. (1997). KAMEDIN: A telemedicine system for computer supported cooperative work and remote image analysis in radiology. *Comput. Methods Programs Biomed.,* 52(3), 175–183.

184. Ngan, S. and Hu, X. (1999). Analysis of functional magnetic resonance imaging data using self-organizing mapping with spatial connectivity. *Magn. Resonance Med.,* 41(5), 939–946.

185. Reddick, W. E., Glass, J. O., Cook, E. N., Elkin, T. D. and Deaton, R. J. (1997). Automated segmentation and classification of multispectral magnetic resonance images of brain using artificial neural networks. *IEEE Trans. Med. Imaging,* 16(6), 911–918.

186. Rout, S., Srivastava, S. and Majumdar, J. (1998). Multi-modal image segmentation using a modified Hopfield neural network. *Pattern Recognition,* 31(6), 743–750.

187. Wang, Y., Adali, T., Kung, S. et al. (1998). Quantification and segmentation of brain tissues from MR images—a probabilistic neural network approach. *IEEE Trans. Image Process.,* 7(8), 1165–1181.

188. Reyes-Aldasoro, C. and Aldeco, A. (2000). Image segmentation and compression using neural networks, *Workshop on Advances in Artificial Perception and Robotics,* Guanajuato, Mexico.

189. Egmont-Petersen, M. and Pelikan, E. (1999). Detection of bone tumours in radiographic images using neural networks, *Pattern Anal. Appl.* 2(2), 172–183.

190. Kobashi, S., Kamiura, N. Hata, Y. and Miyawaki, F. (2001). Volume-quantization-based neural network approach to 3D MR angiography image segmentation. *Image Vision Comput.,* 19(4), 185–193.

191. Dybowski, R. (2000). Neural computation in medicine: Perspective and prospects. In *Proceedings of the ANNIMAB-1 Conference (Artificial Neural Networks in Medicine and Biology),* H. Malmgren, M. Borga and L. Niklasson (Eds), Springer, Goteborg, 13–16 May, pp. 26–36.

192. Dybowski, D. and Gant, V. (2001). *Clinical Applications of Artificial Neural Networks.* Cambridge University Press, London.

193. Rutenberg, T. (1992). Papnet: A neural net based cytological screening system, in *Advances in Neural Information Processing Systems 4*, J. Moody, S. Hanson and R. Lippmann (Eds), Morgan Kaufmann, San Mateo, CA.

194. Blekas, K., Stafylopatis, A., Likas, D. and Karakitsos, P. (1998). Cytological diagnosis based on fuzzy neural networks. *J. Intelligent Syst.*, 8, 55–79.

195. Alonge, F., Ardizzone, E. and Pirrone, R. (2001). Neural classification of multiple sclerosis lesions in MR images. *Int. J. Knowledge-Based Intelligent Eng. Syst.*, 5(4), 228–233.

196. Lee, C., Chung, P. and Tsai, H. (2003). Identifying multiple abdominal organs from CT image series using a multimodule contextual neural network and spatial fuzzy rules. *IEEE Trans. Inf. Technol. Biomed.*, 7(3), 208–217.

197. http://www.hoise.com/vmw/99/articles/vmw/LV-VM-12-99-17.html, December 1999.

198. Cherrak, I., Paul, J. F., Jaulent, M. C., Chatellier, G., Plouin, P. F., Gaux, J. C. and Degoulet, P. (1997). Automatic stenosis detection and quantification in renal arteriography, *AMIA Annual Fall Symposium*, Nashville.

199. Dutch, W., Grabczewski, K. and Adamczak, R. (2001). Rules for melanoma skin cancer diagnosis, *KOSYR*, Wroclaw, pp. 59–68.

200. Clark, M., Hall, L., Dmitry, B., Goldgof, B., Velthuizen, R., Murtaugh, F. and Silbiger, M. S. (1999). Unsupervised brain tumor segmentation using knowledge-based and fuzzy techniques, In *Fuzzy and Neuro-Fuzzy Systems in Medicine*, H.-N. Theodorescu, A. Kandel and L. C. Jain (Eds), CRC Press, Boca Raton, pp. 137–172.

201. Zhang, G. and Berardi, V. (1998). An investigation of neural networks in thyroid function diagnosis. *Health Care Manage. Sci.*, 1, 29–37.

202. Pesonen, E. (1998). Studies on the experimental construction of a neural network-based decision support system for acute abdominal pain, PhD Thesis, Department of Computer Science and Applied Mathematics, University of Kuopio.

203. Smith, B. and Arabshahi, P. (1996). A fuzzy decision system for ultrasonic prenatal examination enhancement. *Proc. 5th IEEE Int. Conf. Fuzzy Syst., FUZZ-IEEE'96*, 3, 1712–1717.

204. Varachiu, N., Karanicolas, C. and Ulieru, M. (2002). Computational intelligence for medical knowledge acquisition with application to glaucoma. *Proceedings of the First IEEE Conference on Cognitive Informatics (ICCI'02)*, Calgary, Canada, August 17–19.

205. Jantzen, J., Axer, H. and Keyserlingk, G. (2002). Diagnosis of aphasia using neural and fuzzy techniques, in *Advances in Computational Intelligence and Learning–Methods and Applications*, H.-J. Zimmermann, G. Tselentis, M. van Someren and G. Dounias (Eds), Kluwer Academic Publishers, Norwell, MA, pp. 461–474.

206. Sordo, M., Buxton, H. and Watson, D. (2001). A hybrid approach to breast cancer diagnosis, in *Practical Applications of Computational Intelligence Techniques*, L. Jain and P. DeWilde (Eds), Kluwer Academic Publishers, Dordrecht.

207. Lin, Y., Tian, J. and He, H. (2002). Image segmentation via object extraction and edge detection and its medical application. *J. X-ray Sci. Technol.*, 10, 95–106.

208. Dokur, Z. and Ölmez, T. (2002). Segmentation of ultrasound images by using a hybrid neural network. *Pattern Recognition Lett.*, 23(14), 1825–1836.

6

The Soft Computing Technique of
Fuzzy Cognitive Maps for Decision
Making in Radiotherapy

Elpiniki Papageorgiou, Chrysostomos Stylios, and Peter Groumpos

CONTENTS

6.1 Introduction

Radiotherapy is the clinical and technological endeavor devoted to cure patients suffering from cancer (and other diseases) using ionizing radiation, alone or combined with other modalities. The aim of radiation therapy is to design and perform a treatment plan for how to deliver a precisely measured dose of radiation to the defined tumor volume with as minimal damage as possible to the surrounding healthy tissue. Successful radiation treatment results in eradication of the tumor, thus high quality of patient's life, and prolongation of survival at a reasonable cost.

The implementation and clinical practice of irradiation is a complex process that involves many professionals who have to take into account a variety of interrelated measurements, tasks, functions, and procedures. Professionals while determining the treatment of a patient have to know how this particular tumor will be destroyed and how the surrounding healthy tissue is likely to be adversely affected by the applied radiation dose. A large number of parameters–factors (medical and technological), which are complementary, similar, and conflicting, are taken into consideration when the radiation treatment procedure is designed. Each factor has a different degree of importance in determining (or influencing) the dose and all factors together determine the success of the therapy [1].

Experts determine the radiation treatment planning taking into consideration a variety of parameters–factors. The number, nature, and characteristics of factors increase the complexity of the procedure and require the implementation of an advanced technique similar to human reasoning, such as the soft-computing modeling technique of fuzzy cognitive maps (FCMs) [2].

Till today, many approaches and methodologies, algorithms, and mathematical tools have been proposed and used for optimizing radiation therapy treatment plans [3,4]. Dose-calculation algorithms [5,6], dose–volume feasibility search algorithms [7], and biological-objective algorithms have been utilized [8]. Dose distributions have been calculated for the treatment planning systems, satisfying objective criteria and dose–volume constraints [4]. Some algorithms have been proposed for optimizing beam weights and beam directions to improve radiotherapy treatment [9]. Moreover, steepest-descent methods and gradient-descent methods have been used to optimize the objective functions, based on biological or physical indices, and have been employed for optimizing intensity distributions [10,11]. Dose–volume histograms analyses the resultant dose distributions, which appears to indicate some merit [12]. Furthermore, methods related to knowledge-based expert systems and neural networks have been proposed for optimizing the treatment variables and developing decision-support systems for radiotherapy planning [13,14]. Much scientific efforts have been made to optimize treatment variables and dose distributions. Toward this direction, there is still a need for a flexible, efficient, and adaptive tool based on an abstract cognitive model, which will be used for clinical practice simulation and decision making [15,16,17].

The complexity and the vagueness of the decision-making process for radiotherapy treatment planning may be handled with soft-computing methods [18]. FCMs is a soft computing technique that incorporates ideas from artificial neural networks (ANNs) and fuzzy logic (FL). Their advantageous modeling features are the flexibility in system design, model, and control; the comprehensive operation; and the abstractive representation of complex systems. We propose the use of FCMs to create a dynamic model for estimating the final dose delivered to the target volume and normal tissues with the ability to evaluate the success of radiotherapy. FCMs have been used to model complex systems that involve discipline and different factors, states, variables, and events. FCMs can integrate and include the partial influence or controversial factors and characteristics in the decision-making problem [18]. The main advantage of implementing FCM in this area is that they can take under consideration causal effect among factors in recalculating the value of concepts that determine the radiation dose, keeping it in a minimum level and at the same time having the best result in destroying tumor with minimum injuries to healthy tissues and organs at risk. This is in accordance with the main goal of any radiation therapy treatment planning [1,19,20].

A decision system based on human knowledge and experience is proposed and developed here, having a two-level hierarchical structure with an FCM in each level, which creates an advanced decision-making system. The lower level FCM models the treatment planning taking into consideration all the factors and treatment variables and their influences. The upper level FCM models the procedure of the treatment execution and calculates the optimal final dose for radiation treatment. The upper level FCM supervises and evaluates the whole radiation therapy process. Thus, the proposed two-level integrated structure for supervising the procedure before treatment execution seems a rather realistic approach to the complex decision-making process in radiation therapy [21].

6.2 Soft Computing Techniques for Decision Making

Soft computing differs from conventional (hard) computing. It is tolerant to imprecision, uncertainty, partial truth, and approximation. In effect, the role model for soft computing is the human mind. The principle of soft computing is to exploit the tolerance for imprecision, uncertainty, partial truth, and approximation to achieve tractability and robustness. Soft computing can be seen as a combination and contribution of FL, neural computing (NC), evolutionary computation (EC), machine learning (ML), and probabilistic reasoning (PR), with the latter subsuming belief networks, chaos theory, and parts of learning theory [22–24]. Soft computing can be seen as a partnership where every partner contributes discipline and different methodologies for addressing problems in its domain. In this perspective, the constituent methodologies of soft computing are complementary rather than competitive.

However, it is widely accepted that complex real-world problems require intelligent methods that combine knowledge, techniques, and methodologies from various sources and areas. Intelligent systems are desired to possess humanlike expertise within a specific domain, adapt themselves and learn to do better in changing environments, and explain how they make decisions or take actions. In confronting real-world computing problems, it is frequently advantageous to use several computing techniques synergistically rather than exclusively, resulting in construction of complementary hybrid intelligent systems.

The synergism allows soft computing to incorporate human knowledge effectively, deal with imprecision and uncertainty, and learns to adapt to unknown or changing environment for better performance. For learning and adapting, soft computing requires extensive computation. In this sense, soft computing shares the same characteristics as computational intelligence. Soft computing applications have been used in different areas: diagnostics in medicine, cluster analysis, discriminant analysis, and pattern recognition [25–29].

6.2.1 Description of Fuzzy Cognitive Maps

FCMs have their roots in graph theory. Axelord first used signed digraphs to represent the assertions of information [30]. He adopted the term "cognitive map" for these graphed causal relationships among variables as defined and described by people. The term "fuzzy cognitive map" was coined by Kosko [2]. An FCM model has two significant characteristics.

Causal relationships between nodes are fuzzy numbers. Instead of only using signs to indicate positive or negative causality, a weight is associated with the relationship to express the degree of relationship between two concepts.

The system is dynamic, permitting feedback, where the effect of change in one concept affects other concepts, which in turn can affect the concept initiating the change; the presence of feedback adds a temporal aspect to the operation of the FCM.

Concepts of FCM model reflect attributes, characteristics, qualities, quantities, and senses of the system. Interconnections among concepts of FCM signify the cause and effect relationships among concepts. These weighted interconnections represent the direction and degree with which concepts influence the value of the interconnected concepts. Figure 6.1 illustrates a graphical representation of an FCM model.

The cause and effect interconnection between two ordered nodes C_j and C_i is described with the weight w_{ji}, which takes a value in the range -1 to 1.

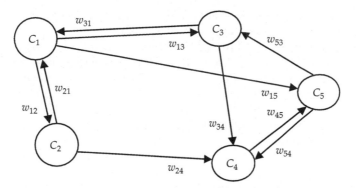

FIGURE 6.1
A simple FCM.

There are three possible types of causal relationships between concepts:

- $w_{ji} > 0$, which indicates positive causality between concepts C_j and C_i. That is, the increase (decrease) in the value of C_j leads to the increase (decrease) in the value of C_i.
- $w_{ji} < 0$, which indicates negative causality between concepts C_j and C_i. That is, the increase (decrease) in the value of C_j leads to the decrease (increase) in the value of C_i.
- $w_{ji} = 0$, which indicates no relationship between C_j and C_i.

Human knowledge and experience on the system are reflected, due to the FCM development procedure on the type and the number of concepts, as well as the initial weights of the FCM. The value A_i of concept C_i expresses the quantity of its corresponding physical value and is derived by the transformation of the fuzzy physical values to numerical ones.

FCM is used to model and simulate the behavior of any system. At each simulation step, the value A_i of a concept is calculated, computing the influence of the interconnected concepts to the specific concept according to the following calculation rule:

$$A_i^{(k+1)} = f\left(A_i^{(k)} + \sum_{\substack{j \neq i \\ j=i}}^{N} A_j^{(k)} \cdot w_{ji}\right) \tag{6.1}$$

where $A_i^{(k+1)}$ is the value of concept C_i at simulation step $k + 1$, $A_j^{(k)}$ is the value of concept C_j at step k, w_{ji} is the weight of the interconnection between concept C_j and concept C_i, and f is a sigmoid threshold function.

The sigmoid function f belongs to the family of squashing functions. Usually the following function is used

$$f(x) = \frac{1}{1 + e^{-\lambda x}} \tag{6.2}$$

which is the unipolar sigmoid function, where $\lambda > 0$ determines the steepness of the continuous function $f(x)$ near $x = 0$.

All the values on the FCM model have fuzzy nature; experts describe FCM characteristics and assign initial values using linguistic notion. These fuzzy variables need to be defuzzified to use mathematical functions and calculate the corresponding results. Thus, values of concepts belong to the interval [0, 1] and values of weights to the interval [−1, 1]. Using Equation 6.1 with the sigmoid function, the calculated values of concepts after each simulation step will belong to the interval [0, 1].

6.2.2   Developing Fuzzy Cognitive Maps

The method that is used to develop and construct the FCM has great importance to sufficiently model a system. The method used depends on the group of experts who operate, monitor, and supervise the system and develop the FCM model. This methodology extracts the knowledge on the system from the experts and exploits their experience on the system's model and behavior [31].

The group of experts determines the number and kind of concepts that comprise the FCM. The expert from his/her experience knows the main factors that describe the behavior of the system; each of these factors is represented by one concept of the FCM. Experts know which elements of the systems influence other elements; for the corresponding concepts, they determine the negative or positive effect of one concept on the others, with a fuzzy degree of causation. In this way, an expert transforms his knowledge in a dynamic weighted graph, the FCM. The methodology of developing an FCM based on fuzzy expressions to describe the interrelationship among concepts is described analytically in Refs. 15 and 31 and is used here. According to the developing methodology, experts are asked to think about and describe the existing relationship between the concepts and so they justify their suggestions. Each expert, in fact, determines the influence of one concept on another as "negative" or "positive" and then evaluates the degree of influence using a linguistic variable, such as "strong influence," "medium influence," and "weak influence."

More specifically, the causal interrelationships among concepts are declared using the variable *influence*, which is interpreted as a linguistic variable taking values in the universe $U = [−1, 1]$. Its term set $T(influence)$ is suggested to comprise nine variables. Using nine linguistic variables, an expert can describe in detail the influence of one concept on another and can discern between different degrees of influence. The nine variables used here are: $T(influence)$ = {negatively very strong, negatively strong, negatively medium, negatively weak, zero, positively weak, positively medium, positively strong, and positively very strong}. The corresponding membership functions for these terms are shown in Figure 6.2 and they are μ_{nvs}, μ_{ns}, μ_{nm}, μ_{nw}, μ_z, μ_{pw}, μ_{pm}, μ_{ps}, and μ_{pvs}.

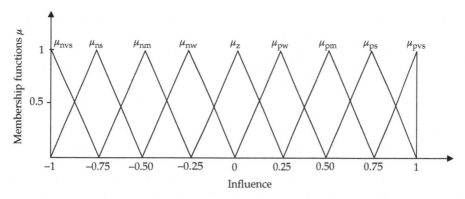

FIGURE 6.2
Membership functions for fuzzy values of FCMs.

Thus, every expert describes each one of the interconnections with a linguistic fuzzy rule; they use IF–THEN rules to justify the cause and effect relationships among concepts, inferring a linguistic weight for each interconnection. Then, the inferred fuzzy weights are integrated using the SUM method, as suggested by experts, and an overall linguistic weight is produced, which with the defuzzification method of center of gravity (CoG) [32] is transformed to a numerical weight w_{ji}, belonging to the interval [−1, 1]. All the weights are gathered into a weight matrix $n \times n$ **W**, where n is the number of concepts.

Every expert describes the relationship between two concepts using the following fuzzy rule with linguistic variables.

IF a change **B** occurs in the value of concept C_j THEN a change **D** in the value of concept C_i is caused.

Infer: The influence from concept C_j to C_i is **E**.

Where **B**, **D**, and **E** are fuzzy linguistic variables that experts use to describe the variance of concept values and the degree of influence from concept C_j to C_i.

6.2.3 Fuzzy Cognitive Map for Decision Support System

Decision support systems (DSS) are widely used in many application areas, from management and operational research sciences to medical applications. DSS are used to suggest solutions and provide advice to people how to conclude to a decision. DSS suggest alternative ways of action based on the advantages, disadvantages, and consequences of each action. DSS are developed utilizing the experience and knowledge of experts in the distinct problem. DSS do not take the decision by themselves but they suggest to human the most appropriate and suitable decision. Especially DSS play a significant role in medical applications, where decisions include humans (patients and doctors), medical equipment, and computers. Medical DSS are used by general practice doctors for specific health problems in order to propose a diagnosis and treatment.

Here, an integrated approach for modeling and medical decision making is presented based on the soft computing technique of FCMs. A decision-making procedure is a complex process that has to consider a variety of interrelated functions. Usually decision making involves many professionals and a variety of interrelated functions.

A generic FCM-DSS model for diagnosis could consist of three kinds of concepts as illustrated in Figure 6.3. There are concepts representing the Factor-concepts, which are either laboratory tests and measurements, or observations of the doctor and other information on patient status. The values of Factor-concepts are taking into consideration to infer the value of Selector-concepts. Selector-concepts represent some intermediate conclusions. The Selector-concepts influence the Output-concepts that conclude the decision. The FCM model can include all the factors and symptoms that can infer a decision along with the existing causal relationships among Factor-concepts, because factors are interdependable and sometimes the existence or lack of a factor requires the existence or lack of another. Moreover, Factor-concepts influence Selector-concepts and the value of each Selector-concept can subsequently influence the degree of the Output-concept of the FCM. This FCM model is an abstract conceptual model of what a doctor does when he makes

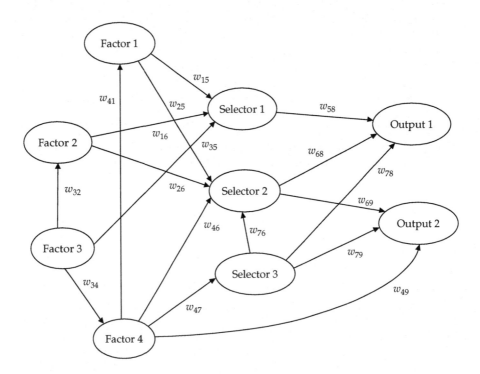

FIGURE 6.3

A generic FCM-DSS model for medical decision making. (From Papageorgiou, E.I., Stylios, C.D., Groumpos, P.P., *IEEE Trans. Biomed. Eng.*, 50(12) 2003. With permission.)

a decision; he reaches some intermediate inferences based on the inputs taking into consideration all the related symptoms, and then according to the intermediate Selector-concepts values, he determines his final decision that in the FCM model are presented as Output-concepts.

6.3 The Nonlinear Hebbian Learning Algorithm

In this section, the nonlinear Hebbian learning (NHL) algorithm that has been proposed to train FCM [33] is described. The NHL algorithm is used to overcome inadequate knowledge of experts and nonacceptable FCM simulation results [33]. The weight adaptation procedure is based on the Hebbian learning rule for nonlinear units [34]. The nonlinear Hebbian-type rule for ANNs learning [35] have been adapted and modified for the FCM case, and the NHL algorithm was proposed [33].

NHL algorithm is based on the premise that all the concepts in FCM model trigger synchronously at each iteration step. During this triggering process, the weight w_{ji} of the causal interconnection of the related concepts is updated and the modified weight $w_{ji}^{(k)}$ is calculated for iteration k.

The value $A_i^{(k+1)}$ of C_i, concept at simulation step $k + 1$, is calculated, computing the influence of interconnected concepts with values A_j to the specific concept C_i due to modified weights $w_{ji}^{(k)}$ at simulation step k, according to Equation 6.1, which takes the form

$$A_i^{(k+1)} = f\left(A_i^{(k)} + \sum_{\substack{j \neq i \\ j=i}}^{N} A_j^{(k)} \cdot w_{ji}^{(k)}\right) \tag{6.3}$$

Furthermore, during the development phase of FCM, experts have defined which concepts of FCM are the decision output concepts (DOCs). These concepts are the outputs of the system that interest us, and we want to estimate their values, which represent the final state of the system. The distinction of FCM concepts as inputs, intermediates, and outputs is determined by the group of experts for each specific problem. Experts select the output concepts and they also define the initial stimulators (Factor-concepts) and the interior concepts (Selector-concepts) of the system.

Taking the advantage of the general nonlinear Hebbian-type learning rule for ANNs, we introduce the mathematical formalism incorporating this learning rule for FCMs, a learning rate parameter and the determination of input and output concepts. This algorithm relates the values of concepts and values of weights in the FCM model.

The proposed learning rule [33] has the general mathematical form for the adjustment of the weights

$$\Delta w_{ji} = \eta_k A_i^{(k-1)} A_j^{(k-1)} - w_{ji}^{(k-1)}(A_i^{(k-1)})^2 \tag{6.4}$$

where the coefficient η_k is a very small positive scalar factor called the "learning parameter" and is determined using an experimental trial and error method to converge the simulation process. $A_j^{(k)}$ is the value of concept C_j, which at next simulation step, $k+1$, triggers the interconnected concepts.

This simple rule states that if $A_i^{(k)}$ is the value of concept C_i at simulation step k and A_j is the value of the concept C_j that triggers the concept C_i, the corresponding weight from concept C_j toward the concept C_i increases proportional to their product multiplied with the learning rate parameter minus the weight decay at simulation step $k-1$, that is multiplied by the value A_j of triggering concept C_j. All the FCM concepts trigger at the same iteration step and their values are updated synchronously.

Equation 6.4 takes the following form of nonlinear weight adaptation algorithm, if we introduce a weight decay parameter:

$$w_{ji}^{(k)} = \gamma \cdot w_{ji}^{(k-1)} + \eta_k A_i^{(k-1)}(A_j^{(k-1)} - \text{sign}(w_{ji}^{(k-1)})w_{ji}^{(k-1)}A_i^{(k-1)}) \tag{6.5}$$

where η_k is the learning rate parameter and γ the weight decay parameter.

The value of each concept of FCM is updated through Equation 6.3, where the value of weight $w_{ji}^{(k)}$ is calculated using Equation 6.5.

When the NHL algorithm is applied, only the initial nonzero weights suggested by experts are updated for each iteration step. All the other weights of weight matrix **W** remain equal to zero, which is their initial value.

For the termination of the proposed algorithm, two termination conditions are proposed. One termination condition is the minimization of function F_1. The termination function F_1 that has been proposed for the NHL examines the values of DOCs. It is supposed that for each DOC_i, experts have defined a target value T_i. This target value can either be the desired value when DOC_i represents a concept, which has to take a value or the mean value when DOC_i represents a concept whose value has to belong to an interval. Thus, the function F_1 is defined as

$$F_1 = \sqrt{\sum_{i=1}^{l}(DOC_i - T_i)^2} \tag{6.6}$$

where l is the number of DOCs.

The second termination condition is the minimization of the variation between two subsequent values of DOCs, represented by the following equation:

$$F_2 = \left| DOC_i^{(k+1)} - DOC_i^{(k)} \right| < e \tag{6.7}$$

This termination condition helps to terminate the iterative process of the learning algorithm. The term e is a tolerance level keeping the variation of values of DOC(s) as low as possible and it is proposed to be equal to $e = 0.001$, satisfying the termination of the iterative process.

Algorithm 1: "Nonlinear Hebbian learning"
Step 1: Read input concept state A^0 and initial weight matrix $\mathbf{W}^0$
Step 2: Repeat for each iteration k
Step 3: Calculate $A_i^{(k)}$ according to Equation 6.3
Step 4: Update the weights: $w_{ji}^{(k)} = \gamma \cdot w_{ji}^{(k-1)} + \eta A_i^{(k-1)}(A_j^{(k-1)} - sgn(w_{ji})w_{ji}^{(k-1)}A_i^{(k-1)})$
Step 5: Calculate the two termination functions
Step 6: Until the termination conditions are met
Step 7: Return the final weights $\mathbf{W}_{NHL}$

FIGURE 6.4
Pseudo code of NHL algorithm.

Through this training process and when both the termination conditions are met, the final weight matrix $\mathbf{W}_{NHL}$ of FCM is derived.

A generic description of the proposed NHL algorithm for FCMs is given in Figure 6.4.

After a number of experiments and implementation the NHL algorithm in different domains, the upper and lower bounds for the learning rate parameters γ and η have been determined [33]. The bounds of learning rate parameter η are determined as $0 < \eta < 0.1$, and for the weight-decay parameter as $0.9 < \gamma < 1$. Larger η values of 0.1, and smaller γ values of 0.9 do not lead the system in convergence for any initial values.

The flowchart of the proposed NHL procedure implemented in FCMs is given in Figure 6.5. It is mentioned that if the learning procedure repeats for over 1000 iteration steps without converging, it stops and experts are asked to reconstruct the FCM.

Training FCM with the NHL algorithm enhances the FCM model and incorporates the expert's knowledge into a proper FCM model of the process or system. This is the case of supervisor-FCM and when the FCM has to converge in desired equilibrium points after simulation results.

6.4 Radiation Therapy Procedure: Background, Issues, and Factors

Radiotherapy is identified as the external application of beams of photons generated by linear accelerator machines to eliminate tumors and treat cancer patients. There are two objectives for "3-dimensional conformal"

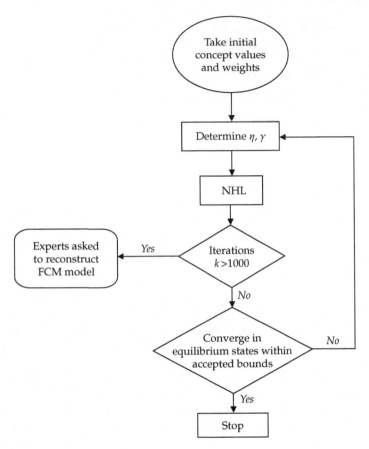

FIGURE 6.5
The flowchart of NHL algorithm.

radiotherapy; the first one is to deliver the highest dose to a volume shaped exactly like the tumor and the second one is to keep the dose level at the minimum for healthy tissues and critical organs. Before the implementation of any beam radiation, a treatment plan is required to be designed. The treatment planning determines how to perform the radiation, which is a complex problem because various complementary of interconnecting conditions and constraints have to be met. The performance criteria for radiation therapy are maximization of dose and dose uniformity within the target region and dose minimization to surrounding critical organs and normal tissues. The process of adjusting radiation variables and displaying the corresponding dose distribution is repeated till the optimizations of these criteria are met.

The depth of the tumor from the skin surface is probably the most important factor in selecting the appropriate radiation therapy machine, but it is definitely not the only one. For treating complex tumors, a variety of factors are taken into consideration to determine the treatment plan [36–38].

An inexhaustive list of factors may include the following:

1. The depth at which the tumor is located from skin surface.
2. The shape (geometrical or irregular) and the size of the tumor.
3. The location of the tumor in part of the body or head and size of cross section treated.
4. The local invasive capacity of the tumor and its potential spread to the regional lymph nodes.
5. The type of tissue within the tumor, as well as the type of tissue that surrounds the tumor. The presence of inhomogeneities within the irradiated volume such as bone, muscle, lung, fat, and air should be taken into consideration.
6. The dose distribution within the target volume should be reasonably uniform (within ±5%).
7. The tumor position regarding the center of the contour cross section.
8. The existence of radiation-sensitive organs within the irradiated volume, such as eyes, bladder, salivary glands, larynx, spinal cord, anus, and skin. These normal critical structures should not receive doses near or beyond tolerance.
9. Damage to the healthy tissue outside the treatment volume (maximum dose ≤95% of prescribed dose).
10. Patient dimension and contour geometry in treatment region.
11. The number of radiation fields that must be used and the daily dosage on the tumor based on the biological damage of the healthy subcutaneous tissue.
12. Cost—this includes cost of equipment, cost of shielding, and cost of usage of space.
13. The length of time required to administer the treatment—it is difficult to keep a patient immobilized for a long period of time. The length of procedure preparation time (both for patient and staff). The time required for obtaining the optimum treatment plan and the time for calculating the distributed doses within the irradiated volume.
14. Amount of secondary (scattered) radiation that can be tolerated by the patient.
15. Matching of beam overlap volume with target volume.
16. Degree of difficulty in repeatability (flexibility) of setup of the patient and treatment geometry.

However, to achieve a good distribution of the radiation on the tumor as well as to protect the healthy tissues, the following should be taken into consideration:

1. Selection of appropriate size of the radiation field
2. Increase of entry points of the beam (more than one radiation field)
3. Selection of appropriate beam directions
4. Selection of weight of each field (dose contribution from individual fields)
5. Selection of appropriate quality, that is, energy and type of radiation (x-rays, γ-rays, and electrons)
6. Modification of field with cerrobend blocks or multileaf collimators, compensating filters or *bolus*, and wedge filters
7. Use of isocentric stationary beam therapy versus isocentric rotation therapy
8. Patient immobilization
9. Use of conformal (3-D) instead of conventional (2-D) radiotherapy

More information with the necessary theoretical justification and detailed description of the concepts mentioned earlier and related terms are provided in Refs. 36–38.

Treatment planning refers to the description and the selection of implementation procedures and how to reach necessary decisions that have to be made before performing radiation treatment. Both physical and clinical procedures are part of the treatment-planning problem. The treatment-planning process comprises several methods for treatment preparation and simulation toward achieving a reproducible and optimal treatment plan for the patient. Irrespective of the temporal order, these procedures include

- Patient fixation, immobilization, and reference point selection
- Dose prescriptions for target volumes and the tolerance level of organ at risk volumes
- Dose distribution calculation
- Treatment simulation
- Selection and optimization of
  - Radiation modality and treatment technique
  - The number of beam portals
  - The directions of incidence of the beams
  - Beam collimation
  - Beam intensity profiles
  - Fractionation schedule

Thus, the treatment planning is a complex process where a great number of treatment variables have to be considered.

6.5 The Clinical Treatment Simulation Tool for Decision Making in Radiotherapy

Radiotherapists and physicists are asked to construct the FCM model according to their knowledge and experience, thus they are using the factors and treatment variables that were briefly presented in Section 6.4. These factors and characteristics will be the concepts of the FCM decision-making model for radiotherapy treatment planning. They are considering the basic beam data from experimental measurements [39] and the information described at American Association of Physicists in Medicine (AAPM) Task Group (TG) 23 test package [40] to retrieve the main Factors-concepts of the FCM model, selectors, and the relationships among them. The AAPM TG 23 test package is useful for the quantitative analysis of treatment planning systems of photon beam radiation [40,41]. Our test package of basic beam dosimetric data has been developed with experimental measurements [39], which are used here for the determination of initial values of concepts and weights.

Radiotherapy experts, following the generic FCM-decision support model presented in Section 6.2.3, identified and divided the concepts, which consist of the FCM model for radiotherapy treatment planning, into three categories: Factor-concepts, Selector-concepts, and Output-concepts. Factors and selectors concepts could be seen as inputs, they represent treatment variables with given, measured, or calculated values, and the corresponding causal weights are identified from experimental data, and data from AAPM TG 23 test package [39–42]. The values of the Selector-concepts are influenced by the value of the Factor-concepts with the corresponding causal weights. The values of the Output-concepts are influenced and determined by the values of the Factor-concepts and the Selector-concepts with the corresponding causal weights. The decision-making procedure is based on the determination of the values of the Output-concepts that lead to the final decision.

The values of concepts in the FCM model can generally take crisp, numeric, and linguistic values. It is considered that the values of concepts in the FCM-DSS model take five positive linguistic variables depending on the characteristics of each particular concept, such as very high, high, medium, weak, and near zero. When concepts represent events or discrete variables, there is a threshold (0.5) that determines if an event is activated or not. All the values of concepts in the FCM belong to the interval [0, 1]. The degree of the influence between concepts is represented by a linguistic variable of the fuzzy set {positive very high, positive high, positive medium, positive weak, zero, negative weak, negative medium, negative low, and negative very low} [15,21].

Experts developed the FCM that models the radiotherapy treatment planning procedure, according to the test packages and experimental data. So, the clinical treatment simulation tool based on fuzzy cognitive map (CTST-FCM) model consists of 26 concepts that are described in Table 6.1.

Concepts F-C1 to F-C13 are the Factor-concepts, concepts S-C1 to S-C10 are the Selector-concepts, and the concepts OUT-C1 to OUT-C3 are the Output concepts. The value of the Output-concept OUT-C1 represents the amount of dose applied to the mean clinical target volume (CTV), which has to be more than 90% of the amount of prescribed dose to the tumor. The value of concept OUT-C2 represents the amount of the surrounding healthy tissues' volume received a dose, which has to be as small as possible, less than 5% of the volume receiving the prescribed dose. The value of concept OUT-C3 represents the volume of organ at risk (OAR) receiving a dose, which should be less than the 10% of volume receiving the prescribed dose. The objective of the FCM model is to keep the values of the OUT-Cs, in the following range:

$$OUT\text{-}C1 \geq 0.90 \tag{6.8}$$

$$OUT\text{-}C2 < 0.05 \tag{6.9}$$

$$OUT\text{-}C3 < 0.10 \tag{6.10}$$

The values of Output-concepts are acceptable when they satisfy the performance criteria in Equations 6.8 through 6.10.

Using the development methodology for FCMs as described in Section 6.2.2, every expert describes each interconnection using a fuzzy rule. Fuzzy rules are evaluated in parallel using fuzzy reasoning and the inferred fuzzy weights are combined so that an aggregated linguistic weight is produced, which is then defuzzified and the result is a crisp value representing the weight of each interconnection. In this way, the weights of interconnections among Factor-concepts and Selector-concepts, Selector-concepts and Output-concepts, and Output-concepts and Factor-concepts, are determined. Five examples are now described to illustrate the determination of the weights for some interconnections.

Example 6.1

One expert describes the influence from the S-C3 toward OUT-C1 representing the amount of dose to target volume using the following fuzzy rule:

IF a small change occurs in the value of S-C3, THEN a small change is caused in the value of OUT-C1.

This means that if a small change occurs in the size of radiation field, then a small change in the value of dose to the target volume is caused,

TABLE 6.1

Concepts of the CTST-FCM: Description and Type of Values

Concepts	Description	Number and Type of Values Scaled
F-C1	Accuracy of depth of tumor	Five fuzzy
F-C2	Size of tumor	Seven fuzzy (very small, small, positive small, medium, negative large, large, and very large)
F-C3	Shape of tumor	Three fuzzy (small, medium, and large)
F-C4	Location of tumor size at cross section	Three fuzzy
F-C5	Regional metastasis of tumor (sites of body)	Five fuzzy
F-C6	Type of irradiated tissues—presence of inhomogeneities	Five fuzzy
F-C7	Dose uniformity (including 90% isodose) within target volume	One fixed
F-C8	Skin sparing—amount of patient skin dose	Three fuzzy (low, medium, and high)
F-C9	Amount of patient thickness irradiated	Five fuzzy
F-C10	Accuracy of patient's contour (taken from CT-scans and portal films)	Five fuzzy
F-C11	Amount of scattered radiation received by patient	Five fuzzy
F-C12	Time required for treatment procedure or preparation	Five fuzzy
F-C13	Amount of perfect match of beam to target volume	Three fuzzy
S-C1	Quality of radiation—four types of machines (orthovoltage, supervoltage, megavoltage and teletherapy)	Four discrete
S-C2	Type of radiation (photons, electrons, protons, and heavy particles)	Four discrete
S-C3	Size of radiation field	Five fuzzy
S-C4	Single or multiple field arrangements	Two discrete
S-C5	Beam direction(s) (angles of beam orientation)	Continuous
S-C6	Weight of each field (percentage of each field)	Continuous
S-C7	Stationery versus rotation—isocentric beam therapy	Continuous
S-C8	Field modification (no field modification, blocks, wedges, filters, and multileaf-collimator shaping blocks)	Five discrete
S-C9	Patient immobilization	Three discrete
S-C10	Use of 2-D or 3-D conformal technique	Two discrete
Out-C1	Dose given to treatment volume (must be within accepted limits)	Five fuzzy
Out-C2	Amount of irradiated volume of healthy tissues	Five fuzzy
Out-C3	Amount of irradiated volume of sensitive organs (OAR)	Five fuzzy

Source: Papageorgiou, E.I., Stylios, C.D., Groumpos, P.P., *IEEE Trans. Biomed. Eng.*, 50, 12, 2003. With permission.

increasing the amount of dose. So, the influence of S-C3 to OUT-C1 is positively small.

The inferred linguistic weight for this interconnection will be aggregated with other linguistic weights proposed by the other experts and an overall linguistic weight will be produced, which will be defuzzified.

Example 6.2

The influence from the F-C2 toward the S-C3 representing the size of radiation field is inferred:

IF a small change occurs in the value of F-C2, THEN a large change is caused in the value of S-C3.

This means that the size of the tumor, determined by the radiotherapist, influences the size of radiation field. If the size of target volume is increased by a small amount, the size of radiation field is increased by a larger amount. The influence of F-C2 to S-C3 is inferred as positively strong.

Example 6.3

The influence from F-C1 toward the OUT-C2 representing the healthy tissues' volume received a prescribed dose is inferred that

IF a large change occurs in the value of F-C1, THEN a very large change is caused in the value of OUT-C2.

This means that if the depth of tumor increases the amount of healthy tissues' volume that received the prescribed dose increases. Thus, the influence is positively very strong.

Example 6.4

The influence from S-C4 toward the F-C13 representing the amount of perfect match of beam to target volume-tumor is inferred as

IF a large change occurs in the value of S-C4, THEN a very large change is caused in the value of F-C13.

This means that if more field arrangements are used, the match of beam to the target volume increases by a very large amount. Thus, the influence is positively very strong.

Example 6.5

The influence from Output-concept OUT-C1 toward the Output-concept OUT-C2 is inferred as

IF a small change occurs in the value of OUT-C1, THEN a large change is caused in the value of OUT-C2.

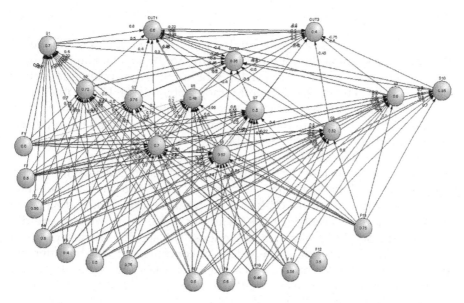

FIGURE 6.6
The CTST-FCM model with 26 concepts and 156 interrelationships. (From Papageorgiou, E.I., Stylios, C.D., Groumpos, P.P., *IEEE Trans. Biomed. Eng.*, 50(12), 2003. With permission.)

This means that if the dose given to the tumor increases, a larger amount of healthy tissues' volume receives the prescribed dose given to the tumor. The influence of OUT-C1 to OUT-C2 is inferred as positively strong.

Analogous is the methodology of determining all the existent influences among Factor-concepts, Selector-concepts, and Output-concepts.

The CTST-FCM model for the decision making in radiotherapy is developed and illustrated in Figure 6.6. It consists of 26 concepts and 156 interconnections. Initial values of concepts are taken from the data set of the AAPM TG 23 [40, 43] and from experimental data [39], and are identified according to each specific treatment technique, then these values are normalized and transformed in the interval [0, 1].

6.5.1 Testing of the Clinical Treatment Simulation Tool for Two Radiotherapy Planning Case Studies

The treatment of localized prostate cancer is commonly treated with the use of radical radiotherapy. In this section, two different treatment cases for prostate cancer therapy have been examined using the CTST-FCM model. In the first case, the 3-D conformal technique consisting of six-field arrangement is suggested and in the second one, the conventional four-field box technique. Radiotherapy physicians and medical physicists choose and specify the initial values of concepts and weights of the proposed CTST-FCM model, for each case.

6.5.1.1 Case Study 1

Conformal radiotherapy allows a smaller amount of rectum and bladder to be treated, by shaping the high dose volume to the prostate and low-dose volume to bladder and rectum [44,45], where the target volume is readily visualized and defined on computed tomography (CT) [46]. Radiotherapists and medical physicists select the treatment variables for the field size, beam direction, beam weights, number of beams, compensating filters, type and quality of radiation, and moreover, they describe and determine the corresponding weights on CTST-FCM.

For the specific therapy technique, a six-field arrangement with gantry angles 0°, 60°, 120°, 180°, 240°, and 300° using a 6 MV photon beam radiation will be considered. Multiple CT-based external contours define the patient anatomy and also isocentric beam therapy is used. Beam weights are different for the six fields, blocks, and wedges. The specific characteristics of conformal therapy determine the values of concepts and weights interconnections of CTST-FCM model. Thus, the S-C4 takes the value of six-field number; S-C3 has the value of "small size" for radiation field, which means that the influence of S-C3 and S-C4 toward OUT-Cs is great. In the same way, the S-C5 and S-C6 have great influence on OUT-Cs because different beam directions and weights of radiation beams are used. S-C8 takes the value for the selected blocks and wedges, influencing the OUT-Cs. The S-C7 takes the discrete value of isocentric beam therapy. The S-C9 takes a value for accurate patient positioning and the S-C10 takes the discrete value of 3-D radiotherapy.

Thus, for the specific technique considering the earlier discussion, the initial values of concepts and weights of interconnections between S-Cs and OUT-Cs are suggested. The value of weights between S-Cs and OUT-Cs is given in Table 6.2. Tables 6.3 and 6.4 contain the weights of interconnections

TABLE 6.2

Weights Representing Relationships among Selector-Concepts and Output-Concepts for First Case Study

Selectors	OUT-C1	OUT-C2	OUT-C3
S-C1	0.6	−0.45	−0.4
S-C2	0.50	−0.6	−0.5
S-C3	0.4	−0.45	−0.4
S-C4	0.3	−0.6	−0.5
S-C5	0.38	−0.40	−0.4
S-C6	0.45	−0.4	−0.4
S-C7	0.30	−0.30	−0.30
S-C8	0.4	−0.5	−0.45
S-C9	0.4	−0.5	−0.45
S-C10	0.6	−0.5	−0.5

Source: Papageorgiou, E.I., Stylios, C.D., Groumpos, P.P., *IEEE Trans. Biomed. Eng.*, 50, 12, 2003. With permission.

TABLE 6.3

Weights of the Interconnections among Factor-Concepts and Selector-Concepts

Factors/ Selectors	S-C1	S-C2	S-C3	S-C4	S-C5	S-C6	S-C7	S-C8	S-C9	S-C10
F-C1	0.7	0.7	0.6	0.62	0.4	0.42	0.6	0.6	0.2	0
F-C2	0.65	0.6	0.7	0.6	0.2	0.53	0.55	0.5	0.6	0.5
F-C3	0.4	0.4	0.6	0.63	0.45	0	0.4	0	0	0.7
F-C4	0.7	0.38	0.3	0.6	0.4	0.52	0.4	0.6	0.7	0
F-C5	0.4	0.78	0.8	0.6	0.7	0.6	0	0.45	0	0
F-C6	0.7	0.75	0.32	0.6	0.5	0.55	0.47	0.5	0	0.6
F-C7	0.62	0.62	0.6	0.7	0.65	0.6	0.2	0.74	0.5	0.4
F-C8	0.52	0.75	0.65	0.67	0.72	0.74	0.45	0.55	0	0.6
F-C9	0.35	0.6	0.5	0.6	0.6	0.6	0.2	0.5	0.5	0
F-C10	0.22	0.5	0	0	0.6	0.58	0.72	0.3	0.7	0.6
F-C11	0.61	0.72	0.75	0.6	0.6	0.55	0.22	0.6	0	0
F-C12	0.33	0	0	0.52	0	0	0	0	0.5	0
F-C13	0.50	0.50	0.7	0.65	0.65	0.7	0.4	0.2	0.6	0.7

Source: Papageorgiou, E.I., Stylios, C.D., Groumpos, P.P., *IEEE Trans. Biomed. Eng.*, 50, 12, 2003. With permission.

TABLE 6.4

Weights of the Interconnections among Output-Concepts

Outputs	OUT-C1	OUT-C2	OUT-C3
OUT-C1	0	0.6	0.6
OUT-C2	−0.7	0	0
OUT-C3	−0.6	0	0

Source: Papageorgiou, E.I., Stylios, C.D., Groumpos, P.P., *IEEE Trans. Biomed. Eng.*, 50, 12, 2003. With permission.

between Factor-concepts and Selector-concepts, and Output-concepts to Output-concepts respectively.

The following matrix is formed with the initial values for this particular treatment technique:

$$A_1^{\text{lower level}} = \begin{matrix} [0.75\ 0.5\ 0.5\ 0.6\ 0\ 0.5\ 0.8\ 0.5\ 0.55\ 0.7\ 0.4 \\ 0.5\ 1\ 0.75\ 0\ 0.4\ 1\ 0.7\ 0.2\ 0.6\ 0.5\ 1\ 1\ 0\ 0\ 0] \end{matrix}$$

where A_i is the value of concept C_i.

When the initial values of concepts have been assigned, the CTST-FCM starts to interact and simulate the radiation procedure. Equation 6.3 calculates the new values of concepts after each simulation step and Figure 6.7 illustrates the values of concepts for eight simulation steps. From Figure 6.7, it is concluded that after the fifth simulation step, the FCM reaches an equilibrium region, where the resulting values of OUT-Cs are OUT-C1 = 0.99, OUT-C2 = 0.025, and OUT-C3 = 0.04.

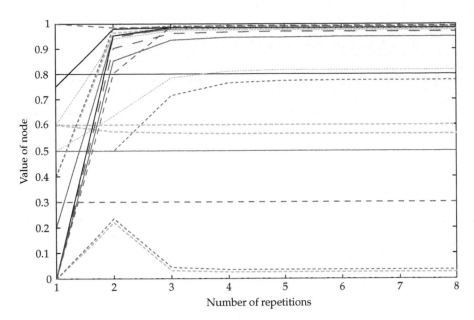

FIGURE 6.7
Variation of values of 26 concepts for the CTST-FCM for the first case for eight simulation steps.

The values of the CTST-FCM concepts at equilibrium region are $A_{leq}^{lower\ level} =$ [0.75 0.5 0.5 0.5564 0 0.5 0.8 0.5 0.55 0.7 0.4 0.779 0.819 0.991 0.99 0.987 0.993 0.99 0.995 0.987 0.99 0.978 0.99 0.025 0.040].

Based on the referred performance criteria in Section 6.5, the calculated values of output concepts are accepted. The calculated value of OUT-C1 is 0.99, which means that the CTV receives the 99% of the amount of the prescribed dose, so it is accepted. The value of OUT-C2 that represents the amount of the surrounding healthy tissues' volume received a dose is found equal to 0.025, so that 2.5% of the volume of healthy tissues receives the prescribed dose of 81 Gy. The value of OUT-C3 that represents the amount of the critical organ's volume (bladder and rectum) is equal to 0.034, which means that the 3.4% of the volume receives the prescribed dose of 81 Gy. The values of OUT-Cs satisfy the performance criteria in Equations 6.8 through 6.10. It is clear that the CTST-FCM model with the initial values of treatment variables and their interconnections which radiotherapists and medical physicists proposed for the specific technique of prostate cancer converged to a set of values that satisfy the performance criteria. Thus, the CTST-FCM suggests that the treatment planning can be executed and there will be acceptable results for the treatment.

6.5.1.2 Case Study 2

For the second case study, the conventional four-field box technique is implemented for the prostate cancer treatment. This radiotherapy technique consists of a four-field box arrangement with gantry angles 0°, 90°, 180°, and 270°.

A single external contour defines the patient anatomy and the isocentric beam therapy is used. Beam weights have the same value for four fields and moreover, no blocks, wedges, collimator settings, and compensating filters are used.

For this case, the CTST-FCM has to be reconstructed, which means that radiotherapists have to reassign weights and interconnections because a different treatment technique is used [45–48]. Data from AAPM TG 23 and experiments determine the treatment variables and their interrelationships, and they modify the CTST-FCM model.

For this case, the Selector-concept S-C4 has the value of four-field number; S-C3 has the value of "large size" of radiation field, which means that the influence of S-C3 and S-C4 toward OUT-Cs is very low. In the same way, the S-C5 and S-C6 have lower influence on OUT-Cs because different beam directions and weights of radiation beams are used. S-C8 has zero influence on OUT-Cs because no blocks and no wedges are selected for this treatment case. The S-C7 takes the discrete value of isocentric beam therapy and has the same influence on OUT-Cs as the conformal treatment case mentioned earlier. The S-C9 takes a low value for no accurate patient positioning and the S-C10 takes the discrete value of 2-D radiotherapy.

The weights between S-Cs and OUT-Cs for this case are given in Table 6.5. If we compare Table 6.5 with Table 6.2, which contains the weights for the first case, we will see that some weighted interconnections have different values.

Using this new CTST-FCM model with the new modified weight matrix, the simulation of the radiotherapy procedure for this case example starts with the following initial values of concepts:

$$A_2^{\text{lower level}} = \begin{matrix} [0.5 \ \ 0.5 \ \ 0.5 \ \ 0.6 \ \ 0 \ \ 0.5 \ \ 0.6 \ \ 0 \ \ 0.5 \ \ 0.3 \ \ 0.6 \ \ 0.5 \\ 0.5 \ \ 0.75 \ \ 0 \ \ 0.2 \ \ 1 \ \ 0.4 \ \ 0.4 \ \ 0.6 \ \ 0 \ \ 0 \ \ 0 \ \ 0 \ \ 0] \end{matrix}$$

The values of concepts are calculated using Equation 6.3 and the variation of values of 26 concepts after eight simulation steps are illustrated in Figure 6.8.

TABLE 6.5

Selector-Concepts–Output-Concepts Weights
for the Second Radiotherapy Case Study

Selectors	OUT-C1	OUT-C2	OUT-C3
S-C1	0.52	−0.45	−0.44
S-C2	0.50	−0.6	−0.48
S-C3	0.27	−0.2	−0.20
S-C4	0.24	−0.4	−0.4
S-C5	0.22	−0.25	−0.2
S-C6	0.25	−0.20	−0.20
S-C7	0.30	−0.30	−0.30
S-C8	0	0	0
S-C9	0.28	−0.30	−0.2
S-C10	0.20	−0.25	−0.20

Source: Papageorgiou, E.I., Stylios, C.D., Groumpos, P.P., *IEEE Trans. Biomed. Eng.*, 50, 12, 2003. With permission.

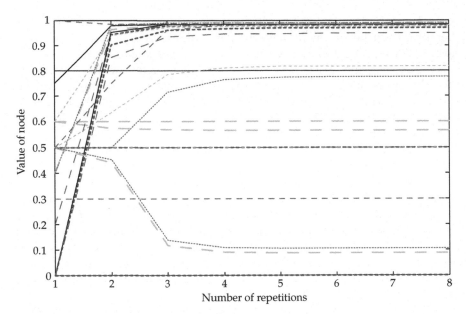

FIGURE 6.8
Variation of values of 26 concepts of CTST-FCM for the second example, with the classical treatment planning case for eight simulation steps.

It shows that the CTST-FCM interacts and reaches an equilibrium region. The values of concepts at equilibrium region are $A_{2eq}^{lower\ level} = [0.5\ 0.5\ 0.5$ $0.566\ 00.5\ 0.6\ 0\ 0.5\ 0.3\ 0.6\ 0.777\ 0.818\ 0.986\ 0.983\ 0.981\ 0.99\ 0.983\ 0.991$ $0.976\ 0.985\ 0.97\ 0.943\ 0.983\ 0.087\ 0.011]$.

At this equilibrium region, the final values of OUT-Cs are OUT-C1 = 0.983, OUT-C2 = 0.087, and OUT-C3 = 0.11. The calculated value of concept OUT-C1 is within the desired limits but the values of concept OUT-C2 and concept OUT-C3 are not accepted. The value of OUT-C2 is equal to 0.087, which means that the 8.7% of the volume of healthy tissues receives a prescribed dose of 81 Gy. The calculated value of OUT-C3 describes that the 11% of volume of organs at risk receives an amount of the prescribed dose. These values for OUT-C2 and OUT-C3 are not accepted according to related protocols [45].

If these suggested values for Output-concepts were adopted, the patient's normal tissues and sensitive organs would receive a larger amount of dose than that desired, which is not accepted. Thus, it is important to examine all the factors and selectors and their cause and effect toward the Output-concepts and suggest new treatment variable values in order to reschedule the planning procedure. This prompts the need for a higher level to lead the rescheduling and supervise the whole treatment planning.

6.5.2 Discussion of the Results on the Two Case Studies

CTST-FCM model integrates different treatment planning procedures where treatment parameters can have different degrees of influence to the treatment execution. CTST-FCM model estimates the final dose, which is actually received by the target volume and the patient. CTST-FCM was modified for some standard treatment techniques that are implemented in clinical practice, and then the CTST-FCM run and advised radiotherapists about the acceptance of the treatment planning technique.

The CTST-FCM model is an efficient and useful tool especially for this case of complex radiotherapy treatment planning problems, where the surrounding normal tissues and organs at risk place severe constrains on the prescription dose as in the case of prostate cancer. In practice, the patient receives a different amount of dose than that determined during the treatment planning due to the presence of some other factors, more general, as machine factors, human factors, and quality processes [37, 49] that influence the treatment execution. In addition to this, there are some factors on the CTST-FCM model, such as tumor localization and patient positioning, which change their values easily and it is necessary to take them into consideration during the final decision-making process with a more generic mode for all the patient cases. Thus, a better solution would be the designing of higher level with a new key-concept named "final dose" (FD). Concept FD would be affected by the parameters referred earlier and the OUT-Cs. The concept of "FD" is an extremely important concept describing the success of radiation treatment and so the prolongation of the patient's life. The purpose of the proposed approach is not to accurately calculate the amount of FD received by the patient, but to describe the success of the radiation therapy process in general and determining the value of FD. This highlights the need to construct a supervisor level.

6.6   Abstract Supervisor-FCM Model for Radiation Therapy Process

The supervisor level is higher than CTST-FCM model and is used for the parameters analyses and the final acceptance of the treatment planning technique. The two-level structure creates an advanced integrated system, which handles abstract information. The supervisor is modeled as an FCM that models, monitors, and evaluates the whole process of radiation therapy.

The supervisor-FCM is developed exploiting and utilizing experts' knowledge (doctors), who actually supervise the process. Radiotherapists usually use the notion and values of tumor localization, patient positioning, and the calculated dose by the treatment planning system to determine the FD [49,50]. They also mentioned that human factors and machine factors play an important role in the determination of the FD and they usually take these values into consideration. Experts, using this method of thinking and concluding, suggested the concepts of the supervisor-FCM.

The suggested supervisor-FCM consists of seven concepts to supervise the decision-making process during the radiation therapy process and it is

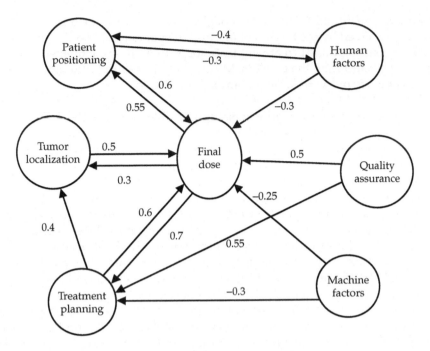

FIGURE 6.9
The Supervisor-FCM model of the radiotherapy process.

depicted in Figure 6.9. This model updates the first introduced supervisor-FCM and now one more concept has been added [21]. This new concept represents the quality assurance (QA) of the whole radiotherapy process. QA refers to the whole range of procedures and technical systems for assuring that the quality parameters of the process are in accordance with the national and international standards (preset) such as the International Standards Organization (ISO standards). Treatment planning systems, imaging devices, simulators, treatment units, checks of beam quality and inhomogeneity, and clinical dose measurements determine the QA process.

The concepts suggested by experts to include in the supervisor-FCM are as follows:

UC_1 *(tumor localization)*. It is dependent on the following three factors concepts of the lower level FCM: patient contour, sensitive critical organs, and tumor volume. It embodies the value and influences these three Factor-concepts.

UC_2 *(dose prescribed from treatment planning [TPD])*. This concept represents the prescribed dose and is dependent on the following concepts of the CTST-FCM model: the delivered dose to target volume, normal tissues, and critical organs that are calculated at the treatment planning model of the lower level FCM.

UC_3 (*machine factors*). This concept represents the equipment character-istics, reliability, efficiency, and maintenance.

UC_4 (*human factors*). A general concept that evaluates the experience and knowledge of medical staff, involved in the treatment.

UC_5 (*patient positioning and immobilization*). This concept describes the cooperation of the patient with the doctors and if the patient accu-rately follows their instructions.

UC_6 (*QA*). This represents and evaluates the qualifications of staff, the efficiency of therapeutic procedures, and the performance of techni-cal systems for complying with the preset standards.

UC_7 (*final dose given to the target volume [FD]*). An estimation of the radi-ation dose received by the target tumor.

The methodology presented in Section 6.2 is used here to construct the supervisor-FCM. The experts are asked to describe the relationships among concepts and they use IF–THEN rules to justify the cause and effect relation-ship among concepts and infer a linguistic weight for each interconnection. The degree of the influence is a linguistic variable, member of the fuzzy set T{*influence*} as illustrated in Figure 6.2.

Experts suggested the following connections among the earlier-described concepts of supervisor-FCM:

Linkage 1. Connects UC_1 with UC_7: it relates the tumor localization with the delivered FD. Higher the value of tumor localization, greater the delivered final dose is.

Linkage 2. Relates concept UC_2 with UC_1: when the dose derived from treatment planning is high, the value of tumor localization increases by a small amount.

Linkage 3. Connects UC_2 with UC_7: when the prescribed dose from treat-ment planning is high, the FD given to the patient will also be high.

Linkage 4. Relates UC_3 with UC_2: when the performance of machine parameters increases, the dose from treatment planning decreases.

Linkage 5. Connects UC_3 with UC_7: any decrease to machine parameters influences negatively, the FD given to target volume.

Linkage 6. Relates UC_4 with UC_7: the human factors cause decrease in the FD.

Linkage 7. Connects UC_4 with UC_5: the presence of human factors causes decrease on the patient positioning.

Linkage 8. Relates UC_5 with UC_4: any decrease on the patient positioning negatively influences the human factors.

Linkage 9. Connects UC_5 with UC_7: when the patient positioning increases, the FD also increases.

Linkage 10. Relates UC_6 with UC_2: any increase on the QA (control) checks, positively influences the treatment planning.

Linkage 11. Connects UC_6 with UC_7: any increase on the QA (control) checks positively influences the FD.

Linkage 12. Relates UC_7 with UC_5: when the FD reaches an upper value, the patient positioning is influenced positively.

Linkage 13. Connects UC_7 with UC_1: any change in FD causes change in tumor localization.

Linkage 14. Connects UC_7 with UC_2: when the FD increases to an acceptable value, the dose from treatment planning increases to a desired one.

After the determination of the relationships and the kind of linkages among concepts, each one of the experts suggests one linguistic weight for each linkage. The linguistic weights for each linkage are aggregated to an overall linguistic weight, which is defuzzified and transformed into crisp weight, corresponding to each linkage. Thus, the following weight matrix for the supervisor-FCM is produced, with the following numerical linkage weights:

$$
W^{second\ level} =
\begin{bmatrix}
0 & 0 & 0 & 0 & 0 & 0 & 0.5 \\
0.4 & 0 & 0 & 0 & 0 & 0 & 0.6 \\
0 & -0.3 & 0 & 0 & 0 & 0 & -0.22 \\
0 & 0 & 0 & 0 & -0.3 & 0 & -0.3 \\
0 & 0 & 0 & -0.4 & 0 & 0 & 0.6 \\
0 & 0.55 & 0 & 0 & 0 & 0 & 0.5 \\
0.3 & 0.7 & 0 & 0 & 0.55 & 0 & 0
\end{bmatrix}
$$

Experts describe the goals of supervisor-FCM and set the objectives. One objective of the supervisor-FCM is to keep the amount of FD, which is delivered to the patient, between some limits, an upper FD_{max} and a lower limit FD_{min}. Another objective is to keep the TPD between maximum value TPD_{max} and minimum value TPD_{min}. These objectives are defined at the related AAPM and International Commission on Radiological Protection (ICRP) protocols [1,11,12], where the accepted dose levels for each organ and region of human body are determined. So, the overall objective for the upper level, the supervisor-FCM, is to keep the values of corresponding concepts, FD given to the target volume and TPD in the range of values:

$$0.90 \leq FD \leq 0.95 \tag{6.11}$$

$$0.80 \leq TPD \leq 0.95 \tag{6.12}$$

The supervisor-FCM evaluates the success or failure of the treatment by checking the value of the FD concept, whether the suggested treatment process is within the accepted limits or not for the specified case of treatment [50].

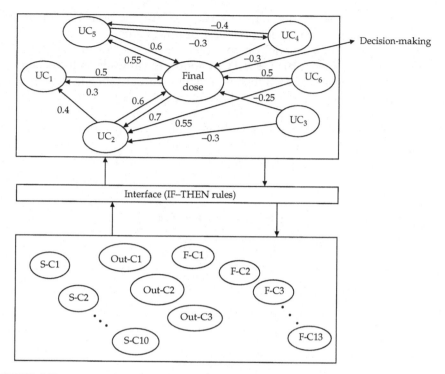

FIGURE 6.10
The integrated two-level hierarchical structure for decision making in radiation therapy.

A two-level hierarchical structure is proposed, which is illustrated in Figure 6.10. The CTST-FCM model for the radiotherapy treatment planning process, which was discussed in Section 6.5, is the lower level, where the 26 concepts of CTST-FCM model the treatment planning and the dose distribution to the target volume and normal tissues. Supervisor-FCM is the upper level, which is used for the determination of acceptance of the treatment therapy.

6.7 An Integrated Hierarchical Structure of Radiotherapy Decision Support System

The integrated hierarchical structure, consisting of the supervisor-FCM and the CTST-FCM, is the advanced DSS, which advises the radiotherapist–doctor on the decisions about the success of treatment therapy and the optimum treatment outcome. The supervisor-FCM aims to plan strategically and to detect and analyze unacceptable treatment before the execution of the treatment procedure. The main supervisory task is the coordination of the whole system, determining the amount of FD given to the target volume.

The proposed two-level hierarchical structure can successfully model the complex radiotherapy treatment planning requirements.

When the CTST-FCM model on the lower level reaches an equilibrium region, information from the CTST-FCM concepts values pass to the supervisor-FCM. Supervisor-FCM interacts, reaches an equilibrium region that determines if the calculated value of FD concept is accepted or not. The flowchart of this procedure is depicted in Figure 6.11.

The FCMs on the hierarchical two levels interact and information must pass from one FCM to the other. An interface is required for the transformation and transmission of information from the CTST-FCM on the lower

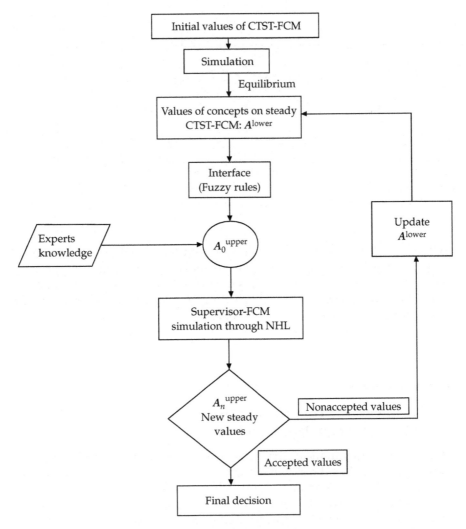

FIGURE 6.11
Schematic representation of the algorithm for supervision execution.

level to the supervisor-FCM on the upper level and vice versa. This interface consists of two parts; one part transmits information from lower level to upper level and the other part from upper level to lower level. Generally, the information from two or more concepts on the lower level CTST-FCM pass through the interface is aggregated and influence one concept in the upper level, and an analogous interface exists for the inverse transformation and transmission of information.

The interface is designed as a set of fuzzy rules. The transformation and transmission of information between concepts of two-level structures are representing using the IF–THEN rules that are embedded into the interface. The fuzzy rules take the values of concept as input from the lower level and infer the value of concepts on the supervisor-FCM. For example, information from the concepts of machine parameters at the lower level (Selector-concepts S-C7 and S-C8) pass through the interface and influence the concept of UC_3 "machine factors" at the upper level. Also, information from the Output-concepts (OUT-C1, OUT-C2, and OUT-C3) influences the UC_2 "dose from the treatment planning system." The following fuzzy rules describe the part of the interface from lower level toward the upper level:

- IF value of OUT-C1 is very high AND values of OUT-C2 AND OUT-C3 are very low THEN value of UC_2 is very high.
- IF value of OUT-C1 is the highest AND values of OUT-C2 AND OUT-C3 are the lowest THEN value of UC_2 is the highest.
- IF value of OUT-C1 is high AND values of OUT-C2 OR OUT-C3 are low THEN value of UC_2 is high.
- IF value of OUT-C1 is very high AND values of OUT-C2 OR OUT-C3 are low THEN value of UC_2 is high.
- IF value of S-C3 is very low AND values of S-C7 AND S-C8 are very high THEN value of UC_3 is high.
- IF value of S-C3 is very low AND values of S-C7 AND S-C8 are the highest THEN value of UC_3 is very high.
- IF value of S-C3 is very low AND values of S-C7 OR S-C8 are very high THEN value of UC_3 is high.
- IF value of S-C3 is medium AND values of S-C7 OR S-C8 are medium THEN value of UC_3 is medium.
- IF value of S-C9 is very high THEN value of concept UC_5 is very high.
- IF value of S-C9 is the highest THEN value of concept UC_5 is the highest.

In the same way, with a corresponding set of fuzzy rules, the interface from the upper level toward the lower level is developed describing analogous influences from the concepts of supervisor-FCM toward the Selector-concepts of the CTST-FCM.

6.7.1 Estimation of the Success or Failure of the Treatment Therapy

The initial values of concepts on supervisor-FCM are determined by the values of concepts of lower level CTST-FCM model, through the interface described earlier, and also there are some external inputs for the values of concepts referred to as UC_5 "human factors" and UC_1 "tumor localization."

6.7.1.1 Case Study 1

Here, the case of Section 6.5 will be discussed under the aspects of the hierarchical two-level structure. The CTST-FCM that was used for the first test case of prostate cancer is the lower level FCM. As presented, this CTST-FCM after the simulation had reached an equilibrium region and the values of Factor-concepts, Selector-concepts, and Output-concepts could be used for the desired treatment planning and calculation of dose on the target volume, normal tissues, and sensitive organs. These values are inputs to the fuzzy rules consisting the interface and so they determine the initial values of concepts on supervisor-FCM that are given in the following matrix:

$$\mathbf{A}_1^0 = [0.75 \ 0.8 \ 0.3 \ 0.6 \ 0.7 \ 0.5 \ 0.65]$$

For these values of concepts, the supervisor-FCM is able to examine if they are within the accepted limits for the radiotherapy execution. The supervisor-FCM simulates through Equation 6.1 using the initial matrix $\mathbf{A}_1^0$ and the initial weight matrix $\mathbf{W}^{upper \ level}$, to find an equilibrium region.

After 10 iteration steps, an equilibrium region is reached and Figure 6.12 gives the subsequent values of calculated concepts. Values of concepts UC_2

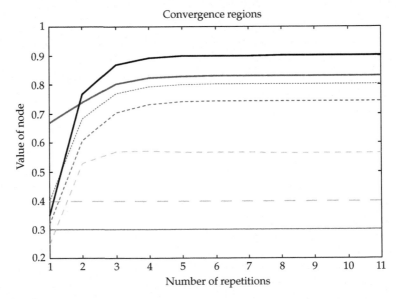

FIGURE 6.12
Equilibrium state for Supervisor-FCM model.

and UC_7, in the equilibrium region, are equal to the values 0.8033 and 0.89 where the value of FD is out of the suggested desired regions in Equations 6.11 and 6.12. Then according to the algorithm of supervision of Figure 6.11, we continue implementing the NHL algorithm.

The supervisor-FCM updates by the implementation of NHL algorithm, which is described in Section 6.3. After trial and error experiments for the specific supervisor-FCM model, the values of learning parameters η and γ have been determined as 0.04 and 0.98, respectively. The desired target values for each of the two DOCs are the mean values of the corresponding Equations 6.11 and 6.12: $T_1 = 0.875$ for the concept TPD and $T_2 = 0.925$ for the FD.

Equation 6.5 is used to modify the weights of supervisor-FCM, and Equation 6.3 is used to calculate the values of concepts after each simulation step. After 11 simulation steps, the supervisor-FCM reaches an equilibrium region, satisfying the criteria of algorithm in Equations 6.6 and 6.7:

$$A_1^{\text{upper level}} = [0.8325\ 0.8462\ 0.3000\ 0.6055\ 0.7693\ 0.5000\ 0.9236]$$

and the new weight matrix derived after training using the NHL algorithm is

$$W_{\text{NHL}}^{\text{supervisor}} = \begin{bmatrix} 0 & 0 & 0 & 0 & 0 & 0 & 0.54 \\ 0.465 & 0 & 0 & 0 & 0 & 0 & 0.61 \\ 0 & -0.1 & 0 & 0 & 0 & 0 & -0.043 \\ 0 & 0 & 0 & 0 & -0.105 & 0 & -0.078 \\ 0 & 0 & 0 & -0.23 & 0 & 0 & 0.611 \\ 0 & 0.52 & 0 & 0 & 0 & 0 & 0.386 \\ 0.409 & 0.681 & 0 & 0 & 0.54 & 0 & 0 \end{bmatrix}$$

The updated weights keep their initial suggested signs and directions, and their values within the initial ranges derived from the fuzzy linguistic variables, as suggested by expert doctors. Protocols and experimental data prescribe the final dose to patient for every treatment case.

In the first example, the calculated values for output concepts are TPD = 0.8462 and FD = 0.9236, respectively, which are within the acceptable values according to Equations 6.11 and 6.12 [50]. Radiotherapists can follow the suggested values and the treatment will be executed with successful results.

6.7.1.2 Case Study 2

To update the values of concepts at lower level, we follow the upper-lower interface and we change the values of the most important Factor-concepts and Selector-concepts. Also, at the same time, for the new employed technique or clinical case, we redefine some of the weights between SCs and OUT-Cs based on experts' suggestions. So, new values are assigned to the size of the radiation field (S-C3), beam direction (S-C5), weight of each field (S-C6), patient immobilization (S-C9), and increase the amount of perfect match of beam to the target volume (F-C15). These values along with the rest of the values of matrix $A_2^{\text{lower level}}$ for the second case study result in producing the following matrix for the lower level:

$$A_{21}^{\text{lower level}} = \begin{matrix}[0.5 \ 0.5 \ 0.5 \ 0.6 \ 1 \ 0.5 \ 1 \ 0 \ 0.5 \ 0.3 \ 0.6 \ 0.5 \ 0.5 \\ 0.75 \ 0 \ 0.8 \ 1 \ 0.8 \ 0.4 \ 0.6 \ 0.5 \ 0.5 \ 0 \ 0.8 \ 0.2 \ 0.4]\end{matrix}$$

The CTST-FCM with the new $A_{21}^{\text{lower level}}$ interacts and new values for the 26 concepts are calculated according to Equation 6.3 and the newly calculated values for Output-concepts are: OUT-C1 is 0.98, OUT-C2 is 0.03, and OUT-C3 is 0.07. These calculated values of Output-concepts are within the accepted limits for the CTST-FCM model. So, these newly updated values of concepts from CTST-FCM model influence the upper concepts of supervisor-FCM through the interface again, determining the next new initial concept values:

$$A_{21}^{\text{upper level}} = [0.87 \ 0.81 \ 0.2 \ 0.4 \ 0.65 \ 0.5 \ 0.86]$$

Then implementing the NHL algorithm for the supervisor-FCM, the following values of concepts on upper level are calculated: $A^{\text{upper level}} = [0.857 \ 0.832 \ 0.2 \ 0.621 \ 0.85 \ 0.5 \ 0.91]$.

Thus, the value of UC_7 is FD = 0.91 and the value of UC_2 is TPD = 0.832, which are accepted for the treatment execution.

If the calculated values of TPD and FD were not accepted, then the procedure mentioned earlier for the final decision could continue until the calculated values of concepts FD and TPD would be accepted. In this way, the supervisor-FCM supervises the treatment for prostate cancer therapy with external beam radiation and more generally the whole procedure.

6.7.2 Evaluation of the Proposed Model

In this research, the FCM modeling methodology is introduced and utilized at lower level to model the process of treatment planning, adjusting the treatment variables and calculating the corresponding dose to the target volumes, organs at risk, and normal tissues. The same modeling methodology is used at upper level to model abstractly and supervise the whole procedure of radiation therapy. For the supervisor-FCM, a novel training algorithm, the NHL is utilized to adjust the interconnections between the generic treatment variables of upper level and calculate the FD.

Weight adaptation and fine-tuning of supervisor-FCM causal links have great importance in updating the model to achieve acceptable results for radiotherapy techniques. This is the reason why we implement the NHL in the supervisor-FCM. Also, we should emphasize that using the NHL algorithm, we combine the human experts' structural knowledge with the data for each specific case. This is exactly the same with the reaction of a human expert who adapts his approach to the input data.

The doctor in charge usually evaluates the value of the FD given to the target volume, and the supervisor-FCM does exactly the same. In the case of unacceptable values for TPD and FD, some concepts on the lower level CTST-FCM have to be influenced; they take new values that causes the lower level CTST-FCM to interact. Then, the new calculated values of lower level, through the interface, determine again the values of upper level supervisor-FCM concepts. Implementing the NHL algorithm, the supervisor-FCM interacts and after some simulation steps, converges to an equilibrium region, which can be accepted or not, according to the related protocols.

Thus, radiotherapists can follow the suggested values and the treatment will be executed with successful results. The proposed approach is efficient and very useful for the FCM-controlled clinical radiotherapy process. The utilization of NHL algorithm recalculates all weights that participate in the simulation process, which enhances the supervisor-FCM model that was initially determined by expert doctors. Its importance to the radiotherapists is underlined by the fact that they will be able to introduce clinical cases based on a range of accepted values for the Output-concepts of the model.

Some requirements and limitations of the proposed approach are

- Experts should have great knowledge and know the proper operation of the whole system to provide useful information on desired values of Output-concepts.
- The proposed training algorithm does not derive new interconnections and there is no influence on the architecture of the FCM model.

6.8 Discussion and Conclusions

The soft computing approach of FCMs is used to determine the success of the radiation therapy process estimating the FD delivered to the target volume. The scope of this research is to advise radiotherapists to find the best treatment or the best dose. Furthermore, a two-level integrated hierarchical structure is proposed to supervise and evaluate the radiotherapy process before treatment execution. The supervisor-FCM determines the treatment variables of cancer therapy and the acceptance level of final radiation dose to the target volume. Two clinical case studies have been used to test the proposed methodology with successful results and demonstrate the efficiency of the CTST-FCM tool.

The proposed CTST-FCM model is evaluated for different treatment cases but it raises the need for an abstract model that will supervise it. An integrated two-level hierarchical structure is proposed, consisting of two-level FCMs to evaluate the radiotherapy planning procedure. The supervisor-FCM represents a second higher level control for prediction, decision analysis, and determination of the FD. The supervisor-FCM model is updated with the implementation of the NHL algorithm that adjusts the weights and ensures the success of the treatment therapy procedure.

The proposed two-level decision model for the radiation treatment procedure considers an extremely large number of factors that are ensured with the use of FCMs. This dynamic decision-making model for the radiotherapy treatment process uses the experts' knowledge and follows a human reasoning similar to that which doctor adopt while deciding on the treatment plan.

This research work was focused on the study of knowledge representation and on the introduction of a two-level hierarchical model based on FCMs. For the radiotherapy planning model, the CTST-FCM model on the lower level was proposed and an abstract generic model to supervise the whole process was suggested, which was enhanced with learning methods to have better convergence results. Furthermore, an interface to transform and transmit information between the levels of hierarchy was described and an algorithm to ensure the flow and exchange of information within the integrated hierarchical system was proposed.

The proposed modeling method based on FCMs could improve the radiotherapist's ability to simulate the treatment procedure and decide whether the treatment execution will or will not be successful by taking into consideration the prescribed dose between the accepted limits. In addition to this, the radiotherapist can simulate the procedure before the treatment process starts. This proposed approach for decision making in radiotherapy was introduced to improve planning efficiency and consistency for treatment cases, selecting the related factors and treatment variables, describing and determining the causal relationships among them.

The proposed hierarchical structure can be easily implemented in clinical practice and thus provides the physicians and medical physicists with a fast, accurate, reliable, and flexible tool for decision making in radiotherapy procedures. The test cases, presented in this work, demonstrate the efficiency of the proposed integrated approach and give very promising results to develop intelligent and adaptive decision support systems for medical applications.

Acknowledgment

The work of Elpiniki I. Papageorgiou was supported by a postdoctoral research grant from Greek State Scholarship Foundation (I.K.Y.).

References

1. F. Khan, *The Physics of Radiation Therapy*, 2nd Edition, Williams & Wilkins, Baltimore, 1994.
2. B. Kosko, *Neural Networks and Fuzzy Systems*, Prentice-Hall, NJ, 1992.

3. A. Brahme, Optimization of radiation therapy and the development of multi-leaf collimation, *Int. J. Radiat. Oncol. Biol. Phys.*, 25(2), 373–375, 1993.
4. A. Brahme, Optimization of radiation therapy, *Int. J. Radiat. Oncol. Biol. Phys.*, 28, 785–787, 1994.
5. J.P. Gibbons, D.N. Mihailidis, and H.A. Alkhatib, A novel method for treatment plan optimisation, *Engineering in Medicine and Biology Society, Proceedings of the 22nd Annual International Conference IEEE*, Vol. 4, pp. 3093–3095, July 2000.
6. G.S. Mageras and R. Mohan, Application of fast simulated annealing to optimization of conformal radiation treatments, *Med. Phys.*, 20, 447–639, 1993.
7. G. Starkschall, A. Pollack, and C.W. Stevens, Treatment planning using dose-volume feasibility search algorithm, *Int. J. Radiat. Oncol. Biol. Phys.*, 49, 1419–1427, 2001.
8. A. Brahme, Treatment optimization using physical and biological objective functions, in: Smith A. (Ed.) *Radiation Therapy Physics*, Springer, Berlin, pp. 209–246, 1995.
9. C.G. Rowbottom, V.S. Khoo, and S. Webb, Simultaneous optimization of beam orientations and beam weights in conformal radiotherapy, *Med. Phys.*, 28, 1696, 2001.
10. S. Soderstrom, Radiobiologically based optimization of external beam radiotherapy techniques using a small number of fields, Thesis, Stockholm University, 1995.
11. G. Kutcher and C. Burman, Calculation of complication probability factors for non-uniform normal tissue irradiation: the effective volume method, *Int. J. Radiat. Oncol. Biol. Phys.*, 16, 1623–1630, 1989.
12. L.J. Beard, M. van den Brink, A.M. Bruce, T. Shouman, L. Gras, A. te Velde, and J.V. Lebesque, Estimation of the incidence of late bladder and rectum complications after high-dose (70–78 Gy) conformal radiotherapy for prostate cancer, using dose-volume histograms, *Int. J. Radiat. Oncol. Biol. Phys.*, 41, 83–99, 1998.
13. T. Willoughby, G. Starkschall, N. Janjan, and I. Rosen, Evaluation and scoring of radiotherapy treatment plans using an artificial neural network, *Int. J. Radiat. Oncol. Biol. Phys.*, 34(4), 923–930, 1996.
14. D. Wells and J. Niederer, A medical expert system approach using artificial neural networks for standardized treatment planning, *Int. J. Radiat. Oncol. Biol. Phys.*, 41(1), 173–182, 1998.
15. C.D. Stylios and P.P. Groumpos, Fuzzy cognitive maps in modeling supervisory control systems, *J. Intell. Fuzzy Systems*, 8, 83–98, 2000.
16. E. Papageorgiou, C.D. Stylios, and P.P. Groumpos, Decision making in external beam radiation therapy based on FCMs, *Proceedings of 1st IEEE International Symposium 'Intelligent Systems' 2002*, 10–12 September, Bulgaria, 2002.
17. V.C. Georgopoulos, G.A. Malandraki, and C.D. Stylios, A fuzzy cognitive map approach to differential diagnosis of specific language impairment, *Artif. Intell. Med.*, 29(3), 261–278, 2003.
18. A. Zadeh, Fuzzy logic, neural networks and soft computing, *Commun. ACM*, 37, 77–84, 1994.
19. ICRU Report 50, Prescribing, recording and reporting photon beam therapy, *International Commission on Radiation Units and Measurements*, Washington, 1993.
20. ICRU Report 24, Determination of absorbed dose in a patient irradiated by beams of x or gamma rays in radiotherapy procedures, *International Commission on Radiation Units and Measurements*, Washington, 1976.

21. E.I. Papageorgiou, C.D. Stylios, and P.P. Groumpos, An integrated two-level hierarchical decision making system based on fuzzy cognitive maps, *IEEE Trans. Biomed. Eng.*, 50(12), 1326–1339, 2003.

22. J.S. Jang, C.T. Sun, and E. Mizutani, *Neuro-Fuzzy and Soft Computing*, Prentice Hall, Upper Saddle River, NJ, 1997.

23. P. Bonissone, Soft computing: the convergence of emerging reasoning technologies, *Soft Computing*, 1, 6–18, 1997.

24. L.A. Zadeh, What is soft computing, *Soft Computing*, 1, 1–2, 1997.

25. P. Mitra, S. Mitra, and S.K. Pal, Staging of cervical cancer with soft computing, *IEEE Trans. Biomed. Eng.*, 47(7), 934–940, 2000.

26. S.K. Pal and S. Mitra, *Neuro-Fuzzy Pattern Recognition: Methods in Soft Computing*, Wiley, New York, 1999.

27. V. Kecman, *Learning and Soft Computing: Support Vector Machines, Neural Networks, and Fuzzy Logic Models (Complex Adaptive Systems)*, Prentice Hall, Upper Saddle River, NJ, 2001.

28. H. Abbass. An evolutionary artificial neural networks approach for breast cancer diagnosis, *Artif. Intell. Med.*, 25(3), 265–281, 2002.

29. C.A. Pena-Reyes and M. Sipper, Evolutionary computation in medicine: an overview, *Artif. Intell. Med.*, 19(1), 1–23, 2000.

30. R. Axelrod, *Structure of Decision, the Cognitive Maps of Political Elites*, Princeton University Press, Princeton, NJ, p. 404, 1976.

31. C.D. Stylios, P.P. Groumpos, and V.C. Georgopoulos, An fuzzy cognitive maps approach to process control systems, *J. Adv. Comput. Intell.*, 3(5), 409–417, 1999.

32. C.T. Lin and C.S.G. Lee, *Neural Fuzzy Systems: A Neuro-Fuzzy Synergism to Intelligent Systems*, Prentice Hall, Upper Saddle River, NJ, 1996.

33. E.I. Papageorgiou, C.D. Stylios, and P.P. Groumpos, Fuzzy cognitive map learning based on nonlinear Hebbian rule, in: Gedeon T.D. and Fung L.C.C. (Eds.) *AI 2003*, *LNAI*, Vol. 2903, Springer-Verlag, Heidelberg, pp. 254–266, 2003.

34. E. Oja, H. Ogawa, and J. Wangviwattana, Learning in nonlinear constrained Hebbian networks, in: Kohonen T., Makisara K., Simula O. and Kangas J. (Eds.) *Artificial Neural Networks*, North-Holland, Amsterdam, pp. 385–390, 1991.

35. M. Hassoun, *Fundamentals of Artificial Neural Networks*, MIT Press, Bradford Book, MA, 1995.

36. ICRU Report 42, *Use of Computers in External Beam Radiotherapy Procedures with High Energy Photons and Electrons*, International Commission on Radiation Units and Measurements (ICRU), Bethesda, MD, 1987.

37. R. Mohan, G.S. Mageras, B. Baldwin, L.J. Brewster, G.J. Kutcher, S. Leibel, C.M. Burman, C.C. Ling, and Z. Fuks, Clinically relevant optimization of 3D-conformal treatments, *Med. Phys.*, 19(4), 933–943, 1992.

38. I. Turesson and A. Brahme, *Clinical Rationale for High Precision Radiotherapy*, ESTRO (European Society of Therapeutic Radiation and Oncology), Malmo, 1992.

39. E. Papageorgiou, A model for dose calculation in treatment planning using pencil beam kernels, Master thesis in Medical Physics, University Medical School of Patras, Patras, Greece, June 2000.

40. AAPM Report 55, *Radiation Treatment Planning Dosimetry Verification*, American Association of Physicists in Medicine, Report of Task Group 23 of the Radiation Therapy Committee, American Institution of Physics, Woodbury, New York, 1995.

41. J. Venselaar and H. Welleweerd, Application of a test package in an intercomparison of the photon dose calculation performance of treatment planning systems used in a clinical setting, *Radioth. Oncol.*, 60, 203–213, 2001.

42. R. Alam, G.S. Ibbott, R. Pourang, and R. Nath, Application of AAPM Radiation Therapy Committee Task Group 23 test package for comparison of two treatment planning systems for photon external beam radiotherapy, *Med. Phys.*, 24, 2043–2054, 1997.

43. F. Dechlich, K. Fumasoni, P. Mangili, G.M. Cattaneo, and M. Iori, Dosimetric evaluation of a commercial 3-D treatment planning system using Report 55 by AAPM Task Group 23, *Radioth. Oncol.*, 52, 69–77, 1999.

44. A. Pollack, G.K. Zagars, G. Starkscall, C.H. Childress, S. Kopplin, A.L. Boyer, and I.I. Rosen, Conventional vs. conformal radiotherapy for prostate cancer: preliminary results of dosimetry and acute toxicity, *Int. J. Radiat. Oncol. Biol. Phys.*, 34(4), 555–564, 1996.

45. Radiation Therapy Oncology Group, *A Phase I/II Dose Escalation Study Using Three-Dimensional Conformal Radiation Therapy for Adenocarcinoma of the Prostate*, Radiation Therapy Oncology Group, Philadelphia, 1996.

46. J. Armstrong, Three-dimensional conformal radiation therapy: evidence-based treatment of prostate cancer, *Radioth. Oncol.*, 64, 235–237, 2002.

47. M.J. Zelefsky, L. Happersett, S.A. Leibel, C.M. Burman, L. Schwartz, A.P. Dicker, G.J. Kutcher, and Z. Fuks, The effect of treatment positioning on normal tissue dose in patients with prostate cancer treated with three-dimensional conformal radiotherapy, *Int. J. Radiat. Oncol. Biol. Phys.*, 37, 13–19, 1997.

48. J. Meyer, A.J. Mills, L.C.O. Haas, J.K. Burnham, and E. Parvin, Accommodation of couch constraints for coplanar intensity modulated radiation therapy, *Radioth. Oncol.*, 61, 23–32, 2001.

49. G. Leunes, J. Verstaete, W. Van de Bogaert, J. Van Dam, A. Dutreix, and E. Van der Schueren, Human errors in data transfer during the preparation and delivery of radiation treatment affecting the final result: garbage in, garbage out, *Radiother. Oncol.*, 23, 217–222, 1992.

50. ICRU Report 29, Dose specification for reporting external beam therapy with photons and electrons, *International Commission on Radiation Units and Measurements*, Washington, 1978.

7

Artificial Neural Networks in Radiation Therapy

John H. Goodband and Olivier C. L. Haas

CONTENTS

7.1   Introduction

7.1.1   A Brief History of Radiotherapy

The past 50 years has seen increasing research into the use of radiation therapy (RT) [1]. This is largely due to the development of high-energy x-ray machines, which allow the treatment of deep-seated tumors without causing serious harm to the patient's epidermis—a phenomenon known as the "skin-sparing effect." The greatest challenge associated with radiotherapy is to ensure that the detrimental effect of radiation on the healthy tissues surrounding the tumors is minimized whilst the curative effect of tumor destruction is maximized. The side effects from overdosing healthy tissue can be as severe as the effect of the tumor itself; conversely, underdosing may allow the tumor to regrow. Brain, bowel, and spinal cord are particularly susceptible to radiation dosage, as these tissues have limited ability to self-repair [2]. When preparing treatment plans, organs that are highly sensitive to irradiation are generically classed as organs at risk (OARs). Overdosage may not be immediately apparent and late reactions to radiation therapy may occur months or sometimes years after the treatment has ceased. It is therefore of paramount importance to accurately differentiate between the OARs and other healthy tissues (OHTs), which can recover from relatively high radiation doses.

The volume irradiated by the combined fields of a treatment plan is called the high dose volume (HDV), and aims to match the planning target volume (PTV) prescribed by an oncologist. Typically, a maximum deviation of 5% from the prescribed dose is allowed [3]. The PTV includes a margin of error that takes into account setup errors and both inter- and intrafractional organ motion.

Introduction of computed tomography (CT) scanners in 1971 revolutionized tumor diagnosis by using hairline-thin x-ray beams and algebraic reconstruction techniques to produce two-dimensional (2-D) image "slices" of the human body, thereby enabling an accurate delineation of tumor contours to be realized [4–6]. Development of the multileaf collimator (MLC) [7] subsequently facilitated the accurate shaping of radiation fields from the beam's eye view—a technique known as conformal radiation therapy (CRT). During the 1980s, with the availability of powerful computers that enabled RT experts to calculate the dose distribution attributable to complex multiple-beam configurations, CRT was recognized as a more accurate method for implementing RT [1]. CRT is now routinely used in most hospitals in the United Kingdom.

The development of intensity-modulated radiation therapy (IMRT) [7] has facilitated modulation of x-ray beam intensity across the field. This has provided the means to improve the targeting of cancerous tissue by delivering radiation dose to convex PTVs, thereby sparing OARs. Each IMRT beam is subdivided into small (typically 0.5–1 cm^2 cross-section at the treatment area) pencil beams or beamlets, the intensities of which are required to be individually modulated. These are referred to in the literature as intensity-modulated beams (IMBs). A combination of IMBs from different angles is then optimized to provide the treatment plan (see Figure 7.1). Compensators and MLCs are used for delivering IMRT [7]. The patient-specific compensator (PSC) [8] is recognized as a relatively inexpensive and practical method, which can be easily verified using standard procedures for quality assurance [9].

IMRT, therefore, has the potential to improve the outcome of treatments and consequently patients' life expectancy and quality of life. In practice, delivery is slow due to the inherent complexities involved in both determining and realizing the shapes required and is expensive as a result of hardware, software, and staff training costs associated with treatment planning. This has resulted in slow uptake for the method, with only the best equipped hospitals worldwide, mainly in the United States, able to deliver IMRT on a routine basis.

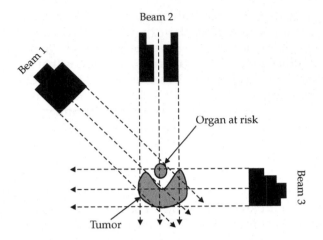

FIGURE 7.1
Schematic of intensity-modulated beams (IMBs).

Until recently, it has been difficult to monitor the accuracy of CRT treatment delivery for quality assurance checks [9]. The development of the electronic portal imaging device (EPID) [10,11] has greatly improved this situation, allowing online detection of treatment errors. Accurate implementation of IMRT is now known to have a profound effect on patient morbidity and quality of life [12,13]. By escalating dose to the tumor while minimizing dose to organs and healthy tissue adjacent to the tumor, control of cancerous cells is improved and side effects are minimized [14]. The first IMRT treatment (using MLC) given in the United Kingdom was in 1996 at the Memorial Sloan Kettering Cancer Institute and Hospital [15].

Although it is desirable to achieve high levels of accuracy when delivering x-rays using IMRT techniques, if exact tumor position relative to an irradiated volume is unknown, theoretical advantage in accuracy compared with less complex techniques may be compromised or even negated. Tumor position varies both between and during treatments (inter- and intrafractional movements, respectively). The movement is caused by a combination of daily changes in patient physiology, for example, bladder size; continuous changes in breathing; and also by setup errors associated with system complexity. Although every effort is made to minimize the cumulative dose error caused by each of these, there is inevitably some residual movement that cannot be prevented during treatment. Adaptive radiation therapy (ART) [16] and image-guided radiation therapy (IGRT) [17] are recent innovations, which attempt to reduce errors by imaging the tumor, either immediately prior to treatment (interfractional) or during (intrafractional) treatment. Ultrasound imaging [18], cone beam CT (CBCT) [19], or CT scans [20] can be used for interfractional tumor positioning. So called 4D imaging techniques (with the fourth dimension being time) such as 4D CT or 4D CBCT are giving volumetric information at different time instant. 4D imaging is starting to be used to assess the organ and target motion prior treatment, enabling clinicians to select margins to accommodate these motions appropriately or plan motion compensation strategies prior treatment delivery. Online intrafractional organ motion detection can be facilitated using either fluoroscopic x-rays for observing internal markers [21] stereoscopic video techniques for tracking external markers [22], magnetic markers [23] or a combination of optical and magnetic markers [24]. This technique is referred to as respiratory tracking and adaptive, real time or online RT because adaptation occurs during, rather than before, the beam delivery.

The development of IMRT and IGRT has significantly increased the complexity of RT planning and execution in recent years. Many decisions required to deliver effective treatment programs are still made by human experts based on their personal experience. The necessity for greater speed and consistency in decision making has led to much research being carried out to replicate this human experience using artificial intelligence. At the same time, ART requires methods that can rapidly adapt to changes in patient breathing characteristics for tracking tumor position online. The artificial neural network (ANN) paradigm is a flexible learning approach that uses external data to synthesize input–output relationships. Among their many applications, ANNs have become popular both for encoding human knowledge and for adaptive online control systems. This has made ANNs the subject of several areas of research in RT.

This chapter commences in Section 7.2 with a brief explanation of ANN theory, followed by a review of contemporary ANN applications for RT in Section 7.3. Sections 7.4 and 7.5 present two case studies that illustrate the process linked with the development and implementation of ANN solutions in RT. The first case study describes a static ANN employed to determine optimal PSC design from a desired dose distribution. The second case study presents an adaptive ANN to predict intrafractional organ movement. Section 7.6 presents discussion and conclusions related to this work and the use of ANNs in RT.

7.2   Artificial Neural Networks

ANN is a method by which a computer can be trained to "learn" a relationship between input and output data using mathematical functions designed to simulate the interactions between neurons and synapses in the human brain. This makes the ANN a very powerful tool in situations where an explicit functional relationship between inputs and outputs cannot be determined analytically. If sufficient input–output data are readily available, the training of an ANN may provide the best combination of speed and accuracy to replicate a desired function without a requirement for sophisticated technical knowledge from the operator. These facets have made ANNs the subject of research in the field of RT, where many complex nonlinear relationships exist, and at present much onus is on human experts to analyze data using knowledge and experience in combination with purely numerical calculations. In many cases, ANNs can accelerate these processes by either replicating human knowledge or carrying out computations that would otherwise be more time-consuming using alternative methods.

7.2.1 Artificial Neural Network Theory

7.2.1.1 McCulloch–Pitts Neuron Model

A diagrammatic representation of the earliest neuron design is illustrated in Figure 7.2. Each element $u_i \in \mathbb{R}, i = 1, \ldots, M$ of an input vector $\mathbf{u}$ is multiplied by a weight, $w_i \in \mathbb{R}$. The sum $\sum_{i=1}^{M} w_i u_i$ is added to a bias $b \in \mathbb{R}$, which acts as an affine transformation upon the decision hyperplane $\sum_{i=1}^{M} w_i u_i = 0$. The total, $v = \sum_{i=1}^{M} w_i u_i + b$, is often called the induced local field* for the neuron. v then becomes the input to an activation or transfer function, $\Phi(\cdot)$, which produces an output, y. This can be written in vector notation as

$$y = \Phi\left(\mathbf{w}^\mathrm{T}\mathbf{u} + b\right) \tag{7.1}$$

where $\mathbf{w} = [w_1, w_2, \ldots, w_M]^\mathrm{T}$ and $\mathbf{u} = [u_1, u_2, \ldots, u_M]^\mathrm{T}$. This design is still the basis for most ANNs used today. In general, Φ is a nonlinear function, although linear functions are used in output layers of many ANNs. The earliest function used in NN design, because it can be implemented with simple on/off circuits, was the Heaviside step function defined as

$$\begin{cases} y = 0, \ v < 0 \\ y = 1, \ v \geq 0 \end{cases} \tag{7.2}$$

Common among contemporary activation functions are the so-called squashing functions, which take an unbounded input and "squash" it to give a bounded output, typically

$$\Phi(\mathbf{w}^\mathrm{T}\mathbf{u} + b) = \Phi(v) \mapsto (0,1) \quad \forall \mathbf{u} \tag{7.3}$$

or

$$\Phi(\mathbf{w}^\mathrm{T}\mathbf{u} + b) = \Phi(v) \mapsto (-1,1) \quad \forall \mathbf{u} \tag{7.4}$$

where Equation 7.3 is realized using the logsig function $y = (1 + e^{-v})^{-1}$ and Equation 7.4 by the tansig function $y = (e^v - e^{-v})(e^v + e^{-v})^{-1}$. In general, multiple inputs are linked to multiple outputs, sometimes *via* complex arrays of weight connections and layer functions (see, e.g., Figure 7.3). The combination of these attributes is called the architecture of the ANN. An ANN consisting of only an input and output layer is called a single-layer perceptron

---

* The term is borrowed from Ernst Ising's model for spin-glass states in a ferromagnetic lattice, which was an early influence on ANN theory.

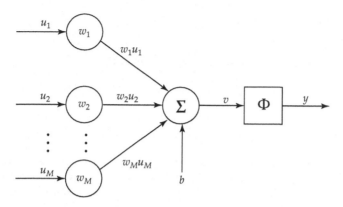

FIGURE 7.2
McCulloch–Pitts neuron model.

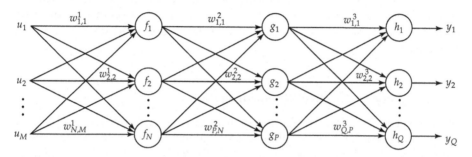

FIGURE 7.3
ANN with two hidden layers. f, g and h represent (possibly) different activation functions and w represents the various connecting weights between layers. The ANN input is $\mathbf{u} = [u_1 u_2 \ldots u_M]^T$, output is $\mathbf{y} = [y_1 y_2 \ldots y_Q]^T$, and hidden layers have N and P neurons, respectively, in each layer.

(SLP). An ANN with at least one hidden layer, that is, a layer between the input and output layers is known as multilayer perceptron (MLP). The size of an MLP is often specified in compact form A-B-L, where A represents the number of input neurons, B the number of neurons in the first layer, etc. Therefore, 3-4-2 represents a MLP with 3 inputs, 4 hidden neurons, and 2 outputs. The MLP has become ubiquitous for modeling nonlinear functions, because it has been proven [25] that any continuous function can be modeled by a MLP with sufficient neurons.

7.2.2 Training Algorithms

When ANNs are required to model an input–output relationship, a set of training data is used, which should in some way encode the mapping. This method is known as supervised learning. Data are presented to an ANN in the form of a set of input and target vectors and a training algorithm is implemented, which aims to reduce the value of a scalar energy function,

F, typically the mean squared error (MSE) given by

$$F = \frac{1}{N} \sum_{i=1}^{N} (\mathbf{t}_i - \mathbf{y}_i)^{\mathrm{T}} (\mathbf{t}_i - \mathbf{y}_i) \tag{7.5}$$

$$= \frac{1}{N} \sum_{i=1}^{N} \mathbf{e}_i^{\mathrm{T}} \mathbf{e}_i \tag{7.6}$$

where $\mathbf{t}_i$ is a target vector, $\mathbf{y}_i$ a training output vector, and N the number of input-target data pairs. This reduction is achieved by iteratively changing the values of weights connecting inputs to neurons until, either the desired value for F is attained or the algorithm is terminated by an alternative criterion, for example, time restriction or cross-validation (see Section 7.4). Each iteration is called an epoch.

There are a large number of algorithms available to train ANNs, see [26] for a good introduction to several of these. A brief description of four training algorithms commonly applied to ANNs in RT is provided in the following sections.

7.2.2.1 Steepest Gradient Backpropagation

Twenty years after its introduction by Rumelhardt and McClelland [27], the steepest descent backpropagation (SDBP) algorithm is still used in many contemporary ANN RT applications. In SDBP, weights are initialized using a random distribution with zero mean. The Nguyen–Widrow layer initialization function [28] is often used. The gradient of F is calculated and a small step is made in the negative direction to this gradient. Weight adjustments are then made throughout the ANN to decrease F—this is the backpropagation aspect of the algorithm. Although reliable, it can be extremely slow to converge on an error minimum for even relatively small networks. A momentum term [26] can be added to stabilize and accelerate learning, using SDBP, but the speed of convergence still cannot compare with the following alternative methods.

7.2.2.2 Conjugate-Gradient Backpropagation

The conjugate-gradient (CG) search method has been shown to guarantee convergence to the minimum of a quadratic function in a finite number of iterations $\leq W$, where W is the number of parameters of the function [29]. Although F is not quadratic, and there is no longer guaranteed convergence, a modification of CG called conjugate-gradient backpropagation (CGBP) can be used to train a MLP. CGBP employs a line search to find successive local minimum values of F. At each local minimum, the algorithm is reset by calculating a new search direction until the algorithm either converges or is terminated by a suitable criterion, for example, time limitation or maximum acceptable error.

7.2.2.3 Levenberg–Marquardt

The Levenberg–Marquardt (LM) algorithm [30,31] is designed specifically to minimize the sum-of-squares error function, using a formula that (partly) assumes that the underlying function modeled by the network is linear. Close to a minimum, this assumption is approximately true and the algorithm can make very rapid progress. Further away it may be a very poor assumption. The LM algorithm therefore compromises between the linear model and a gradient-descent approach. A move is only accepted if it improves the error, and if necessary the gradient-descent model is used with a sufficiently small step to guarantee downhill movement. The LM algorithm requires the inversion of $[\mathbf{J}_k^T \mathbf{J}_k + \lambda_k \mathbf{I}]$ at the kth iteration, where $\mathbf{J}$ is the Jacobian matrix and λ is an adjustable scalar parameter. The size of this matrix is dependent on the number of training pairs, N, presented to the ANN and the computational cost of the required matrix inversion scales as $O(N^3)$ [32]. Therefore, there is a practical limit to the amount of data that can be used to train any one ANN using the LM algorithm.

7.2.2.4 Bayesian Regularization

Bayesian regularization (BR) is a useful method for limiting the complexity of a model where some *a priori* knowledge, for example, smoothness or monotonicity is available. The training performance function defined by Equation 7.5 is modified using a term that is the mean-square of the total number of weights and biases, w_j, where $j = 1, 2, \ldots, W$, in the ANN and is given by

$$\text{MSW} = \frac{1}{W} \sum_{j=1}^{W} w_j^2 \tag{7.7}$$

This is added to MSE to give a cost function defined as

$$\text{CF} = F + \text{MSW} \tag{7.8}$$

$$= \frac{1}{N} \sum_{i=1}^{N} (\mathbf{t}_i - \mathbf{y}_i)^T (\mathbf{t}_i - \mathbf{y}_i) + \frac{1}{W} \sum_{j=1}^{W} w_j^2 \tag{7.9}$$

The additional term defined by Equation 7.7 serves to minimize both the magnitude and number of weights and biases in the network, thereby creating the smoothest function by allowing the elimination of redundant weights, that is, those with negligible values.

7.2.3 Time-Series Prediction Neural Network

A time-series prediction (TSP), $\hat{x}(t + T)$, T seconds ahead of the present time t is given by

$$\hat{x}(t + T) = G\left\{(x(t), x(t - \tau), x(t - 2\tau), \ldots,\right.$$

$$\left. x(t - (H - 1)\tau), x(t - H\tau) + \sum_{h=0}^{H} \varepsilon(t - h\tau)\right\}$$

$$= G\left\{\mathbf{x}^T + \sum_{h=0}^{H} \varepsilon(t - h\tau)\right\} \qquad (7.10)$$

where G is the nonlinear function described by the ANN, τ the sampling period, H the number of historical samples used in combination with the present sample $x(t)$, ε observation noise, usually assumed to be Gaussian, and $\mathbf{x} = [x(t), x(t - \tau), x(t - 2\tau), \ldots, x(t - (H - 1)\tau), x(t - H\tau)]^T$. The "black-box" ability of an ANN to model input–output mappings with little or no requirement for *a priori* knowledge has made ANNs popular for both modeling and predicting time-series data. Consequently, they are frequently used in predictive control [33–35]. Each of the following two sections describes the ANN design, which has been demonstrated as suitable for TSP.

7.2.3.1 Time-Series Prediction Multilayer Perceptron

The MLP architecture to predict time-series data constitutes a tapped delay line input, hidden layer, the size of which should ideally be optimized using training data, and one output (Figure 7.4). This type of MLP is known as a TSP MLP [36,37]. Although the output need not be scalar, it has been shown that more precise results are obtained using individual networks for each prediction output, that is, MLPs with scalar outputs [37]. A TSP MLP is trained by using a sequence of vectors $\{\mathbf{x}_n\}$, $\mathbf{x}_n \in \mathbb{R}^{H+1}$, $n = 1, 2, \ldots, N$ as inputs,

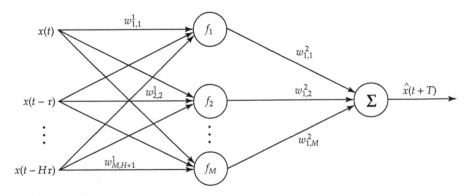

FIGURE 7.4
MLP prediction network with one hidden layer. The MLP illustrated represents a 1-D (scalar) predictor with $H + 1$ inputs, one output, and a hidden layer with M neurons.

and $\{t_n\} \equiv \{x_n(t + T)\}$ as targets. The accuracy of a TSP MLP is dependent on a number of factors:

1. Signal characteristics
2. Desired prediction horizon
3. Sampling frequency
4. Frequency of online updating (if any)
5. Number of inputs ≡ number of historical points used
6. Number of hidden neurons (if any) in network
7. Type of training algorithm (see Section 7.2) implemented

In general, points 1–3 are beyond the control of the ANN designer. It should be noted that for periodic signals to avoid aliasing, the sampling frequency should be at least twice the bandwidth of the signal [38]. Online updating needs to be carried out frequently if the signal is nonstationary, but this frequency may be restricted by the speed of the training algorithm used. Finding optimal solutions for points 5–7 can be very time-consuming. Optimization of architecture and type of training algorithm is, therefore, generally carried out offline, with online updating (where implemented) limited to weight changes facilitated by short periods of retraining. As a result, it may be difficult or even impossible for the ANN to give an accurate response at points where there is a significant change in the signal properties.

7.2.3.2 Generalized Regression Neural Network for Time-Series Prediction

The generalized regression NN (GRNN) is a modification of the standard Gaussian radial basis function (GRBF) ANN [33] and is designed specifically to facilitate function approximation for system modeling and prediction. The architecture is shown in Figure 7.5 for a scalar TSP output. The center, c_n, $n = 1, 2, \ldots, N$ of each GRBF is an input vector from the training set and

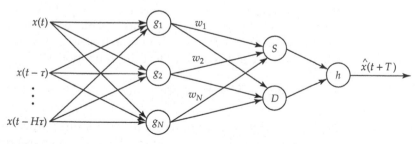

FIGURE 7.5
A GRNN with one output. g represents a Gaussian activation function, S an S-summation neuron, $S = \sum_{n=1}^{N} w_n g(\mathbf{x})$, and D a D-summation neuron, $D = \sum_{n=1}^{N} g(\mathbf{x})$. h carries out the computation $\hat{x}(t + T) = S/D$.

each weight w_n is the target associated with the respective input vector, that is, $w_n = t_n$. The output, $\hat{x}(t + T)$ of the GRNN is given by

$$\hat{x}(t + T) = \frac{\sum_{n=1}^{N} w_n g(\mathbf{x})}{\sum_{n=1}^{N} g(\mathbf{x})} \qquad (7.11)$$

where

$$g(\mathbf{x}) = \exp(-\| \mathbf{c}_1 - \mathbf{x}_n \|/\sigma^2)$$

The value assigned to σ controls the generalization capability of the RBF ANN. As a rule of thumb, σ should be greater than the minimum Euclidean distance between pairs of data and smaller than the maximum distance between pairs. A value commonly used is

$$\sigma = \frac{d_{max}}{\sqrt{C}} \qquad (7.12)$$

where d_{max} is the maximum Euclidean distance between input vectors and C is the number of GRBFs chosen for the network [33]. This value can produce a good function approximation where data points are evenly distributed throughout the input space. Clearly, the magnitude of the GRNN output is bounded by the size of the maximum value of $w_n \equiv t_n$, that is,

$$|\hat{x}(t + T)| \leq \max_n |w_n| \qquad (7.13)$$

This contrasts with MLP training, where there is no guaranteed *a priori* bound on outputs due to the nature of the learning process.

7.2.4   Early Stopping

Care must be taken to ensure that good generalization is built into an ANN, so that when presented with an earlier unseen input, it will produce an appropriate output. An ANN trained to a very small error may have effectively "rote-memorized" training data without learning the true underlying function. To prevent this, it is sometimes better to interrupt training at some point before the predetermined MSE target is reached. This is achieved by splitting data into a training subset and a validation subset. The first subset is used to train the ANN, whereas the second is used to test the validation MSE,

$$\text{MSE}_{val} = \sum_{i=1}^{N} (\mathbf{t}_i^{val} - \mathbf{y}_i)^T (\mathbf{t}_i^{val} - \mathbf{y}_i) \qquad (7.14)$$

of the temporarily "frozen" network at intervals in the training procedure, where $\mathbf{t}^{val}$ is a target vector from the validation set. This is known as cross-validation. If a point is reached where the error of the validation set begins to rise (typically, a few epochs of consecutive increasing error is necessary if MSE_{val} is not monotonically decreasing), the network training is stopped.

In the context of an adaptive TSP MLP, the use of online validation data may be impractical. Following are the three options to be considered:

1. Train using the most recent data and then validate using older data
2. Train using older data and then validate using the most recent data
3. Train using a selection of recent data points and then validate using the data unused for training

The first two methods may give poor results for rapidly changing nonstationary signals because there is no guarantee that the validation set is within the domain of the training data. The third method can sometimes produce an "average" network that may miss a rapid change occurring in the most recent data samples. This is, however, the only practical method for implementing online, where a signal is nonstationary, and is effective where data are noisy. In this work, early stopping is implemented by keeping 1 in 5 data points for validation purposes, whereas training the ANN on the remaining 4 in 5 points. This ratio is considered optimal for most ANN training applications [39]. The validation points are sampled at uniform intervals to avoid the possibility of bias (see Figure 7.6).

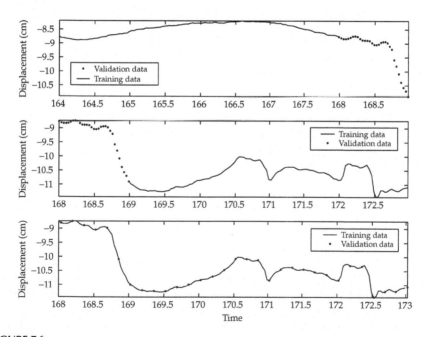

FIGURE 7.6
Illustrating the importance of uniform sampling for validation data. In the top plot the most recent 1 s of data are used for validation, although in the center plot the most recent 4 s of data are used for training. In both cases training ceases immediately because of the domain difference between the training and validation data. In the bottom plot uniform sampling is applied, enabling the ANN to "learn" information from the domain of the whole 5 s of data.

Having introduced the necessary ANN structures and training schemes, the following section provides a review of ANN applications with respect to treatment planning and prediction of tumor position in RT.

7.3   Review of Artificial Neural Networks in Radiation Therapy

7.3.1   Classification

Various authors have investigated the use of ANNs in RT treatment planning. The following studies utilize ANNs to perform classification problems, usually assigned to human experts, attempting to replicate the human experience by the ANN training process. The major advantage cited is time saving, because very experienced treatment planners also spend much time trying out alternative plans to find an optimal plan. Improved consistency is another benefit of using computerised systems which output do not depend on the fatigue of the operator or on the level and type of experience possessed by each human planner.

7.3.1.1   Evaluation of Treatment Set-Up Plans

Leszcynski et al. [40] used an ANN to evaluate treatment set-up plans. An MLP with seven inputs, two hidden layers (no indication of the number of neurons in each layer is given), and one output is trained using SDBP with data obtained from 328 portal images of tangential breast irradiations. Inputs consist of parameters describing the plans and the respective targets are an oncologist's ratings for each plan on a scale of 0–10. The MLP is found to produce an excellent level of agreement with the oncologist's decisions, with an accuracy greater than 92% for classification as "acceptable" and 100% agreement for "unacceptable."

7.3.1.2   Treatment Plan Rating Based on Dose–Volume Histogram Data

Lennernas et al. [41] used an ANN to accept or reject radiation dose–volume histogram data for rectum and urinary bladder. An MLP with 50 inputs, 1 hidden layer (no indication of the number of neurons is given) and 2 outputs is trained using SDBP with data obtained from 27 treatment plans for external radiotherapy of prostate adenocarcinoma. Inputs represent relative dose–volume histogram data and target acceptance or rejection based on the consensus of opinions given by three radiation oncologists. Acknowledging that the amount of data used for training the size of network is less than optimal, positive results are obtained, with the ANN accurately replicating the consensus of opinions given by the radiation oncologists.

Dose–volume information is again used as training data by Willoughby et al. [42]. An ANN is implemented to score treatment plans for patients receiving external-beam RT for soft tissue sarcomas of the lower extremity. An MLP with 13 inputs, 1 hidden layer of unspecified size and 4 output nodes is trained using SDBP with 150 sets of data. Inputs represent dose–volume information and target the score given by a single oncologist. The resulting ANN produces outputs comparable with the scoring decisions of the oncologist 90% of the time on training data and 83% of the time on validation data. It is noted that the oncologist is consistent to within one point 88% of the time.

7.3.2 Nonlinear Regression

ANNs can also be designed to produce continuous functional mappings from inputs to outputs where complex nonlinear relationships need to be modeled. The following studies have investigated the utilization of ANNs as an alternative to existing algorithms in calculating such relationships.

7.3.2.1 Optimal Treatment Plan Selection

Hosseini-Ashrafi et al. [43] used an MLP with one hidden layer to select the best plan from a preoptimized ANN database. The plans are from three separate treatment groups classified by tumor position. Group A has six subcategories: (1) tonsillor tumor, (2) small anterior floor of mouth tumor, (3) small primary tumor of the tonsillar fossa, (4) lateral tumor of the tongue, (5) tumor of the retromolar trigone, and (6) small tumor of the buccal mucosa. Group B includes five treatment groups: (1) tumor of the right vocal cord, (2) larynx, (3) supraglottic carcinoma, (4) pyriform fossa tumor including ipislateral lymph nodes, and (5) larynx of patients with rounded neck. Group C: (1) central mediastinal lesion, (2) peripheral lung lesion, (3) minimizing contra-lateral lung dose, (4) peripheral lung lesion minimizing dose to spinal cord, (5) middle section esophageal cancer minimizing lung and spinal cord irradiation, (6) middle section esophageal cancer where no lung tissue is present, (7) mid-level cervical esophageal cancer, and (8) vertebral body tumor. The optimal treatment plan for each subcategory has been decided earlier by human experts. Each of the three ANNs has 12 inputs, each input representing a geometric feature of each scanned tumor contour. ANN outputs are in terms of gantry and wedge angles and field weights for either two or three beam treatment plans. A "leave-one-out" cross-validation method [44] is used, because the number of training pairs is as low as seven for the third category of Group C. Optimization of hidden neurons is carried out for each of the three categories. Training produces classification successes of 83%, 98%, and 93% for Groups A, B, and C, respectively, using the medical experts' opinions as the benchmark. The Group A ANN uses five hidden neurons, B uses six hidden neurons, and C uses four. The ANN approach is recommended as being able to decrease the number of steps required to produce an optimal treatment plan.

7.3.2.2   Treatment Plan Design

Wells and Niederer [45] compared a set of MLPs, each with one hidden layer with a best contour match (BCM) method in a medical expert system used to design optimal treatment plans for prostate tumors. A simplified four-field box technique for prostate treatment is used to assess the viability of both approaches. Six individual MLPs are trained on specific plans using SDBP: without wedge, with physical wedge, and with dynamic wedge, for both small and large field sizes. Fourteen hidden neurons are used, although no justification is given for this. To obtain the required tolerance level 1,000 –10,000 epochs are required. Data sets vary in size from 12 to 68 for each class. Input data are in the form of seven weighted radial distances to a contour point from the centroid of either the prostate organ or the proximal lymph nodes, depending on the field size. The centroid position is determined by a radiation oncologist based on CT data. Each MLP output represents one of three treatment variables, that is, two for beam weights (the lateral beams are assumed to have the same weight) and the third for wedge angle for the lateral beams (either 0° or 15°). Using a tolerance level of 10% to prevent overtraining, the ANNs consistently perform better than the BCM system when compared with a human dosimetrist.

Wu et al. [46] compared an ANN approach with linear programming (LP) to produce a near-optimal combination of beam angles and weights for treating prostate tumors. Beam angles, θ, are constrained to $30° \leq \theta \leq 330°$ for 5° increments. A treatment plan is produced using a SLP with SDBP training in 27 min, compared with 45 min for the LP method and several hours by conventional methods. By comparing isodose curves, the ANN-generated plan is shown to be superior to the LP plan.

Rowbottom et al. [47] used an ANN to produce customized three beam orientations based on the size and relative positions of organs at risk. An MLP with 12 inputs, 2 hidden layers (no details of the number of hidden neurons is given), and 2 outputs is trained using SDBP. The geometry of patients suffering from cancer of the prostate is modeled by reducing the external contour, PTV, and OARs to a set of cuboids. The MLP inputs are the coordinates and size of the cuboids. Each output represents the orientation of a beam relative to the 0° gantry angle. Only a small (by accepted ANN theory) number of data pairs are used, 45 for training and 12 for validation. Results are inconclusive, as the ANN is only able to produce beam orientations within 5% accuracy of those produced by a previously developed beam-orientation constrained-customization scheme in 62.5% of cases.

Knowles et al. [48] used an MLP with two hidden layers to replicate beam weights devised by medical experts using CT scan data as inputs. The CT input data are reduced to six data points with a seventh point added to indicate the type of cancer. The three MLP outputs represent the relative weight of beam 1 and the relative weight of the motorized wedge for beams 2 and 3. Twenty-five prostate plans are used: 20 for training and 5 for validation. The SDBP algorithm with momentum is initially used for training and progressing to simple adaptive momentum to reduce training times. A second training method using an evolutionary algorithm is assessed and found to give superior generalization capability, although the training process is approximately 10 times longer. Acceptable results (better than 93%) are produced, occasionally bettering those produced by trained radiotherapists.

7.3.2.3 Tumor Delineation

Kaspari et al. [49] used ANNs to predict desired coefficients for polynomials to describe PTVs for glioblastoma multiforme tumors. Polynomial coefficients from measured tumor data are the ANN inputs. The polynomials are both functions of θ and φ in a spherical coordinate system with its center at the tumor isocenter. Although no explicit details of ANN architecture or training are given, it can be inferred from the information given that an MLP is used;* the polynomials used for input data are of third order. An assertion is made that more complicated tumor shapes need higher order polynomials and therefore a larger quantity of hidden neurons. Three radiologists assess the accuracy of predicted surfaces. The results show that 87% of PTVs predicted by the ANN are accepted.

7.3.2.4 Modeling Dose Distribution

Blake [50] models dose distribution from a Varian 2100C linac using an MLP with good agreement between measured data and the trained MLP. Inputs to the MLP consist of field size, fractional distance off axis and depth, and a sigmoidal "template" function, which makes use of *a priori* knowledge, that is, the Bentley–Milan interpolation algorithm [51] for depth-dose profiles. The MLP output is percentage depth dose. No attempt is made to model intensity-modulated beams.

Mathieu et al. [52] used an ANN to model 2-D RT dose distributions. Much of the article is taken up in describing general ANN theory without describing the actual ANN design used by the authors. It can only be inferred that an MLP is used. Two inputs are made to the ANN, representing radial distance from the linac central axis and depth relative to the surface of a water phantom. ANN output gives relative dose. No indication is made of the number of hidden neurons. SDBP is implemented for training, although no justification is given and no attempt is made to use a faster algorithm. More than 10^6 epochs are used to reach a validation error of less than 2%. The time required for training is not mentioned. The authors comment that, compared with "painfully slow" (several hours) Monte Carlo computations, ANNs provide almost instant results, giving 61 profiles of 201 points each in less than 30 s.

Beavis et al. [53] used an ANN for quality assurance (QA) verification of IMBs. The dose distribution attributable to MLCs delivering simple wedge-shaped distributions is calculated by the ANN and compared with the required prescription.

---

* In many academic papers dating from the 1990s, it would seem that the authors assume all ANNs are MLPs.

7.3.2.5 Predicting Beam Compensation

Gulliford et al. [54] trained an MLP with one hidden layer to model the relationship between portal imaging data and fluence maps for treating breast cancer. Inputs to the MLP are pixel densities from a portal image and outputs are fluence maps designed for multiple-static-field MLC delivery. A 4343-4059-2121 MLP (see Section 7.2.1.1) is used, inputs corresponding to the number of pixels in each image and outputs to the number of pixels in the compensation map. Eighty sets of patient data are used as input data. Fluence maps calculated with an in-house algorithm are used as targets. The size of training data is much smaller than those ideally required for the size of MLP; and it is acknowledged that the MLP cannot be more accurate than the algorithm used to create the data, because by definition, this algorithm is used as the benchmark. The MLP is trained four times using SDBP with momentum, 60 pairs of data for training and 20 for validation, with a different validation set being used for each training run. This ensures that each pair is used for both training and validation. An average of the MSE_{val} produced by each run is used to represent output error. Results show the MLP to be 7.6% less accurate than the algorithm. However, the study points out that "there may be a role for ANN solutions in producing intensity modulated fields that are currently produced using time-consuming iterative techniques". Interestingly, Gulliford says, "As the physical basis of the algorithm is known, the ANN will not improve in any way on the current clinical method but may provide an insight into the wider issue of designing intensity-modulated fields."

7.3.2.6 Prediction of Biological Outcomes

Gulliford et al. [55] used ANNs to correlate IMRT dose distributions with radiobiological effect for patients receiving RT treatment. Three different single hidden-layer MLPs are used to model biochemical control, rectal complications, and bladder complications. The MLP architectures are 62-80-2, 37-45-3, and 36-45-3, respectively. Historical clinical data from patients who received radical prostate RT are used to train the MLPs. Inputs to the MLP are parameters associated with treatment plan prescription and dose distribution with known biological outcomes as targets. It is shown that the ANNs are able to predict biochemical control and specific bladder and rectum complications with sensitivity and specificity of greater than 55%. The effects of individual treatment parameters are also analyzed using the ANN results. ANNs are shown to be able to learn some aspects of the complex relationships between treatment parameters and biological outcomes.

7.3.2.7 Predicting Compensator Dimensions for Intensity-Modulated Radiation Therapy

Goodband et al. [56] introduced a novel approach using an ANN to determine the optimal configuration of a novel discrete liquid metal compensating device for delivering IMRT. Mercury and indium gallium have been investigated as appropriate attenuating materials, although the approach is applicable to any suitable material. Initial investigations show that an 86-68 SLP, with Boolean, that is, 1 or 0 output, can be trained to accurately predict simple 2-D compensator profiles with four depth increments. Virtual training data are produced using an algorithm derived from empirically determined parameters with prediction errors of less than 1% reported. An extension of this work is presented in Ref. 57, where SDBP is used to train an 86-17 SLP with continuous outputs, which are then converted into the nearest-integer (NI) values, giving negligible prediction error. A more complex 3-D PSC design is predicted by a 12-17-12 MLP in Ref. 58. Although a high resolution (25 depth increments) is implemented by designing a near-optimal ANN architecture and using the LM training algorithm, prediction error is reduced to less than 1%.

7.3.3   Predictive Control

7.3.3.1   *Tumor Tracking Prediction Methods Using Off-Line Training*

Krell et al. [59] used an ANN to track patient movement. A TSP ANN with associative memory is trained off-line using data from an EPID, which is subsequently incorporated into online data collected by a stereoscopic system that has been developed by the authors to track landmarks on the body surface of patients during treatment. The method is presented as a new in-treatment verification and no metrical comparison with existing methods is given. No details of the ANN architecture are provided, although a Kohonen network [26] seems to be most likely.

Sharp et al. [60] investigated different prediction methods for tracking organ movement. Tracking is facilitated through the use of implanted 2 mm diameter gold fiducial markers imaged fluoroscopically by diagnostic x-rays in real time. Linear filters, TSP MLPs, and a Kalman filter are compared for various imaging rates and system time delays or latencies. Best results are achieved using a linear model for 33 and 200 ms latency, and an MLP with one hidden layer for 1 s latency. Best root-mean squared error (RMSE) ranges from 1 mm for an imaging rate of 30 Hz and 33 ms latency to 6 mm for 1 Hz and 1 s latency. The TSP MLP is trained for exactly 2 s for each record using 15 s of data and the CG algorithm. Note that only static forms of the predictors are used—no online adaptive parameter adjustment is attempted.

Kakar et al. [61] used an adaptive neurofuzzy inference system (ANFIS) for prediction of organ movement for 11 breast cancer patients. Mean RMSE is reduced to submillimetre accuracy over a period of 20 s for a 0.24 s latency. Average RMSE is found to be 35% of respiratory amplitude for patients adopting free breathing and 6% for those coached earlier to produce consistent breathing characteristics.

7.3.3.2 *Tumor Tracking Prediction Methods Using Online Training*

Isaksson et al. [62] used an adaptive TSP MLP to predict lung tumor movement through correlating external and internal markers. Updating is carried out at 1 and 5 s intervals using 2 s of data sampled at 10 Hz. The LM training algorithm is used. All of the data points are used as inputs to a TSP MLP with two hidden neurons and one output neuron. All three neurons are sigmoidal. No indication is given as to whether they are logsig or tansig, and no justification is given for this architecture. From the explanation given, the training is therefore carried out using only one input data vector. Validation is carried out using a second 2 s sequence, although whether this is the previous or following sequence is not explained. The tumor motion of three patients is analyzed, with the conclusion that an adaptive filter is superior to a stationary filter when the tumor motion is nonstationary.

Yan et al. [63] presented a method to correlate internal and external marker motion using a linear ANN. SD is used for training the ANN. Unfortunately, details of the sampling frequency used are not given; this makes comparison with other methods difficult. A prediction error of 23% between external and internal marker position is reported.

A common trait (with the single exception of Ref. 58) of the investigations where static MLPs are used is a reliance on SDBP for training. In all cases where SDBP is implemented, there is no explicit justification for the use of this algorithm, and given the size of the MLPs used, it would appear that CGBP could have been at least examined as a faster alternative. In some instances, the MLPs utilized have been small enough to use LM. Frequently, there is no systematic optimization of network size before training, and the number of hidden layer neurons is either omitted from the work completely or appears to have been arrived at with no logical justification. Where more than one hidden layer is used, no reason is given for this.

The following two sections illustrate the importance of optimizing ANN architecture and selecting appropriate training algorithms. Section 7.4 presents a new method for calculating 3-D PSCs using an ensemble of MLPs called a committee machine (CM). In Section 7.5 a comparison is made between three training algorithms to update adaptive TSP MLPs for predicting future tumor positions.

7.4 Prediction of 3-D Physical Compensator Profiles

7.4.1 Introduction

Traditionally, a radiation therapy physics model can accurately calculate the dose produced by a physical compensator. The inverse problem of determining a suitable compensator profile from a desired dose distribution still, however, presents a challenge [64]. As mentioned in Section 7.3, Goodband et al. [56–58] trained ANNs to determine the appropriate configuration of novel physical compensators, which incorporate mercury as an attenuating material. Ideally, empirically measured data should be used for training an ANN. However, in RT physics it is impractical, both on the grounds of time and financial cost, to collect the amount of data required to accurately train an ANN. Making use of an approach that employs virtual data derived using an algorithm, it is theoretically possible to produce an indefinitely large set of training data. The resulting virtual dose distributions and their corresponding compensator profiles can then respectively be used as input and target vectors to train an ANN. The work presented here extends this concept that was originally introduced in Ref. 58. The architecture for an MLP is optimized using the LM training algorithm and early stopping. However, due to restrictions imposed by the memory requirements of the LM algorithm, it is necessary in the present case to limit the data set for a single ANN to about 1900 pairs. A CM approach is therefore investigated, whereby the averaged output of several ANNs trained on randomized subsets of the total data are used. An extension of an existing algorithm is implemented to optimize the size of the CM. The CM is then applied to a new method for producing PSC dimensions. The 3-D PSC model is discretized into a series of 2-D models, arranged in parallel "slices." A PSC profile is calculated by splitting a desired dose matrix into a corresponding series of slices. Each dose distribution slice is converted into an input vector for the CM. The output of the CM is then the vector of predicted PSC dimensions for that slice. The process is repeated for each slice of the dose matrix. If the process is stopped at this stage, the resulting compensator would only take into account scatter parallel to and within each slice. To simulate scatter between slices, the dose matrix is rotated by 90° and the process is repeated. This new approach allows the CM to approximate 2-D photon scatter. Because at present scatter is ignored completely when calculating compensator dimensions, this does not represent a degradation compared with existing methods.

7.4.2 Physical Compensator Model

Elements of compensating material are modeled as discrete, identical units (see Figure 7.7). Each element represents 6.7 mm length and 1 mm height of the PSC. A 3-D profile is built up using a series of parallel 2-D "slices" 6.7 mm apart. It is hypothesized that compensator elements are focused into the beam to minimize the effect of beam divergence, which is not modeled. A similar approach is used in Ref. 65 for a binary attenuation device. For treatment purposes, a compensator can be mounted on a linac tray at a distance of 670 mm from the beam source and 330 mm from the isocenter of the prescribed treatment volume, that is, the center of the tumor. This results in a dose resolution of 10 mm at the patient level. Each vector, $c = [c_1 c_2 \ldots c_{12}]^T$, $c_i \in \mathbb{N}$, $0 \le c_i \le 25$, represents a slice of the compensator, equal to 80.4 mm width on the linac tray. Each depth increment is equivalent to 1 mm of mercury, which is comparable with the degree of resolution achieved by commercial milling

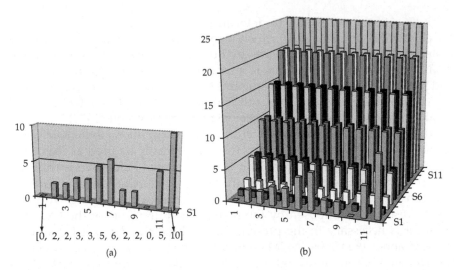

FIGURE 7.7
Schematic representation of compensator dimensions. (a) 2-D vector representation of one slice within the 3-D compensator, (b) 3-D representation of the compensator dimension. Note that in practice there is no gap between columns.

machines. 25 mm of mercury attenuates approximately 80% (depending on the field size) of the dose from a 6 MeV x-ray beam. By comparison, 10 depth increments are used for validation purposes in Ref. 65.

7.4.3 Network Training Data

A set of training data of possible compensator profiles ranging from flat to graded and asymmetric is produced. These represent a generic set encompassing typical breast or prostate treatment fields for IMRT. 3714 different 12-element compensator vector "slices" are generated. To replicate measured dose distributions, the resolution of the PSC is increased by a factor of 20 to 240 mm × 0.33 mm elements. This facilitates a more accurate representation of photon scatter [66] within the material. The intensity distribution attributable to each profile is calculated in two parts. A primary beam calculation for the depth, d, of material is made using

$$I(d) = I_0 e^{p(d)} \tag{7.15}$$

where $I(d)$ is the residual x-ray intensity due to an incident beam with intensity I_0, and $p(d)$ is a polynomial in d. A Gaussian filter is then applied to the resultant profile to approximate the effect of photon scatter. Model parameters are identified from experimental measurements. The Nelder–Mead Simplex algorithm [67] is used to calculate the best fit value for $p(d)$ with respect to a set of eight patient specific compensator (PSC) profiles and resultant dose distributions; this produces a value

$$p(d) = 0.1542d^4 - 0.7344d^3 + 1.0987d^2 - 1.2020d \tag{7.16}$$

Using an iterative approach, the following Gaussian filter is adopted:

$$y(x(k)) = \alpha_3 x(k - 3) + \alpha_2 x(k - 2) + \alpha_1 x(k - 1) + \alpha_0 x(k) + \alpha_1 x(k + 1)$$

$$+ \alpha_2 x(k - 2) + \alpha_3 x(k + 3) \tag{7.17}$$

where

$$\alpha_n = \frac{\exp(-n^2/162)}{\sum_{n=0}^{3} \exp(-n^2/162)}$$

Gaussian filters are also used to model scatter in Ref. 68. The model is accurate to within 4.55% relative to the maximum dose level. Although more accurate models, for example, Monte Carlo [69] are available, these are considerably slower, generally taking several hours to simulate dose distribution for a typical profile. On the basis of practicality, the present algorithm is therefore used to demonstrate the ANN training method. The resulting dose vectors are subsequently rescaled to 12 elements for the MLP training process. The labeled sets of dose and dimension vectors are used, respectively, as the input and target data set for training the MLPs. The set of fluence vectors is normalized with zero mean and elements in the range $[-1, 1]$. This accelerates training by preventing weight oscillation [33]. Target vectors **s** are rescaled using a linear mapping $\mathbb{N}^{12} \mapsto \mathbb{R}^{12}$ given by

$$\mathbf{s} = [s_1, s_2, \ldots, s_{12}]^T, s_k \in [0, 25], (k = 1, 2, \ldots, 12) \mapsto$$

$$\mathbf{s}^l = [s_1^l, s_2^l, \ldots, s_{12}^l]^T, s_k^l \in [0.25, 0.75], s_k^l = \frac{s_k + 12.5}{50}$$

This restricts outputs to be well within the asymptotic bounds of the logsig neurons used on the output layer of the MLPs, which constitute the ensemble members. The requirement for an additional linear layer and the associated increase in training time is thus avoided [57].

7.4.4 Ensemble Averaging Model

An ensemble averaging model (EAM) uses an average, possibly weighted, of the outputs of a group of two or more ANNs to provide the required response to a single input applied to the ensemble. The work that first formalized the relationship between network ensembles and improved performance was that by Krogh and Vedlesby [70]. In this paper, dating from 1995, the output error of an ensemble was presented in terms of the errors of individual members of a committee and a measure introduced by the authors called the "ambiguity" of the ensemble. A schematic of an EAM is shown in Figure 7.8. The output vector **a** of the EAM is the weighted sum $\sum_{i=1}^{N} w_i \mathbf{y}_i$ of the outputs from each MLP. The values of $\{w_i\}$ are usually fixed at $w_i = 1/N$, that is, **a** is the mean value of $\mathbf{y}_i$.

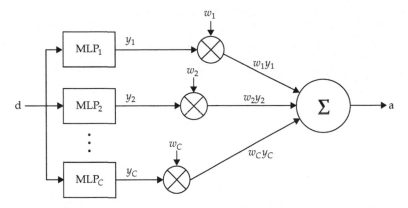

FIGURE 7.8
Schematic of ensemble averaging model d represents a dose input vector and a, a (possibly weighted), average output vector of PSC dimensions.

7.4.5 Optimizing Multilayer Perceptron Architecture

Although genetic algorithms (GAs) have been successfully implemented for optimizing ANN architectures [71–73], it is known that GAs cannot be guaranteed to converge to the global minimum of a cost function. Given the training limitations of the LM method, it is also not practical to produce an initial population of ANNs from a wide selection of architectures. A typical initial GA population would require about 50 members [74]. The deterministic enumeration approach described in the following text searches through $(23 \times 19) - 437$ different combinations of ANN architecture and training sets. This is equivalent to 9 generations of a GA with 49 members in the initial population. A typical GA would take several times this number of generations to converge. Clearly, this would be more time-consuming than an exhaustive search, and several runs would be needed to fine-tune parameters. The present study therefore examines a deterministic approach based on increasing the size of data sets and growing the ANN architecture incrementally.

Training is carried out for a sequence of ANN architectures commencing with one hidden neuron and 100 data pairs chosen at random from the total set. The optimal number of training pairs depends on the size of ANN architecture and the nature of the functional relationship and cannot be established with certainty before training commences. ANN theory gives generalized worst-case lower bounds for data size, but these predictions are often impractical to implement [33]. In the present study, 20 training runs are made for each data set with reinitialization of random weights for each run. This is consistent with standard ensemble averaging in ANN training [33], allowing training to commence from different points in the weight space that increases the chance of at least one of the training runs to converge on an error minimum. The number of data pairs is increased in 100 s to 1900 with 20 training runs being carried out for each data set. For the particular PSC used in training (a Pentium® 4 2.0 MHz processor with 261 MB RAM), 1900 is found to be the limiting data size. Three different methods for optimizing the ANN architecture are compared. In each case, after 20 training runs are executed for all sizes of data set, the following criteria for each architecture are calculated:

i. The minimum MSE.

ii. The smallest mean for each set of training runs.

iii. The smallest mean of each set of training runs excluding the worst single MSE from each set. This eliminates some of the variance introduced by poor results obtained through training routines, which get trapped in local minima.

A discussion of these different techniques can be found in Ref. 75. In each case, if there is an improvement on the validation error performance of the previous architecture, an additional neuron is added and the process is repeated. This is continued until two consecutive architectures show no improvement in performance, at which point the algorithm is terminated. This termination criterion is arbitrary and is based solely on the training time available. Training and validation is carried out using the MATLAB® 6.5 Neural Network toolbox.

7.4.6 Optimizing Ensemble Membership: Genetic Algorithm-Based Selective Artificial Neural Network Ensembles

The most popular methods for implementing EAMs are simple ensemble [76], Adaboost [77] and Bagging [78], each of which use combined outputs from all the ANNs available. However, an ANN with poor generalization error is likely to degrade the performance of the ensemble. Conversely, it cannot simply be stated that a combination of only the best performing ANNs from the population will automatically produce the best committee performance. In Ref. 79, an incremental addition approach is used, where a committee member is added if it improves generalization performance of the committee. A simple average of the outputs is made, with no investigation into altering the weighting of members. Although this approach outperforms simple averaging of all members of a population, no attempt is made to improve the committee by *removing* some of its members. A method proposed in Ref. 80 uses a GA to select an optimal subset of ANNs for committee membership. The method, called GA based Selective Ensemble (GASEN), shows that a small subset of ANNs from a population can outperform a simple average of the whole population. This method includes weighting the output of EAM members to optimize performance, although the criterion for selecting networks for membership is rather vague. The authors express some surprise at the frequency with which the GA search produces the same result as the enumerating process.

For the present study, each of the 21 MLPs are trained for 500 epochs without early stopping. They are divided into three groups of seven, each group being trained on one-third of the total data, that is, 1238 data pairs. This allows the LM algorithm to reach several local minima commencing from different points in weight space. An optimal combination of these MLPs is then found for the EAM. For a population of 21 MLPs, the number of combinations is $2^{21} - 1 = 2{,}097{,}151$. Given that a calculation to find the MSE for an EAM composed of all 21 ANNs takes approximately 0.04 s, an exhaustive search can be undertaken in approximately 16 h.

The exhaustive search is carried out by first calculating the individual MLP responses to all 3714 data pairs. These are stored in a $21 \times 12 \times 3714$ matrix. The simple average MSE for each combination is then systematically calculated for every combination. The GA search uses the following algorithm:

i. Initial population $\mathbf{p}$, of C randomly generated chromosomes, is created. Each chromosome is a 21-point binary string representing inclusion (1) or exclusion (0) from EAM membership of each labeled MLP. Position 1 in a string represents the MLP producing the best MSE response, with position 21 producing the worst.

ii. A check is made to ensure that every potential member is represented throughout the initial population for crossover to work effectively.

iii. The fitness function for each chromosome is calculated. This is simply the MSE output for each EAM.

iv. EAMs are ranked in order of fitness, that is, lowest MSE = highest rank.

v. A vector, $\mathbf{r}$, of random numbers $r_i \in (0,1)$, $i = 1, \ldots, C$ is generated. k members are extracted for the new population $\mathbf{p}'$ using the formula

$$\mathbf{p}' = \mathbf{p}\left(\text{round}\left[\frac{-(2\alpha + \beta) + \sqrt{(2\alpha + \beta)^2 + 8\beta\mathbf{r}}}{2\beta}\right]\right) \tag{7.18}$$

where α and β are scalar parameters related to the number of chromosomes in the initial population and the selection pressure, ϕ, by the equations

$$\alpha = \frac{2C - \phi(C + 1)}{C(C - 1)} \tag{7.19}$$

and

$$\beta = \frac{2(\phi - 1)}{C(C - 1)} \tag{7.20}$$

$\mathbf{p}'$, therefore, contains members of $\mathbf{p}$ in random order. A selection pressure equal to 2 is used, which favors selection of fitter members of the population, making duplication of members very likely [74].

vi. Uniform crossover is carried out on all adjacent pairs of chromosomes in $\mathbf{p}'$ except for the highest ranked in each generation. This is saved and carried over to the next generation, a process known as elitism [81].

vii. The GA search is implemented both with and without a mutation operator. The mutation rates investigated are 0.0476 ($1/l$, where l is the length of a chromosome, a commonly used value) and 0.01, genes being randomly changed in each generation according to a Bernoulli, that is, binomial distribution.

viii. If the average fitness of the newly created population is less than that of the previous population, it is rejected. This ensures that the algorithm will converge to some error minimum (although not necessarily the global minimum).

In addition to finding the error produced by a simple average of the optimal EAM membership, optimal values for the set of weights $\{w_i\}$ (Section 7.4) are calculated using the simplex search method [67].

7.4.7 Results and Discussion

7.4.7.1 *Optimizing Multilayer Perceptron Architecture*

A plot of the root-mean-squared validation error (RMSE^{val}) against the number of hidden neurons for the various architectures is shown in Figure 7.9 with lowest values given in Table 7.1. It can be seen that the three methods for assessing

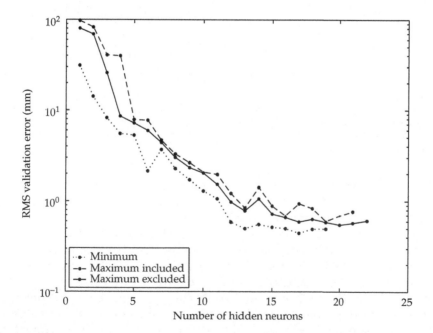

FIGURE 7.9
Plot of validation MSE versus number of hidden neurons in ANN architecture for methods 1, 2, and 3. (From Goodband, J.H., Haas, O.C.L., Mills, J.A., A mixture of experts committee machine to design compensators for intensity modulated radiation therapy, *Pattern Recognition*, 39(9), 1704–1714, 2006.)

TABLE 7.1

Results for Optimizing MLP Architecture

Method	1	2	3
Hidden neurons	17	19	20
RMSE (mm)	0.444	0.481	0.549

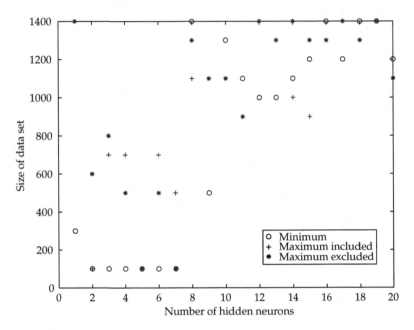

FIGURE 7.10

Plot of number of hidden neurons against size of training data set for methods 1, 2, and 3. (From Goodband, J.H., Haas, O.C.L., Mills, J.A., A mixture of experts committee machine to design compensators for intensity modulated radiation therapy, *Pattern Recognition*, 39(9), 1704–1714, 2006.)

optimal architecture produce results of similar magnitude. Because the lowest value of RMSE[val] (0.444 mm) is achieved by an MLP with 17 hidden neurons, this architecture is adopted for all MEM members.

Figure 7.10 shows the size of the data sets, which produce the best results for each method. Methods 1 and 3 produce the best 7 hidden neuron result using only 100 data pairs, that is, 80 for training and 20 for validation. The number of free parameters for an architecture of this size is $(12 \times 7) + (7 \times 12) + 7 + 12 = 187$. Accepted ANN theory states that the number of data pairs used in training should exceed the number of free parameters to produce reasonable results. For eight hidden neurons $\equiv 212$ free parameters, best results are obtained by data sets >1000 in size, which would normally be expected. The results show that it may be possible to use small, experimentally obtained data sets to produce reasonable, although not optimal, training results.

TABLE 7.2

Optimal Weights for MLP Outputs in EAM

ANN (in Performance Order)	1	2	3	5
Weighting coefficients (a_i)	0.3261	0.3363	0.1978	0.1398

7.4.7.2 Ensemble Averaging

The exhaustive search takes approximately 16 hours and produces an optimal combination of four ANNs, comprising the first, second, third, and fifth ANNs ranked in order of MSE performance. The optimal EAM performance using simple averaging is a RMSE^{val} of 0.392 mm across the whole training set. (Note that the simple averaged output from all 21 ANNs gives a RMSE^{val} of 0.552 mm.) Using a GA approach, the same result is obtained in $\approx$1 min in 12 out of 20 runs using a mutation rate of 0.01. The mutation rate of 0.0476 produces inferior results, with no instance of predicting the optimal combination in 20 runs.

Further refinement is achieved by optimizing the weighting coefficients for the four ANNs. The optimal coefficients are shown in Table 7.2. These give an RMSE^{val} = 0.389 mm. This represents an 8.25% improvement in performance compared with the best single MLP.

Compared with existing methods using Monte Carlo techniques, the EAM is at least three orders of magnitude faster in computational time and therefore produces significant improvements in optimization programs where iterative techniques may be required. The complete process to predict a required PSC profile and compare simulated dose distribution with desired dose distribution for error produced by photon scatter is illustrated in Figure 7.11.

7.4.8 Conclusions

In this section, the performance of a single MLP has been compared with that of an EAM committee machine using virtual data, which has been generated from empirically derived parameters. A restriction has been made on the size of input–output vectors, making it possible to use the LM algorithm that has facilitated fast and accurate training of MLPs. In contrast with existing ensemble methods, emphasis has been on optimizing both the number of hidden neurons for the component MLP architecture as well as the number of MLPs constituting the ensemble. An optimal combination of MLPs to construct an EAM has been selected using a GA and demonstrated to provide an 8.25% improvement in prediction error compared with a single MLP. RMSE between a desired dose distribution and the distribution simulated for a predicted compensator is 1.63%. Note that such an error is calculated assuming that the desired distribution is achievable. In the simulation studies carried out, see Figure 7.11 the 'desired distribution' does not take into account the effect of scatter. The latter causes the 'blurring' in the 'simulated distribution.' Such distribution is the 'best' achievable solution. The trained MEM is capable of calculating the inverse of a 12 × 12 dose matrix in about 0.02 s. Clearly, the

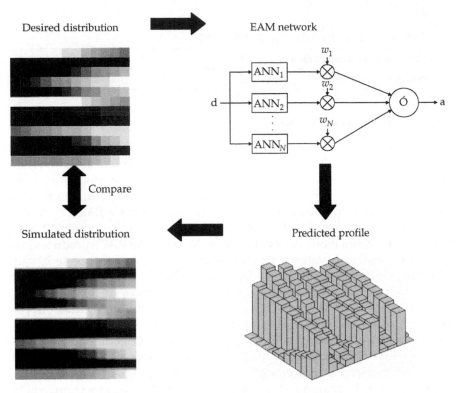

FIGURE 7.11
Illustration of physical compensator prediction and comparison process.

prediction method can be extended, with appropriate change in attenuation parameters, to solid materials possessing different attenuation characteristics.

The aim of the first case study has been to demonstrate the applicability of the ANN approach to model the type of nonlinear input–output relationships found in RT physics. Although earlier ANN applications in RT have made use of individual ANNs, this study has shown that improvements can be achieved by adopting a CM approach. Although the training of ANNs and CMs may take several hours, once they have been trained, calculations can be executed in a fraction of the time required by traditional medical physics models.

7.5 Lung Tumor Tracking

7.5.1 Introduction

Section 7.4 describes a static modeling problem, where the ANNs, once trained, maintain fixed weights throughout the modeling phase, whereas Section 7.5 introduces a problem where ANNs are required to adapt to model data, which change with time.

Delivering a prescribed radiation dose to a moving lung tumor is one of the biggest challenges in RT [21]. The conventional CRT approach is to use an extended PTV that takes into account tumor movement. This inevitably leads to extensive overdosing of healthy tissue. As an attempt to keep the tumor in a fixed position, both breath holding and breath training have been attempted at some RT treatment centers [82,83]. It is, however, impractical to expect controlled breathing from a patient suffering from lung cancer. Indeed Berbeco et al. [22] comment that there is no evidence that breath training provides any significant improvement in clinical outcome. ART techniques are therefore being pursued to actively track tumor movement during treatment. A static breathing model devised by Lujan et al. [84] has become popular, but relies on training to achieve consistency. If breathing is inconsistent, an adaptive, that is, dynamic predictor of movement is required.

The aim of the work presented in this section is to identify the most suitable method for predicting breathing/tumor movement. A comparison is made of three different algorithms for adaptive training of TSP MLPs as well as an assessment of the performance of a GRNN. This work extends the methods introduced in Refs. 58 and 60, neither of which justifies the use of its specific training algorithms (CGBP and LM, respectively). The data used for training the MLPs are introduced in Section 7.5.2, followed in Section 7.5.3 by an explanation of the properties necessary for a predictor to be used in tracking tumor motion. Section 7.5.4 explains the process used to derive the MLP architecture and methodology adopted for online retraining. Section 7.5.5 describes the error measures used and introduces new metrics for assessing prediction error, which take into account maximum error. Results and discussion are present in Section 7.5.6.

7.5.2 Tumor Motion Data

The data used for this study have been acquired from a breathing training database collated by George (RGeorge@mcvh-vcu.edu) at Virginia Commonwealth University. Twenty-four adult patients suffering from lung cancer were observed over a 1-year period; a record being made of 331 4-min breathing traces of respiratory motion using three types of training. None of the patients were oxygen-dependant and all were capable of lying in a supine position without feeling the pain. A marker block resting on the chest of each patient between umbilicus and xyphoid allowed tracking of respiratory movement in the anterior–posterior direction using a real-time position management system developed by Varian Medical Systems. The marker position was sampled at 30 Hz.

Tumor trajectory No.1 from the Virginia Data set (hereafter abbreviated V1) is shown in Figure 7.12. Note that from 75 s into the observation, breathing becomes more irregular, culminating in a disruptive exhalation at 170 s. Clearly, in this case, a sinusoidal model as proposed by Lujan et al. [84] cannot be used for the whole of the period under observation. The probability density function (PDF) for organ displacement over the period of 1–165 s

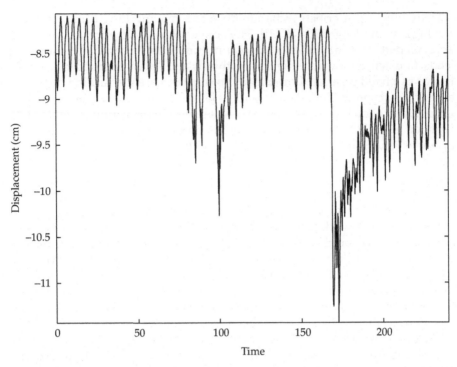

FIGURE 7.12
Four minutes of respiratory motion (Virginia data set No.1) tracked in real time using an external
fiducial marker block.

compared with the PDF for 165–240 s shows a change in the character of
marker movement, both in terms of mean position and amplitude (Figure
7.13). Frequency of breathing cycles also changes. Figure 7.14 illustrates two
10-s periods of V1 sampled at different times (0–10 s and 190–200 s) in the
4 min observation period. Although many of the trajectories are more consistent than this particular example, a robust prediction system needs to be
capable of adapting to most of the breathing changes that may occur during
treatment.

7.5.3 Adaptive Predictor Model

An ideal tumor motion predictor needs to be

1. Patient specific—the predictor should be designed using patient data
   collected either immediately before treatment or preferably during
   treatment.

2. Rapid—A linac takes a finite period of time, known as the latency
   [60,85] of the system (typically 0.2–0.4 s), to respond to an input signal.
   The predictor should be capable of offering an accurate prediction
   for future tumor position at least this far ahead, thereby enabling

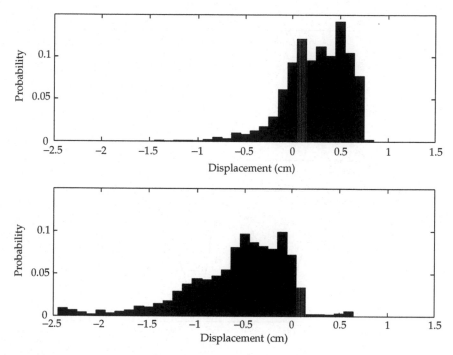

FIGURE 7.13
Probability density functions for V1. Upper plot is for 1–165 s, lower plot for 166–240 s. Zero displacement is the mean position of the total, that is, 1–240 s data.

rapid action to be taken to prevent overdosing of healthy tissue if the tumor position changed significantly. Updating of predictor parameters therefore needs to be carried out in the shortest possible time to facilitate this.

3. Robust—Spurious predictions of tumor position can lead to overdosing of healthy tissue or underdosing of tumor volume. If a rapid updating method leads to a significant degradation in prediction accuracy, there is little point in using it.

In the present study, the ability of an adaptive TSP ANN (Section 7.3) (MLP and GRNN) to predict tumor motion is investigated using the CG, LM, and BR training algorithms for MLP implementation.

7.5.4 Time-Series Prediction Artificial Neural Network Parameterization and Training

Section 7.4 shows that it is beneficial to fine-tune ANN architecture. Ideally, therefore, a different architecture should be used for each breathing trace, that is, each treatment session. The present study is, however, concerned with comparing training algorithms, therefore to reduce the number of

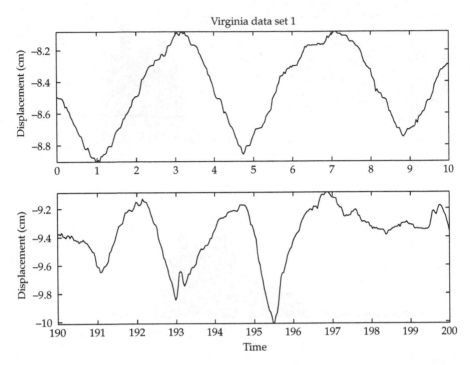

FIGURE 7.14
Change in breathing characteristics for V1. Note a change in both frequency and range of displacement.

variables in the problem, no attempt is made to find an optimal architecture for every breathing trace. Instead, a "reasonable" MLP architecture is created by training various MLP architectures using the first 20 s of data, 16 s for training, 4 s for validation, from a random 30 breathing traces, and then averaging the results. This approach is also taken by Sharp et al. [60]. The number of MLP inputs and hidden neurons are increased starting with one of each. This procedure is carried out using the LM algorithm with early stopping. Subsequently, training is implemented using the following protocol:

1. Train the TSP MLP using the first 20 s of data. CGBP and LM use 16 s of data for training and 4 s, that is, every fifth point for validation. BR uses all 20 s for training.
2. Subsequently update online every 1 s using the latest 5 s of historical data. This ensures that at least one complete breathing cycle is used for training at each update. Training time is limited to 0.1 s. Owing to the longer time required for BR convergence, only CGBP and the LM algorithm are used for online updating.

To assess the effect of filtering, training results are compared for both raw data and filtered data.

7.5.4.1 Generalized Regression Neural Network

The GRNN is constructed using centers selected at every fifth point in the training data $\equiv 6$ Hz, that is, using the same data points as those used for validation purposes with the TSP MLP. The fastest human breathing cycle frequency is approximately 2 Hz [86]. During RT treatment, frequencies are generally maintained between 0.25 and 0.5 Hz. Sampling at 6 Hz ($>2 \times 2$ Hz), therefore, ensures that aliasing (Section 7.1) does not occur. As with the TSP MLP, training is carried out using both unfiltered data and filtered data. The number of GRNN inputs is first assessed using an averaging procedure similar to that used with the TSP MLP, that is, using the first 20 s of data from 30 randomly sampled data sets. The number of inputs is increased from one until there is no improvement in mean-squared validation error (MSE^{val}). The width of the GRBF neurons constituting the hidden layer is calculated for each change in the number of inputs using $\sigma = d_{max}/\sqrt{30}$ (Section 7.2). Because the speed of design for the GRNN (≈ 0.2 s) is much faster than a comparable architecture of MLP, after the number of inputs has been deduced, it is possible to determine near-optimal GRBF widths for each data set by iteratively calculating the GRNN response to validation data for several width values. A time of 20 s equivalent to 100 calculations is allotted for this procedure. Training commences with the initial value $\sigma_0 = d_{max}/\sqrt{30}/20$ and the value of σ is increased in steps of $\sigma_0/20$, again until there is no improvement in MSE^{val}.

7.5.5 Assessment Criteria

The RMSE between predicted and observed marker positions is a standard metric used in IGRT [60,61,85]. It is defined for N observations by

$$\text{RMSE} = \sqrt{\frac{1}{N} \sum_{n=1}^{N} (\hat{x}_n(t + T) - x_n(t + T))^2} \tag{7.21}$$

where $\hat{x}(t + T)$ is the predicted value of $x(t + T)$ at time t and $x(t + T)$ the observed marker position at time $t + T$. When using filtered data for training, RMSE is calculated using the observed, that is, unfiltered observation so that the value is not artificially reduced. The average RMSE for all 331 breathing traces

$$\overline{\text{RMSE}} = \frac{1}{331} \sum_{j=1}^{331} (\text{RMSE})_j \tag{7.22}$$

is then used as the final RMSE measure. Error is also presented in terms of the magnitude of the maximum single error over all breathing traces, given by

$$|e^{max}|^* = \arg \max_j |e_j^{max}|, j = 1, 2, ..., 331 \tag{7.23}$$

and the average value of the maximum errors

$$\overline{|e^{max}|} = \frac{1}{331} \sum_{j=1}^{331} |e_j^{max}|$$ (7.24)

In addition to the errors, the maximum ANN weight and bias values on the output layer of the ANN are recorded for each trace. $|\text{weight}^{max}|*$, $\overline{|\text{weight}^{max}|}$, standard deviation of $|\text{weight}^{max}|$, $|\text{bias}^{max}|*$, $\overline{|\text{bias}^{max}|}$ and standard deviation of $|\text{bias}^{max}|$ are calculated and evaluated. All training is carried out using the MATLAB 6.5 Neural Network toolbox version 4.0.2.

7.5.6   Results and Discussion

To emphasize the magnitude of extrapolation errors that are possible when using a static ANN, Figure 7.15 illustrates a possible predicted trajectory produced by a static TSP MLP using CGBP training, compared with that produced by the BR + CG adaptive algorithm. Although some prediction errors occur with the adaptive algorithm, these are relatively small when compared with those of the static ANN. What is very noticeable is that when significant changes occur in the observed trajectory, the positions predicted by the static

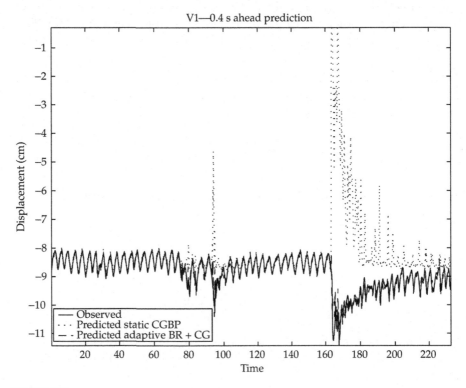

FIGURE 7.15
Plot comparing 0.4 s ahead prediction of static CGBP with adaptive BR + CG algorithm for data set No.1 from the Virginia Commonwealth University database.

ANN may even lead to prediction in the opposite direction to those observed. The stochastic nature of MLP training means that each time the TSP MLP is trained, a different set of network weights is produced and consequently, the direction and magnitude of prediction errors change. This example serves to illustrate why the use of an adaptive system is essential when applied to IGRT. Note that in practice any algorithm implemented for a safety critical application would be combined with a fail safe mode. In this particular example, the predicted position can be constantly monitored against expected changes and if the predicted position is outside pre-calculated thresholds then alternative position, such as the last known 'good' position, would be use in its place. Another alternative currently pursued within the Control theory and Applications Centre is to have a bank of predictors working in parallel with the final predicted solution being selected from the set of possible solution based on expected prediction error.

Table 7.3 gives the results for nonfiltered data and Table 7.4 presents the results for filtered data. The inferior results for the LM algorithm are attributable to much higher maximum errors in all four categories. This gives an indication of the unsuitability of LM as a robust algorithm for online updating. Based on the RMSE criterion, both the CGBP and BR + CG algorithms are significantly superior to using either the LM algorithm or no prediction method ($p < 0.006$). Best RMSE results for 0.2 and 0.4 s latency are obtained using BR ($p < 0.0071$)

TABLE 7.3

Optimization Results for TSP MLPs—Unfiltered Data

Training Algorithm	CGBP	LM	BR + CG	GRNN	No Prediction
0.2 s Latency					
RMSE (mm)	0.727	1.608	0.765	0.780	1.196
Standard deviation of RMSE (mm)	0.547	10.857	0.715	0.166	0.326
$|e^{max}|$* (mm)	15.91	321.6	15.80	8.339	6.256
$\overline{|e^{max}|}$ (mm)	3.283	84.26	4.279	5.353	3.676
No. of inputs	4	3	3	2	—
σ	—	—	—	0.031	—
No. of hidden neurons	8	7	8	30	—
0.4 s Latency					
RMSE (mm)	1.092	2.267	1.341	1.696	2.276
Standard deviation of RMSE (mm)	0.700	23.87	0.879	0.505	0.652
$|e^{max}|$* (mm)	18.58	1750	20.78	15.83	10.90
$\overline{|e^{max}|}$ (mm)	4.241	158.8	6.852	4.377	5.750
No. of inputs	4	4	3	2	—
σ	—	—	—	0.031	—
No. of hidden neurons	10	10	9	30	—

TABLE 7.4

Optimization Results for TSP MLPs—Filtered Data

Training Algorithm	CGBP	LM	BR + CG	GRNN	No Prediction
0.2 s Latency					
$\overline{\text{RMSE}}$ (mm)	0.626	0.755	0.484	0.773	1.182
Standard deviation of RMSE (mm)	0.538	5.229	0.542	0.164	0.327
$|e^{max}|^*$(mm)	15.62	228.3	14.81	8.297	5.584
$\overline{|e^{max}|}$ (mm)	3.178	40.82	3.082	5.253	3.152
No. of inputs	4	3	3	2	—
σ	—	—	—	0.023	—
No. of hidden neurons	8	7	8	30	—
0.4 s Latency					
$\overline{\text{RMSE}}$ (mm)	1.202	1.614	0.970	1.688	2.264
Standard deviation of RMSE (mm)	0.931	8.402	0.783	0.502	0.652
$|e^{max}|*$ (mm)	18.40	340.6	19.26	15.83	10.45
$\overline{|e^{max}|}$ (mm)	5.027	66.56	5.661	4.346	5.310
No. of inputs	4	4	3	2	—
σ	—	—	—	0.023	—
No. of hidden neurons	10	10	9	30	—

with filtered input. This illustrates the need to pre-process data and filter them on line to remove noise and outliers.

Analysis of maximum error results shows that using no prediction minimizes the worst prediction errors. The GRNN consistently produces best maximum absolute error results among the ANN prediction methods. From a quality assurance standpoint, the GRNN also has an advantage over the MLP methods in having a bounded output. The radiobiological effect caused by any tracking errors is patient specific that may determine which of the prediction methods, if any, should be used. It is clear, however, that the use of the LM algorithm as the sole method for training and updating may not be justified for this specific application. Table 7.5 presents the maximum magnitude of layer weights and biases associated with each algorithm. An investigation of these values highlights the considerably larger magnitude of weights induced by training using the LM algorithm. This explains the proportionately greater errors, which occur when using the LM algorithm, even for relatively small changes in breathing characteristics.

7.5.7 Conclusions on ANN for Respiratory Tracking

A study of four different algorithms used to train a TSP MLP for tracking tumor movement by external markers has been made. It has been found that assessing the performance of an algorithm solely based on RMSE can potentially lead to

TABLE 7.5

Analysis of Weights and Biases—Filtered Data

Training Algorithm	CGBP	LM	BR + CG	GRNN
0.2 s Latency				
|weight^{max}|*	1.884	28.71	9.316	12.33
|weight^{max}|	1.253	7.346	4.372	7.030
Standard deviation of |weight^{max}|	0.274	6.528	2.173	3.280
|bias^{max}|*	66.38	354.8	11.81	—
|bias^{max}|	30.72	37.58	2.848	—
Standard deviation of |bias^{max}|	20.03	62.48	2.536	—
0.4 s Latency				
|weight^{max}|*	2.238	44.41	10.94	12.33
|weight^{max}|	1.213	10.92	4.638	7.030
Standard deviation of |weight|^{max}	0.357	9.826	2.657	3.280
|bias^{max}|*	93.28	216.7	28.00	—
|bias^{max}|	28.67	41.97	5.855	—
Standard deviation of |bias^{max}|	20.83	46.52	6.418	—

unsafe conclusions in terms of the relative advantage of the ANN approach. New error criteria have therefore been devised to highlight error maxima. A novel, hybrid algorithm combining BR and CGBP produces best average RMSE results. By minimizing the magnitude of network weights in addition to minimizing output error, the algorithm significantly reduces mean tracking error compared with MLP training algorithms previously used in IGRT. Further reduction in RMSE is achieved by implementing an averaging filter on inputs. A comparison of filters applied to the training data would be advantageous to find an optimal filter to minimize error for this specific application. Although isolated, relatively high error values still occur; these are of an order of magnitude lower than those observed when using the LM algorithm. An analysis of the magnitude of MLP layer weights produced using each algorithm, provides some insight into why the LM algorithm maximum error results are considerably higher than those produced by the other algorithms. Overall results demonstrate that mean and maximum error reduction can be achieved by using a combination of training algorithms that are appropriate for IGRT application.

7.6 Overall Conclusions

This chapter has presented an overview of ANN applications in RT together with latest investigations carried out by the authors in two specific areas.

A novel method has been presented for predicting 3-D PSC dimensions using an EAM composed of MLPs. In contrast to existing ensemble methods, emphasis is on optimizing both the number of hidden neurons for the individual MLPs as well as the number of MLPs constituting the ensemble.

This is shown to significantly reduce prediction error compared with a single MLP. Each MLP within the EAM is trained on a subset of virtual data that have been created using experimentally determined parameters and incorporate scatter simulation. A restriction is made on the size of input–output vectors, making it possible to use the LM algorithm that facilitates fast training of MLPs. Each compensator profile is calculated by splitting a required dose matrix into a series of parallel slices. The slices are converted into input vectors for the EAM. The response of the EAM to each slice is then the vector of predicted compensator dimensions for that slice. The process is repeated for each slice of the dose matrix. To simulate scatter among slices, the dose matrix is rotated by 90° and the process is repeated. This new approach allows the EAM to approximate 2-D photon scatter. Results show that this novel application to IMRT produces a 12.4% reduction in validation error over that achieved using a single ANN. Clearly, the method can be extended to materials possessing different attenuation properties.

An investigation of training algorithms for use in adaptive TSP MLPs for tumor tracking has highlighted the shortcomings of the LM algorithm for online updating. It has also shown that the CGBP and BR algorithms can significantly reduce average tracking error compared to using no prediction method. It has been highlighted that occasionally the MLP is capable of predictions that are inferior in accuracy to using no prediction method, and that the decision as to which of the two methods to use must be based on possible clinical outcome.

A general requirement in relation to any ANN applications in RT (indeed any health- and safety-related applications) is that of robustness and optimization. Many researchers in this area have employed ANNs without making any attempt to optimize the ANN architectures. This creates inaccurate, undersized ANNs and unwieldy, oversized ANNs, which take too long to train. Where training sets are limited in size and some *a priori* knowledge of the underlying function exists, a recommendation is made to supplement existing data by creating model based virtual data. Otherwise, there is a danger of overfitting sparse data using ANNs that are too large.

With these points in mind, however, the immediate future of ANNs seems assured. Increased computer power brings decreased training times, which makes adaptive ANNs more attractive. Hybrid networks such as those using neurofuzzy techniques are increasingly being developed and proven to be robust for control systems. There is no doubt that ANNs are a powerful tool and that their use in RT will improve delivery of life-saving treatment.

Acknowledgments

This work was sponsored by the Engineering and Physical Sciences Research Council, Industrial Case Training award no. 02303507 and supported by the Framework 6 European integrated project Methods and

Advanced Equipment for Simulation and Treatment in Radiation Oncology (MAESTRO) CE LSHC CT 2004 503564. We also thank Rohini George from the Virginia Commonwealth University and Richmond, Virginia, 23298 for supplying the breathing training database.

References

1. Webb, S, *The Physics of Conformal Radiotherapy—Advances in Technology*, Institute of Physics Publishing, Bristol and Philadelphia, 1997.
2. Hope, R A, Longmore, J M, McManus, S K, Wood-Allum, C A, *Oxford Handbook of Clinical Medicine*, Oxford University Press, Oxford, 1999.
3. International Commission on Radiological Units and Measurements (ICRU), Prescribing, recording and reporting photon beam therapy, *ICRU Report 50* (ICRU, Bethesda, MD), Washington, 1993.
4. Hounsfield, G N, Computed transverse axial scanning (tomography): Part 1 description of system, *British Journal of Radiology*, 1016–1022, 1973.
5. Gordon, R, Herman, G T, Johnson, S A, Image reconstruction from projections, *Scientific American*, 56–68, October 1975.
6. Anton, H, Rorres, C, *Elementary Linear Algebra, Applications Version*, 8th ed., Wiley, New York, 2000.
7. Webb, S, *Intensity-Modulated Radiation Therapy*, Institute of Physics Publishing, Bristol and Philadelphia, 2001.
8. Chang, S X, Cullip, T J, Deschesne, K M, Miller, E P, Rosenman, J G, Compensators: An alternative IMRT delivery technique, *Journal of Applied Clinical Medical Physics*, 3, 15–35, 2004.
9. Bakai, A, Laub, W U, Nüsslein, F, Compensators for IMRT—An investigation in quality assurance, *Medical Physics*, 15–22, 2001.
10. Shalev, S, Progress in the evaluation of electronic portal imaging—taking one step at a time, *International Journal of Radiation Oncology Biology Physics*, 1043–1045, 1994.
11. Shalev, S, The design and clinical application of digital portal imaging systems, (Proceedings of 36th ASTRO Meeting), *International Journal of Radiation Oncology Biology Physics*, 138(Supplement 1), 1996.
12. Derycke, S, Van Duyse, B, De Gersem, W, De Wagter, C, De Neve, W, Non-coplanar beam intensity modulation allows large dose escalation in Stage III lung cancer, *Radiotherapy Oncology*, 253–261, 1997.
13. Meeks, S L, Buatti, J M, Bova, F J, Friedman, W A, Mendelhall, W M, Zlotecki, R A, Potential clinical efficacy of intensity-modulated conformal therapy, *International Journal of Radiation Oncology Biology Physics*, 2, 483–495, 1998.
14. Brahme, A, Development of radiation therapy optimization, *Acta Oncologica*, 5, 579–595, 2000.
15. Ling, C C, Burman, C, Chui, C S, LoSasso, T, Mohan, R, Spirou, S, Implementation of photon IMRT with dynamic MLC for the treatment of prostate cancer, In Sternick, E S, ed., *Intensity Modulated Radiation Therapy*. Advanced Medical Publishing, Madison, WI, 1997, pp. 219–228.
16. Yan, D, Vicini, F, Wong, J, Martinez, A A, Adaptive radiation therapy, *Physics in Medicine and Biology*, 123–132, 1997.

17. Jaffray, D A, *Emergent Technologies for 3-Dimensional Image-Guided Radiation Delivery*, Seminars in Radiation Oncology, Elsevier, Amsterdam, 2005.
18. Fuss, M, Salter, B J, Cavanaugh, S X, Fuss, C, Sadeghi, A, Fuller, C D, Amerduri, A, Hevezi, J M, Herman, T S, Thomas, C R, Jr, Daily ultrasound-based image-guided targeting for radiotherapy of upper abdominal malignancies, *International Journal of Radiation Oncology Biology Physics*, 1245–1256, 2004.
19. Matsinos, E, Endo, M, Kohno, R, Minohara, S, Kohno, K, Asakura, H, Fujiwara, H, Murase, K, Current status of the CBCT project at Varian Medical Systems, *Progress in Biomedical Optics and Imaging—Proceedings of SPIE, Medical Imaging 2005—Physics of Medical Imaging*, 340–351, 2005.
20. Mori, S, Respiratory-gated segment reconstruction for radiation treatment planning using 256-slice CT-scanner during free breathing, *Progress in Biomedical Optics and Imaging—Proceedings of SPIE, Medical Imaging 2005—Physics of Medical Imaging*, 711–721, 2005.
21. Shirato, H, Oita, M, Fujita, K, Watanabe, Y, Miyasaka, K, Feasibility of synchronization of real-time tumor-tracking radiotherapy and intensity-modulated radiotherapy from viewpoint of excessive dose from fluoroscopy, *International Journal of Radiation Oncology Biology Physics*, 335–341, 2004.
22. Berbeco, R I, Nishioka, S, Shirato, H, Chen, G T Y, Jiang, S B, Residual motion of lung tumours in gated radiotherapy with external respiratory surrogates, *Physics in Medicine and Biology*, 3655–3667, 2005.
23. Seiler, P G, Blattmann, H, Kirsch, S, Muench, R K, Schilling, C, A novel tracking technique for the continuous precise measurement of tumour positions in conformal radiotherapy, *Physics in Medicine and Biology*, 45, 103–110, 2000.
24. Muench, R K, Blattmann, H, Kaser-Hotz, B, Bley, C R, Seiler, P G, Sumova, A, Verwey, J, Combining magnetic and optical tracking for computer aided therapy, *Zeitschrift fur Medizinische Physik*, 14, 189–194.
25. Hertz, J, Krogh, A, Palmer, R G, *Introduction to the Theory of Neural Computation*, Addison-Wesley, Reading, MA, 1991.
26. Hagen, M T, Demuth, H B, Beale, M, *Neural Network Design*, PWS Publishing Company, Boston, MA, 1996.
27. Rumelhardt, D E, McClelland, J L, *Parallel Distributed Processing: Explorations in the Microstructure of Cognition*, MIT Press, Cambridge, MA, 1986.
28. Nguyen, D, Widrow, B, Improving the learning speed of 2-layer neural networks by choosing initial values of the adaptive weights, *Proceedings of the IJCNN*, 21–26, 1990.
29. Gill, P E, Murray, W, Wright, M H, *Practical Optimization*, Academic Press, New York, 1981.
30. Levenberg, K, A method for the solution of certain problems in least squares, *Quarterly Applied Mathematics*, 164–168, 1944.
31. Marquardt, D, An algorithm for least-squares estimation of nonlinear parameters, *SIAM Journal of Applied Mathematics*, 431–441, 1963.
32. Tresp, V, Committee Machines, In Hu, Y H, Hwang, J-N, eds., *Handbook for Neural Network Signal Processing*, CRC Press, Boca Raton, FL, 2001.
33. Haykin, S, *Neural Networks: A Comprehensive Foundation*, 2nd ed., Prentice-Hall Inc, New York, 1999.
34. Liu, G P, *Nonlinear Identification and Control*, Advances in industrial control monograph, Springer-Verlag, London, 2001.
35. Tsoukalas, L H, Uhrig, R E, *Fuzzy and Neural Approaches in Engineering*, Wiley, New York, 1997.

36. Hines, J W, *MATLAB Supplement to Fuzzy and Neural Approaches in Engineering*, Wiley, New York, 1997.
37. Tsoukalas L H, Uhrig R E, Fuzzy and Neural Approaches in Engineering, New York: Wiley, 1997.
38. Beauchamp, K G, Yuen, C K, *Digital Methods for Signal Analysis*, George Allen & Unwin Ltd, London, 1979.
39. Kearns, M, Mansour, Y, On the boosting ability of top-down decision tree learning algorithms, Proceedings of the 28th Annual ACM Symposium on Theory of Computing, Philadelphia, PA, 459–468, 1997.
40. Leszcynski, K, Provost, D, Bissett, R, Cosby, S, Boyko, S, Computer-assisted decision making in portal verification - optimisation of the neural network approach, *International Journal of Radiation Oncology Biology Physics*, 1, 215–225, 1999.
41. Lennernas, B, Isaksson, U, Nilsson, S, The use of artificial intelligence neural networks in the evaluation of treatment plans for external beam radiotherapy, *Oncology Reports*, 863–869, 1995.
42. Willoughby, T R, Starkschall, G, Janjan, N A, Rosen, I I, Evaluation and scoring of radiotherapy treatment plans using an artificial neural network, *International Journal of Radiation Oncology Biology Physics*, 4, 923–930, 1996.
43. Hosseini-Ashrafi, M E, Bagherebadian, H, Yahaqi, E, Pre-optimisation of radiotherapy treatment planning: an artificial neural network classification aided technique, *Physics in Medicine and Biology*, 1513–1528, 1999.
44. Martinez, W L, Martinez, A R, *Computational Statistics Handbook with MATLAB®*, Chapman & Hall/CRC, London 2002.
45. Wells, D M, Niederer, J, A Medical expert System approach using artificial neural networks for standardized treatment planning, *International Journal of Radiation Oncology Biology Physics*, 1, 173–182, 1998.
46. Wu, X, Zhu, Y, Luo, L, Linear programming based on neural networks for radiotherapy treatment planning, *Physics in Medicine and Biology*, 719–728, 2000.
47. Rowbottom, C G, Webb, S, Oldham, M, Beam-orientation customisation using an artificial neural network, *Physics in Medicine and Biology*, 2251–2262, 1999.
48. Knowles, J, Corne, D, Bishop, M, Evolutionary training of artificial neural networks for radiotherapy treatment of cancers, *1998-IEEE-International Conference on Evolutionary Computation Proceedings-Cat No.98 TH 8360*, 398–403, 1998.
49. Kaspari, N, Michaelis, B, Gademann, G, Using an artificial neural network to define the planning target volume in radiotherapy, *Proceedings of the 10th Conference on Computer based Medical Systems Table of Contents*, Maribor, Slovenia, 389–401, 1997.
50. Blake, S W, Artificial neural network modeling of megavoltage photon dose distributions, *Physics in Medicine and Biology*, 2515–2526, 2004.
51. Milan, J, Bentley, R E, The storage and manipulation of radiation dose data in a small digital computer, *British Journal of Radiology*, 115–121, 1974.
52. Mathieu, R, Martin, E, Gschwind, R, Makovicka, L, Contassot-Vivier, S, Bahi, J, Calculations of dose distributions using a neural network model, *Physics in Medicine and Biology*, 1019–1028, 2005.
53. Beavis, A W, Kennerley, J, Whitton, V J, Virtual dosimetry: quality control of dynamic intensity modulated radiation therapy, *British Journal of Radiology*, 9, 1998.
54. Gulliford, S, Corne, D, Rowbottom, C, Webb, S, Generating compensation design for tangential breast irradiation with artificial neural networks, *Physics in Medicine and Biology*, 2, 277–288, 2003.

55. Gulliford, S L, Webb, S, Rowbottom, C G, Corne, D W, Dearnaley, D P, Use of artificial neural networks to predict biological outcomes for patients receiving radical radiotherapy of the prostate, *Radiotherapy and Oncology*, 3–12, 2004.

56. Goodband, J H, Haas, O C L, Mills, J A, A reuseable attenuation device for intensity modulated radiation therapy, *Proceedings of Controlo 2004*, 636–641, 2004.

57. Goodband, J H, Haas, O C L, Mills, J A, Neural network approaches to predicting attenuator profiles for radiation therapy, *Proceedings of Control 2004*, 87 (full copy available on CD ROM only), 2004.

58. Goodband, J H, Haas, O C L, Mills, J A, Optimizing neural network architectures for compensator design, *Proceedings of IFAC 2005*, CD ROM only, 2005.

59. Krell, G, Michaelis, B, Gademann, G, Using pre-treatment images for evaluation of on-line data in radiotherapy with neural networks, *European Medical and Biological Engineering Conference EMBEC*, Vienna, 1999.

60. Sharp, G C, Jiang, S B, Shimizu, S, Shirato, H, Prediction of respiratory tumour motion for real-time image-guided radiotherapy, *Physics in Medicine and Biology*, 425–440, 2004.

61. Kakar, M, Nystrom, H, Aarup, L R, Nøttrup, T J, Olsen, D R, Respiratory motion prediction by using the adaptive neuro fuzzy inference system (ANFIS), *Physics in Medicine and Biology*, 4721–4728, 2005.

62. Isaksson, M, Jalden, J, Murphy, M J, On using an adaptive neural network to predict lung tumor motion during respiration for radiotherapy applications, *Medical Physics*, 3801–3809, 2005.

63. Yan, H, Yin, F-F, Zhu, G-P, Ajlouni, M, Kim, J H, Adaptive prediction of internal target motion using external marker motion: a technical study, *Physics in Medicine and Biology*, 31–44, 2006.

64. Webb, S, The physical basis of IMRT and inverse planning, *British Journal of Radiology*, 678–689, 2003.

65. Webb, S, IMRT using only jaws and a mask, *Physics in Medicine and Biology*, 257–275, 2002.

66. Hendee, W R, Ibbott, G S, *Radiation Therapy Physics*, 2nd ed., Mosby-Year Book Medical Publishers, Chicago, 1996.

67. Lagarias, J C, Reeds, J A, Wright, M H, Wright, P E, Convergence properties of the Nelder-Mead simplex method in low dimensions, *SIAM Journal of Optimization*, 1, 112–147, 1998.

68. Mejaddem, Y, Hyödynmaa, S, Svensson, R, Brahme, A, Photon scatter kernels for intensity modulating therapy filters, *Physics in Medicine and Biology*, 3215–3228, 2001.

69. van der Zee, W, Hogenbirk, A, van der Marck, S C, ORANGE: A Monte Carlo dose engine for radiotherapy, *Physics in Medicine Biology*, 625–641, 2005.

70. Krogh, A, Vedelsby, J, Neural network ensembles, cross validation and active learning, In Tesauro, G, Touretsky, D S, Leen, T K, eds., *Advances in Neural Information Processing Systems*, MIT Press, Cambridge, MA, 1995, pp. 231–238.

71. Reeves, C R, Steele, N C, Genetic algorithms and the design of artificial neural networks, *IEEE 'MICROARCH'*, 15–20, 1991.

72. Wu, J-X, Zhou, Z-H, Chen, Z-Q, Ensemble of GA based Selective Neural Network Ensembles, *Proceedings of the 8th International Conference on Neural Information Processing*, Shanghai, China, 1477–1482, 2001.

73. Ozturk, N, Use of genetic algorithm to design optimal neural network structure, *Engineering Computations*, 7, 979–997, 2001.

74. Reeves, R C, Rowe, J E, *Genetic Algorithms—Principles and Perspectives: A Guide to GA Theory*, Kluwer Academic Publishers, Dordrecht, 2003.
75. Fahlman, S E, Fast-learning variations on back propagation: an empirical study, *Proceedings of the 1988 Connectionist Models Summer School*, Pittsburg, 1988, pp. 38–51.
76. Solich, P, Krogh, A, Learning with ensembles: how over-fitting can be used, *Advances in Neural Information Processing Systems*, 190–196, 1996.
77. Freund, Y, Schapire, R E, A decision-theoretic generalization of on-line learning and an application to boosting, *Journal of Computer and Systems Sciences*, 1, 119–139, 1997.
78. Breiman, L, *Bagging Predictors Machine Learning*, 2, 123–140, 1996.
79. Navone, H D, Granitto, P M, Verdes, P F, Ceccatto, H A, A learning algorithm for neural network ensembles, inteligencia artificial, *Revista Iberoamericana de Inteligencia Artificial*, 70–74, 2001.
80. Zhou, Z-H, Wu, J-X, Jiang, Y, Chen, S-F, Genetic algorithm based selective neural network ensemble, *Proceedings of IJCAI'01*, Seattle, 2001.
81. De Jong, K A, An analysis of the behavior of a class of genetic adaptive systems, Doctoral dissertation, University of Michigan, Ann Arbor, MI, 1975. Available as PDF files from http://cs.gmu.edu/eclab/kdj_thesis.html.
82. Kim, D J, Murray, B R, Halperin, R, Roa, W H, Held-breath self-gating technique for radiotherapy of non- small-cell lung cancer: a feasibility study, *International Journal of Radiation Oncology Biology Physics*, 43–49, 2001.
83. Remouchamps, V M, Vicini, F A, Sharpe, M B, Kestin, L L, Martinez, A A, Wong, J W, Significant reductions in heart and lung doses using deep inspiration breath hold with active breathing control and intensity-modulated radiation therapy for patients treated with locoregional breast irradiation, *International Journal of Radiation Oncology Biology Physics*, 392–406, 2003.
84. Lujan, A E, Larsen, E W, Balter, J M, Ten Haken, R K, A method for incorporating organ motion due to breathing into 3D dose calculations, *Medical Physics*, 715–720, 1999.
85. Vedam, S, Docef, A, Fix, M, Murphy, M, Keall, P, Dosimetric impact of geometric errors due to respiratory motion prediction on dynamic multileaf collimator-based four-dimensional radiation delivery, *Medical Physics*, 1607–1620, 2005.
86. Vander, A, Sherman, J, Luciano, D, *Human Physiology—The Mechanisms of Body Function*, 7th ed., McGraw-Hill, New York, 1998.
87. Fletcher, R, *Practical Methods of Optimization*, 2nd ed., Wiley, New York, 1987.
88. Webb, A, *Statistical Pattern Recognition*, Arnold, London, 1999.
89. George, R, Vedam, S S, Chung, T D, Ramakrishnan, V, Keall, P J, The application of the sinusoidal model to lung cancer patient respiratory motion, *Medical Physics*, 2850–2861, 2005.

8

Neural Networks for Estimating Probability Distributions for Survival Analysis

Charles Johnston and Colin R. Reeves

CONTENTS

8.1 Basics of Survival Analysis

Survival analysis is concerned with the time that elapses till the occurrence of some event of interest in the study (e.g., death of a patient who has been followed up since diagnosis of a particular illness). Usually a number of cases are monitored or followed up. Some cases may drop out of the follow-up regime for various reasons. The result is that for these cases, the time till an event of interest is unknown and the time when they are lost from scrutiny is referred to as a "censoring time." Times of events are "event times." The probability distribution of time-to-event is usually skewed. The existence of both censoring times and skewness make for difficulties in the estimation of the underlying probability distribution of time-to-event. Basic equations of survival analysis are as follows:

Probability density function of time-to-event, $f(t)$ is defined as

$f(t)dt$ = probability of an event occurring in time interval $(t, t + dt)$

Survival function, $S(t)$, is defined as

$S(t)$ = probability of no event in time interval $(0, t)$

Hazard function, $h(t)$ is defined as

$h(t)dt$ = conditional probability of an event occurring in the time interval $(t, t + dt)$, given that no event has occurred in $(0, t)$

$$h(t)dt = \frac{f(t)dt}{S(t)}$$

from which we get

$$S(t) = \exp\{-H(t)\} \tag{8.1}$$

where $H(t)$ is the cumulative hazard function given by

$$H(t) = \int_0^t h(x)dx$$

In most situations of interest, the probability distributions are conditioned on some covariates or factors such as patient characteristics, clinical variables or factors, and the type of treatment given.

Conventional methods of survival analysis are either nonparametric, parametric model-fitting, or quasi-parametric modeling. The first and third of these are the most popular in medical applications, for example, modeling the probability of time-to-death or other events in patient experience. The first and second methods are often used in engineering studies, for example, time-to-failure of equipment in service or life test. The following outlines the three groups of survival analysis methods.

8.1.1   Nonparametric Methods

The Kaplan–Meier method estimates the probability of survival in the successive intervals between events, Δt_i. The survival probability, $S(t_i)$, is

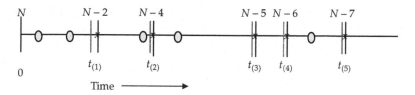

x represents an event on the time scale and o represents a
censored individual on the time scale

FIGURE 8.1
Schematic view of survival data.

assumed to be constant until a new event occurs. The survival probability
at any event time t_i is calculated as the product of all earlier interval values,
$S(\Delta t_j)$, $j = 1$ to i, with $S(\Delta t_0)$ equal to 1, and $S(\Delta t_j) = (n_j - d_j)/n_j$, where n_j is
the number of cases at risk just before the jth interval, and d_j is the number
of events at the jth event time. The number of cases at risk just before the
jth event time is the number at risk after the $(j - 1)$th event time less the
number censored between this and the jth event time. This is depicted in
Figure 8.1.

The Nelson–Aalen method estimates the cumulative hazard function, $H(t)$
in the intervals between events and $h(t)$ is assumed constant until an event
occurs. The cumulative hazard function is calculated as the sum of all the
interval values of hazard function till the latest event time.

These two methods are approximately equivalent, the results being related
as in Equation 8.1.

8.1.2   Parametric Modeling

A parametric model is fitted to the data by maximizing the profile likelihood
of the data sample, including the censoring times. Useful models are the
Weibull, lognormal (or the log-logistic), and gamma distributions. Variations
of these may provide better fit to the data but the models are very flexible
and there is often little to choose between them in practice. They may also be
fit piecemeal to time-sections of the data by including location parameters.
However, each distribution has a characteristic shape to its hazard function
that distinguishes it. The Weibull hazard function is monotonic, the lognormal is unimodal, and the gamma hazard function is asymptotic to a constant
value. These distinguishing characteristics may be useful and may have
practical implications for the cases involved.

8.1.3   Quasi-Parametric Modeling

The method of choice for medical statisticians is usually proportional hazards
modeling (Cox, 1972; Cox and Oakes, 1984). This assumes that the hazard
functions of cases differ only by a factor so that the ratio of the hazards for

case *i*, $h_i(t)$, to the hazard for a baseline case, $h_o(t)$, does not vary with time but can be expressed as

$$\frac{h_i(t)}{h_o(t)} = \exp(\beta_1 x_{1i} + \beta_2 x_{2i} + \cdots + \beta_{p-1} x_{p-1,i}) \tag{8.2}$$

where $\beta_1 \ldots \beta_{p-1}$ are the regression coefficients of $(p-1)$ covariates, $x_1, x_2, \ldots, x_{p-1}$ for the *i*th case.

The baseline hazard function has to be estimated. It relates to the case where all the covariates are zero. Its functional form is arbitrary. Thus, the only parameters in proportional hazards modeling are the covariates. However, the assumption that the hazard ratios are not time-dependent can be restrictive and various methods may be used to detect departures from the assumption and to build time dependence into the regression.

The methodology for conventional survival analysis is well developed and includes variations in the model formulations as well as an armory of diagnostic tools for assessing the adequacy of the fitted models. A good introduction to survival analysis is given by Collett (1994) and a more advanced treatment by Klein and Moeschberger (2003). The book by Therneau and Grambsch (2000) deals with extensions to the Cox proportional hazards model.

8.2   Artificial Neural Networks for Survival Analysis

Artificial neural networks (ANNs) have attracted a lot of attention during the past 20 years and they have a record of very successful applications in modeling nonlinear processes for regression and classification problems. The books by Bishop (1995) and Ripley (1996) provide a statistical perspective on the interpretation of ANNs that is becoming increasingly normative in discussing them. The textbook by Haykin (1999) gives a comprehensive introduction to many different forms of ANN.

The most popular and pervasive ANN model is probably still the *multilayer perceptron* (MLP), which has produced impressive results in regression and classification problems. This is the form of ANN that we have adopted for survival analysis. Although the MLP is the most common, there are many other ANN models. However, these are not discussed here.

The learning process for the MLP is based on a set of data, which can be used to train the network to predict (or estimate) the probability of survival. The data consists of a set of cases defined by their clinical and other characteristics along with event or censoring times. Once trained, the network can then be used to produce a survival function (i.e., survival probability as a function of time) for a new case whose characteristics are known in the same terms as those on which the network was trained.

The MLP is depicted in Figure 8.2. It consists of three layers (this structure may be altered if the complexity of the data requires it). Each layer consists

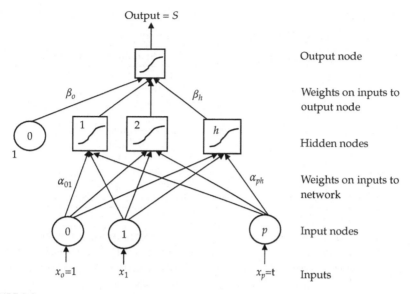

FIGURE 8.2
Multilayer perceptron for survival analysis: in this example of an MLP, the output is a scalar y and the input x is a feature vector of $p + 1$ components. A dummy input with $x_o = 1$ is included as a "bias" or constant term. A similar bias node is included in the hidden layer. The output S is a survival probability for the input feature vector.

of a set of nodes. At the input layer, a feature vector **x** (for each) case is presented to the network, each node in the layer receiving one component of **x**. An extra node is included whose input is always unity; this is a bias node. This augmented feature vector is passed to the hidden layer as a weighted sum of its components and presented at the input side of the hidden layer nodes; see Figure 8.2. Each hidden node then applies a nonlinear transformation or activation function, the result, including a bias of unity, being passed to the next layer as a weighted sum of the nodal outputs, where the process is repeated. The next layer may be another hidden layer or it may be the output layer. It can be shown that a single hidden layer is sufficient to represent virtually any relationship between inputs and outputs arbitrarily closely; however, the learning or network optimization process may be eased by including a skip layer (for the ease of presentation, this is not dealt with here).

At the output, there is a target y_i for each case i (assumed to be a scalar here but it could be a vector y_i, whose components would correspond with the required number of output nodes). The output produced by the network as a prediction can be compared with the target. The errors between target and predicted values can then be minimized by altering the values of the weights; this is the learning, or optimization, process. A number of error measures have been suggested: the usual criterion in most applications is mean squared error averaged over all input/output pairs (and over the number of components in the output, should this be a vector).

TABLE 8.1

Comparison of ANN and Statistical Terminology

Neural Network Terminology	Statistical Equivalent
Network	Model
Training/learning/optimization	Fitting/estimation
Weight	Coefficient
Bias	Constant
Feature	Variable/covariate
Pattern	Observation/case
Input	Independent/explanatory variable
Output	Dependent/response variable
Supervised learning	Prediction of dependent variable
Unsupervised learning	No dependent variable/classification

Many algorithms can be used for error minimization; a common one is called "back propagation," which has led to the MLP being referred to as a "back-propagation network." However, more sophisticated gradient search methods tend to be more efficient. This is discussed in detail later and in fact, for present purposes, much more can be achieved by using the maximization of the profile likelihood than error minimization.

8.2.1   Terminology

From a statistical perspective, some of the terminology used by ANN aficionados is different from that commonly used by statisticians. For convenience, a comparison is shown in Table 8.1.

8.3   Mathematical Formulation

The data for N cases may be represented by a matrix $\mathbf{X}$ with N rows and $p + 1$ columns. The columns contain the values of the covariates or features of the cases and each row is in ANN terms and a pattern/case denoted by $\mathbf{X}_i$ a row vector consisting of components x_{ij}. The weights between the input patterns and the hidden layer are α_{jk} denoting the weight connecting the jth input to the kth hidden node. These weights can be arranged in a matrix $\mathbf{A}$, whose dimensions are $(p + 1) \times h$. Similarly, the weights between the hidden layer and the output node (assuming a scalar output) are $\beta_k(k = 0, ..., h)$ and can be arranged in a vector $\mathbf{b}$.

The mathematical model that maps inputs to outputs for the MLP may be expressed as follows:

$$S_i = G_2\left\{\beta_o + \sum_{k=1}^{h} \beta_k G_1\left(\sum_{j=0}^{p} \alpha_{ij}\, x_{ij}\right)\right\} \tag{8.3}$$

or, in matrix notation,

$$S_i = G_2\{\beta_o + G_1(\mathbf{X}_i^T \mathbf{A})\mathbf{b}_s\} \tag{8.4}$$

where the vector $\mathbf{b}_s$ is $\mathbf{b}$ shortened to just h components by the extraction of β_o. G_1 and G_2 denote the nonlinear transformations applied to the hidden nodes and the output node, respectively, sometimes referred to as "squashing" functions. The logistic transformation is the most commonly used "squashing" function and is expressed as

$$G(y) = \frac{1}{1 + \frac{1}{b}\exp\left(-\frac{y}{a}\right)} \tag{8.5}$$

where a and b are the parameters. The shape of the logistic transformation is shown in Figure 8.3, illustrating the effect of the parameter a. The effect of the parameter b is to shift the curve up or down; for fitting survival functions b is set to unity. Sometimes, there is an advantage in varying the parameter a between the hidden and output layers.

The differential of $G(y)$ is given by

$$G'(y) = \frac{b\exp\left(-\frac{y}{a}\right)}{a\left(b + \exp\left(-\frac{y}{a}\right)\right)^2} = \frac{1}{a}G(y)[1 - G(y)] \tag{8.6}$$

and has a bell-shaped appearance that approximates to a standard normal (Gaussian) curve with mean zero and variance $(\pi a)^2/3$. The maximum difference between the standard normal and logistic distribution (with $a = 1$) is only 0.0228 (Johnson et al., 1995).

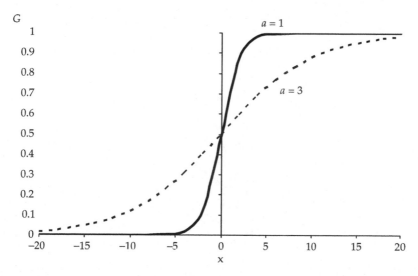

FIGURE 8.3
Logistic function illustrating the effect of changing the value of parameter a.

8.3.1 Constraints That Ensure $S(t)$ Is Monotonic

For $S(t)$ to be a proper distribution function, it must not only be limited to the range (0, 1), which is achieved by the use of the squashing function, but it must also be monotonic. In other words, the differential of $S(t)$ must be negative for all t, because $-S'(t)$ is the probability density function of time-to-event. This may be expressed as

$$f(t) = -S'(t) = -G'_2(t) \left[\sum_{k=1}^{h} \beta_k G'_1(t) \, \alpha_{pk} \right] \qquad (8.7)$$

Because $G'(t)$ is always positive, to make $f(t)$ positive, it is sufficient for α_{pk} and β_k to be of opposite signs. In the optimization process, it is sufficient to apply a constraint that α_{pk} is positive and that β_k is negative for all $k = 1$ to h (where $h + 1$ is the number of nodes in the hidden layer, including the bias node). The weights that are to be constrained to be of opposite signs are shown in Figure 8.4 as dashed (β_k) and dotted lines (α_{pk}), respectively.

This strategy ensures a monotonic survival function and is suitable with Excel's solver tool, which accepts linear constraints in the parameters to be

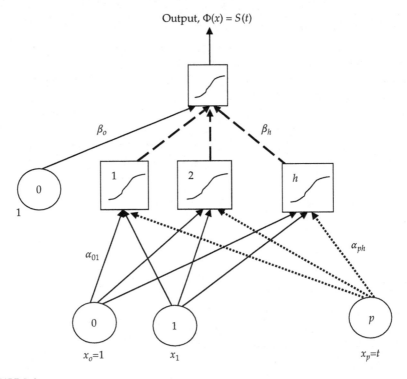

FIGURE 8.4

To ensure a monotonic survival function, α and β weights for the connections (shown by dashed and dotted lines) must have opposite signs.

changed to obtain the optimization. However, this may result in an optimization where one or more constraints are binding and the gradient of the objective function is not zeroed. A better approach is to apply a nonlinear constraint with a more sophisticated optimization tool. This can be achieved, for example, in MATLAB® using the optimization toolbox. The nonlinear constraint is

$$\alpha_{pk}\beta_k \leq 0, \qquad k = 1 \text{ to } h$$

Even then, these constraints are not all individually necessary: note that it is the sum that has to be negative in Equation 8.7, so that the weights may be overconstrained. Nevertheless, the approach described earlier seemed to have enough flexibility to provide a plausible model in the work reported in following sections.

8.4 Optimization: The Learning Process

8.4.1 Minimizing the Error and Maximizing the Log-Likelihood

A common problem in statistics is to use a sample of N observations to estimate the probability distribution of the population from which the sample values are drawn. A simple approach is to use ANNs to fit the empirical distribution function (EDF), and given the rapidly spreading availability of software for fitting ANNs, this may be an attractive option. Thus, the output values are given as $(i - 0.5)/N$, where i is the order number of the individual data point x_i, approximating the cumulative probabilities from 0 to 1, whereas the corresponding input values are the order statistics x_i. Reeves and Johnston (2001) showed that, for the case of a logistic squashing function, this is in fact equivalent to fitting a mixture of logistic densities.

Clearly, another area where such a technique can be applied is in survival analysis. Owing to censoring, the order statistics are not available, but the Kaplan–Meier estimate of the survival function, $S(t)$ is nonparametric (or model-free) and may be taken to be an empirical estimate, which we will denote as $\hat{S}(t_i)$. This may be used as the target output from the ANN shown in Figure 8.3.

Conventionally, the objective of the learning process is to adjust the weights in the network to minimize the sum of squares of the error, SSE, between a desired (target) output and the actual output.

$$\text{SSE} = \sum_{i=1}^{N} \{S(t_i) - \hat{S}(t_i)\}^2 \tag{8.8}$$

where $\hat{S}(t_i)$ is the empirical estimate of the survival probability at event/censoring time t_i given by a Kaplan–Meier analysis of the data.

The back-propagation algorithm can be used for minimizing the SSE as expressed in Equation 8.8 and second-order derivatives are then also readily available. But to compute the Kaplan–Meier estimate, it is necessary to arrange the data into subsets of cases with the same values or ranges of covariates, which is cumbersome and the most unsatisfactory if the subsets become small due to a large number of covariates.

A better approach is to maximize the sample likelihood expressed as

$$L = \prod_{i=1}^{N} \{ f(t_i) \}^{d_i} \{ S(t_i) \}^{1-d_i} \tag{8.9}$$

where d_i is an indicator variable such that

$d_i = 1$ if t_i is an event time and
$d_i = 0$ if t_i is a censoring time

It is much easier, of course, to maximize the log-likelihood, log L, expressed as

$$L = \log L = \sum_{i=1}^{n} d_i \log \{ f(t_i) \} + (1 - d_i) \log \{ S(t_i) \} \tag{8.10}$$

The values of $S(t_i)$ are given directly by the output of the network and the values of $f(t_i)$ may be computed by applying Equation 8.7. It is useful to note that the statistic $-2\log L$ provides a measure of how well a particular model fits the data (smaller the $-2\log L$, better is the fit). Further, the difference between the (minimized) values of $-2\log L$ for two models fitted to the same data takes a chi-squared distribution for which the number of degrees of freedom is equal to the difference in the number of parameters included in the two models. This provides a quantitative measure of significance when parameters are dropped or added to the model.

The ANN was initially implemented in Excel using matrix multiplication and optimization (i.e., maximizing the log-likelihood) was achieved using the Excel Solver tool. Solver proved amenable to the inclusion of the constraints on α_{pk} and β_k for $k = 1$ to h, given by Equation 8.7, that is, to ensure that $f(t)$ is positive.

8.4.2 Weight Regularization

A difficulty in the optimization process is that there is no natural end point. As described till now, experience shows that the process continues with both the weights and the log-likelihood increasing. The result is overfitting. This occurs when the model fits the data so well that it "explains" the noise in the data as well as the structure. If this happens, we cannot expect the model to provide a good generalization to unseen data, and its usefulness is severely affected. Measures to combat this problem are generally known as "regularization," and many techniques have been proposed.

One of the simplest techniques, which often works very well in the context of MLP fitting, is to use some form of weight penalty. The idea is to penalize large weights to encourage small ones. Weights whose value is close to zero will not have a great effect on predictions made from the model, and can effectively be discounted. Weight penalties thus encourage the generation of a parsimonious model, which is likely to be a good generalizer.

A straightforward method, then, is to add a penalty term to the log-likelihood. This is analogous to Akaike's information criterion used in conventional survival analysis (see Collett, 1994). The objective function to be maximized is redefined as

$$\Phi = \log L - \lambda \sum w^2 \tag{8.11}$$

where $\sum w^2$ denotes the sum of squared weights in both hidden and output layers of the network and λ is a regularization parameter.

This method is sometimes referred to as "weight decay." The choice of a value for λ is a nontrivial problem and may depend on the sample data. For the data used in Section 8.5, values of the order of 0.1 gave satisfactory results. But the problem is to find a compromise between overfitting (when the network is too influenced by the noise in the data) and excessive smoothness in the fit to the sample data. One way of testing for the optimum value of λ is to minimize the prediction error when the trained network is applied to a validation data set, but this is not usually a very practical approach because of the implied need to iterate the learning process.

An important aspect of network regularization is that it can control network complexity as it is, in effect, a complexity penalty. Experience shows that if the hidden layer contains too many nodes, the penalty term has the effect of reducing the weights to zero on unnecessary nodes. The dimensionality of the input data is also relevant, because the complexity of the network increases exponentially with increasing input data dimensionality, unless the smoothness of the fit is also increased (Haykin, 1999, pp. 211–212, 291).

8.5 Illustrative Analysis of a Data Set

The data to be analyzed are provided with the statistical software package, SPSS. The provenance is unknown, but it has characteristics that are useful for illustrating certain aspects of the application of ANNs to survival analysis. These data relate to breast cancer but lack any treatment history, thus they are used here simply as a convenient vehicle for illustrating the methodology. All cases with missing values of covariates have been eliminated, leaving 660 cases for present purposes.

8.5.1   Brief Exploratory Analysis of the Data

To understand the nature of the data set, it is necessary to do an exploratory analysis of the data. The following is by no means complete but will serve present purposes. These are the covariates provided in the SPSS data set.

Covariate	Explanation
Age	Patient age, in years, at the time of diagnosis
Pathsize	Pathological tumor size in centimeters
lnpos	Number of positive auxiliary lymph nodes
Histgrad	Histological grade
ER	Estrogen receptor status
PR	Progesterone receptor status

The following is by no means a thorough exploratory analysis of the data but serves to provide an overall appreciation of its characteristics. First, the distribution of pathsize is skewed and can be approximately normalized by a cube root transformation. The resulting new variable is named "curtpath" and this has been used in place of pathsize as inputs to the model-fitting processes. The distributions are shown in Figures 8.5a and 8.5b.

The relationship between age at diagnosis and histological grade and also between age at diagnosis and whether lymph nodes are involved are shown in Figure 8.6 and 8.7. The median age for patients with grade 1 cancer is slightly higher than for those with grade 2, and those with grade 3 have the lowest median age. The same kind of conclusion is reached with respect to lymph node involvement. This might lead us to expect that, in this data set, younger patients will be at a greater hazard of death than older patients. Of course, in reality, treatment has to be taken into account and as the data set is old, the present situation is likely to be different.

The correlation between estrogen and progesterone receptor status is clearly shown in Figure 8.8. The significance of these factors relates to the likely reaction to hormone treatment; if ER or PR is positive then treatment response is more likely to be positive.

8.5.2   Conventional Survival Analysis of the SPSS Breast Cancer Data

As a benchmark, it is worthwhile considering the results of a conventional analysis of the data. In medical survival analysis, the method that has won favor is proportional hazards analysis; first proposed by Cox (1972) and is often referred to as Cox regression. The hazard function for unembellished Cox regression is stated in Equation 8.2. The statistically significant coefficients of covariates from the data are seen in Table 8.2 (note that only the cases with missing values of covariates involved in the model have been eliminated).

The values of "exp(coef)" are the hazard ratios corresponding to the covariates. The values in the column headed "p value" indicate the computed probabilities that the corresponding coefficients are zero.

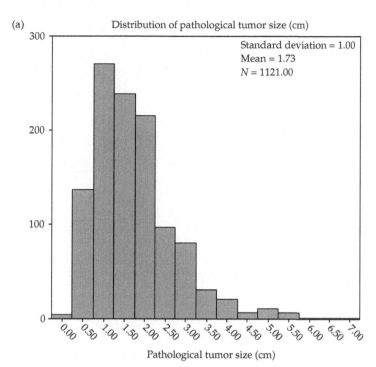

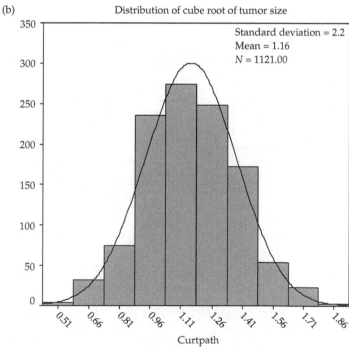

FIGURE 8.5
(a) Histogram of pathsize. (b) Histogram of curtpath.

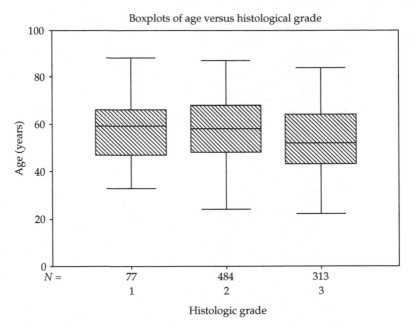

FIGURE 8.6
Cases with higher grades have a younger age distribution (in this data set).

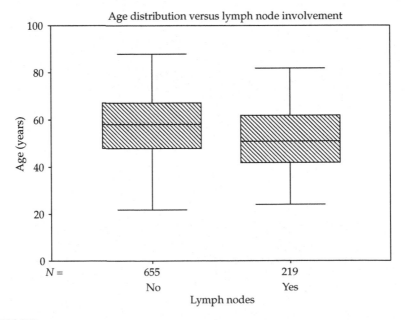

FIGURE 8.7
Cases with lymph node involvement have a younger age distribution (in this data set).

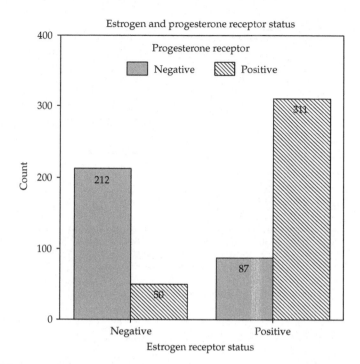

FIGURE 8.8
Association between estrogen and progesterone receptor status.

TABLE 8.2

Proportional Hazards Model (S-PLUS Analysis)

	Coef	Exp(Coef)	Se(Coef)	*p* Value
LNPOS	0.1469	1.158	0.0367	0.000064
PR	−0.5444	0.580	0.3065	0.076000
Curtpath	2.2037	9.058	0.6974	0.001600
Age	−0.0253	0.975	0.0123	0.041000

Note: $n = 819$ (388 observations deleted due to missing values).

Interactions between covariates do not appear to play a part in the hazard. A number of factors that are undoubtedly significant from a clinical perspective have been rejected as not being statistically significant. Part of the reason for this may be their correlations with covariates or factors that are included in the regression. This may mean that some cases are not represented in the final hazard function, for example, cases with positive progesterone receptor and negative estrogen receptor status.

The adequacy of the Cox model could be investigated using various methods such as the inclusion of time-dependent covariates or by plotting Shoenfeld residuals (see Klein and Moeschberger, 2003). This is not the main object here so we may be content with a comparative fit to Kaplan–Meier survival estimates, when an ANN fit is also available (Figure 8.9).

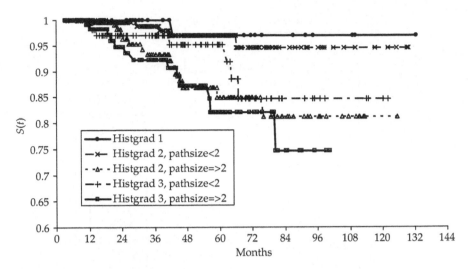

FIGURE 8.9
Kaplan–Meier estimates of $S(t)$ for subsets of data (SPSS breast cancer data).

8.5.3 Training the Artificial Neural Network

As shown in the brief exploratory analysis of the covariates (and factors), the input variables are correlated. This has an effect on conventional survival analysis and conditions the model that results in making it less applicable across the spectrum of case types in the data. It also made for difficulties in getting convergence of the training/optimization process for the ANN. To deal with this, principal components were used as covariates in place of the original covariates. One result of this is that all the clinical variables left out as not statistically significant in conventional analysis are automatically included in the principal components; another is, of course, that the principal components are uncorrelated. An introduction to principal components can be found in any textbook on multivariate statistics, for example, Everitt and Dunn (2001). The advantages of principal component analysis as a method of preprocessing the input to an ANN is discussed by Azoff (1993).

Applying weight decay as the regularization method, the weights applied to some of the principal components are much larger than those applied to others. This is an automatic method of assessing the relative significance of the input variables (the principal components), but it is not satisfactory to use this method to simplify the network by eliminating the input nodes with the lowest weights. It is better to supplement it by comparing the values of $-2\log L$ calculated with and without particular inputs (using the chi-squared approximation). An additional tool for simplifying the network would be to calculate the standard errors on the values of the weights produced by the optimization process. It can then be assumed that any weight whose absolute value was less than twice the standard error can be eliminated; this sometimes results in the elimination of an input feature and the corresponding nodes in the input layer. However, this is beyond the scope of an introduction to the method but will appear in a subsequent publication.

Other methods of network pruning are available, see Haykin (1999), Bishop (1995), and Ripley (1996).

8.5.4 Choice of Regularization Parameter

With the objective function as defined in Equation 8.11, the problem is to find the value of regularization parameter, λ, which generalizes the resulting model, that is, it enables predictions that are good fits to unseen data.

As the data set is too small to allow splitting it into learning and validation subsets, we need to explore other approaches. The two possible optimization objectives are to maximize the value of Φ or to minimize the SSE as given in Equation 8.8 between predictions and a nonparametric analysis.

Kaplan–Meier plots of $\hat{S}(t)$ are shown in Figure 8.9 for subsets of the data defined in Table 8.3.

The rest of the covariates vary within each subset. When we come to compare the Kaplan–Meier plots with predictions from the ANN or the proportional hazards model, the covariates not specified in Table 8.2 are set at representative values from the subsets.

The values of $\hat{S}(t)$ given by these plots are compared with the values of $S(t)$ fitted by the ANN varying λ over a range so that SSE can be plotted against λ. On the same plot we show the values of log-likelihood derived from Equation 8.11 when Φ has been maximized. The results are shown in Figure 8.10. A distinct minimum squared error is obtained at about $\lambda = 0.15$ while the log-likelihood continues to increase as λ is reduced.

It is not at all clear how to select the "best" value of λ from this result. Should the value that minimizes the SSE ($\lambda = 0.14$), be chosen or is there a critical point on the log-likelihood curve? For example, the log-likelihood increases more steeply for $\lambda < 0.005$ (i.e., $\log \lambda = -5.3$). It is interesting to note that for $\lambda < 0.005$, the value of SSE increases rapidly as λ decreases.

If we compare plots of Kaplan–Meier estimates with ANN predictions for different values of λ, we may get some indication of their effects. Such plots are shown in Figure 8.10 for subset 3 as defined in Table 8.2. (Values of the

TABLE 8.3

Subsets of SPSS Breast Cancer Data for Kaplan–Meier Analysis

Subset of Data	Histological Grade (Histgrad)	Pathological Tumor Size (Pathsize)
1	1	All
2	2	< 2 cm
3	2	≥ 2 cm
4	3	< 2 cm
5	3	≥ 2 cm

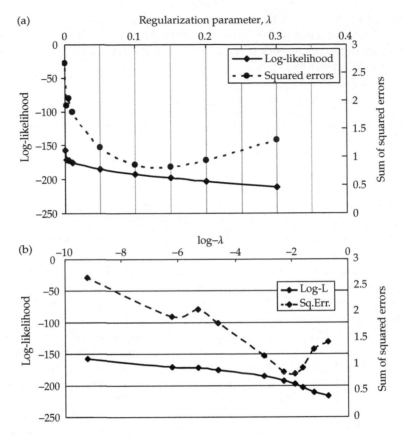

FIGURE 8.10
Effect of regularization parameter.

other covariates are age 54, tumor size 2.3 cm, one lymph node is positive, and the case is estrogen receptor positive and progesterone receptor negative.)

In Figure 8.11, the regularization parameter, λ, is denoted by w. The ANN curve with $\lambda = 0.14$ (minimized SSE) fails to follow the Kaplan–Meier plot (nevertheless, the SSE is minimized over the whole data set) whereas the curve with $\lambda = 0.0001$ tries to follow too many of the kinks in the Kaplan–Meier and makes erroneous long-term predictions. So far the best compromise is achieved with $\lambda = 0.005$. This suggests that minimizing the sum of squared errors, SSE, is a poor criterion for choice of λ. However, maximum likelihood is inherently a biased estimation procedure, and if minimizing SSE had been used as an optimization objective, the result would have been different. But then the ANN would be "learning" a best fit for these five subsets of the data, and modeling over the rest of the covariates may be suboptimal. This area would bear further research, and other methods of weight control should be investigated.

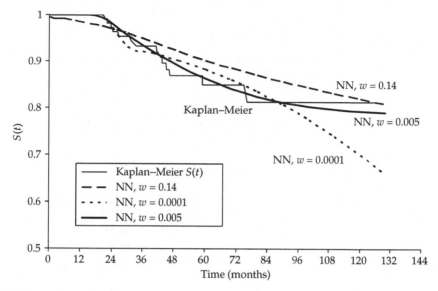

FIGURE 8.11
NN and Kaplan–Meier estimates for subgroup 3.

It is recommended that a small value of λ be used in practice (e.g., 0.01), and some exploration of sensitivity of predictions be undertaken when using the ANN approach to survival analysis. The effect is certainly different for different data sets. If the data can be subdivided into learning and validation sets then the value of λ that gives best predictions for the validation set should be chosen, but this could be a tedious approach.

8.5.5 Comparison between Artificial Neural Network and Proportional Hazards Analysis

Figure 8.12 shows a plot of the predictions made by the proportional hazards model as detailed in Table 8.1 and uses the same covariate values as in the ANN model. The fit is good, so it may be argued that the ANN provides at least as useful a prediction tool as the Cox model.

8.5.6 Investigation of Prediction Errors from Artificial Neural Network

A method that is commonly used by statisticians to examine the validity of a model is to compare the predictions made by the model and the sample estimates. We can use the differences between the Kaplan–Meier survival probabilities (the sample estimates) and the predictions from the ANN for comparison. Figure 8.13 shows a plot of these "errors" for the whole data set. The pattern of these errors is not surprising: the errors are fairly and randomly scattered around zero; they are low for small time values because the estimates all start at $S(t) = 1$ and then decrease. However, there is a scatter of negative errors, which

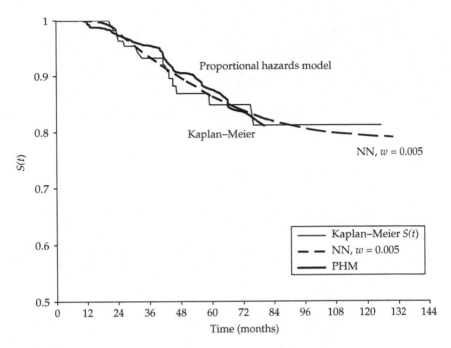

FIGURE 8.12
ANN, proportional hazards model (PHM), and Kaplan–Meier estimates for subgroup 3.

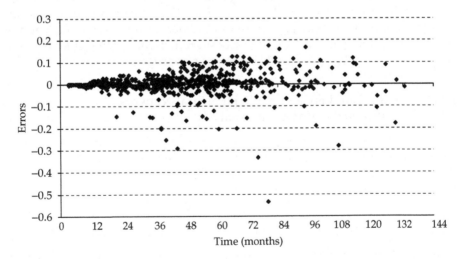

FIGURE 8.13
Plot of errors = ANN predictions – Kaplan–Meier estimates of $S(t)$ for whole data set.

indicates that for some cases the ANN prediction is pessimistic. It is worthwhile investigating this. The most negative error is −0.53 at 80 months. The age of this case is 40 years, tumor size is 5.5 cm (which is very large), 7 lymph nodes are involved, and the cancer is histological grade 3. By any standard, this case has a low probability of survival and all that the data tells us is that the patient was still surviving at 80 months after diagnosis. The ANN predicts a survival probability of 0.29 and the Kaplan–Meier estimate relating to this case was 0.82 (note that Kaplan–Meier does not give a prediction; the estimate depends on the number of cases in the relevant subgroup, sub5, that have died or been censored before 80 months). In fact this case shows that the ANN model is working and this particular case has survived against all odds.

8.6 Conclusion

A methodology for the use of ANNs in survival analysis has been shown to be viable and to be at least as effective as Cox's proportional hazards analysis in the context of a well-known data set. The network is trained with the maximization of log-likelihood and penalized by the sum of squares of the weights multiplied by a regularization parameter. The main difficulty is in choosing the most appropriate value of regularization parameter for the network. This may well depend on the data, but in general is not too critical and very often, good predictions can be obtained from a network trained with a low value of the parameter (e.g., 0.01).

The potential for the method is good, depending on further refinement and the provision of software to automate the fitting process and to facilitate network structuring. The most important refinement is to calculate confidence intervals on the predictions of survival probability and work on this is well advanced. A MATLAB m-file is available for the implementation of the ANN.

References

Azoff E M, 1993, Neural network principal components pre-processing and diffraction tomography. *Journal of Neural Computing and Applications*, **1**, 107–114.

Bishop C M, 1995, *Neural Networks for Pattern Recognition*, Oxford University Press, Oxford.

Collett D, 1994, *Modeling Survival Data in Medical Research*, Chapman & Hall, London.

Cox D R, 1972, Regression models and life tables (with discussion). *Journal of the Royal Statistical Society, B*, **74**, 187–220.

Cox D R and Oakes D, 1984, *Analysis of Survival Data*, Chapman & Hall, London.

Everitt B and Dunn G, 2001, *Applied Multivariate Data Analysis*, Edward Arnold (a division of Hodder and Stoughton), London.

Haykin S, 1999, *Neural Networks—A Comprehensive Foundation*, 2nd Edition, Prentice Hall, Upper Saddle River, NJ.

Johnson N L, Kotz S, and Balakrishnan N, 1995, *Continuous Univariate Distributions, Vol. II*, Wiley, Chichester.

Klein J P and Moeschberger M L, 2003, *Survival Analysis—Techniques for Censored and Truncated Data*, 2nd Edition, Springer-Verlag, New York.

Reeves C R and Johnston C, 2001, Fitting densities and hazard functions with neural networks. In V Kurková, N C Steele, R Neruda and M Kárný (Eds.) *Proceedings of 5th International Conference on Artificial Neural Nets and Genetic Algorithms*, Springer-Verlag, Vienna, pp. 47–50.

Ripley B D, 1996, *Pattern Recognition and Neural Networks*, Cambridge University Press, Cambridge.

Therneau T M and Grambsch P M, 2000, *Modeling Survival Data—Extending the Cox Model*, Springer-Verlag, New York.

Part III

Intelligent and Adaptive Techniques in Medical Imaging

9

Some Applications of Intelligent Systems in Cancer Treatment: A Review

Mark Fisher, Yu Su, and Richard Aldridge

CONTENTS

9.1 Introduction

Since the term was first coined in 1955 [1], artificial intelligence (AI) has become a rapidly expanding area of research, popularized by numerous books, both fiction and nonfiction. However, although it may be safe to assume that all readers will be familiar with the term, we need to establish a working definition for the purpose of setting the context for this chapter. Although several definitions of AI systems have been proposed [2–4], the question of how to quantify AI remains an open research topic within the AI research community. For the purpose of this review we will use the relatively loose definition that an artificially intelligent system (IS) is one that includes a knowledge component and is capable of learning such that it is able to sense and adapt to its environment. Well-known examples of such systems include expert systems, fuzzy systems, artificial neural networks (ANN), and models of statistical inference such as Bayesian networks.

Cancer is currently the cause of 12% of all deaths worldwide and the estimated number of new cancer cases every year is expected to rise to 15 million by the year 2020 [5]. Although most members of the medical and scientific

TABLE 9.1

Estimated New Cases and Deaths from Cancer

Site	New Cases (%)	Estimated Deaths
Digestive system	19	134,840
Prostate	17	29,900
Breast	16	40,580
Respiratory system	14	165,130

Source: Adapted from American Cancer Society, Estimated new cancer cases and deaths by sex for all sites US, 2004. http://www.cancer.org/downloads/STT/CAFF–finalPWSecured.pdf, 2005.

community agree that the best hope for a future cancer cure lies with systemic treatments such as immunotherapy and gene therapy, the surgical removal of cancer tissue supplemented by radiotherapy and chemotherapy currently remains the most effective treatment. Presently, in Europe, 45% of all cancer patients treated are cured (i.e., have a symptom-free survival period exceeding 5 years) following radiotherapy and surgery [6,7]. There are about 100 different types of cancer and significant variations exist between the methods used for diagnosis and treatment; since there are too many to review in detail here, we survey some applications of ISs in cancer treatment focusing particularly on prostate and breast cancer—these are responsible for a high proportion of deaths among men and women (Table 9.1). The organization of the remaining sections reflects the approach used for treatment of solid organ tumors as divided into three stages: diagnosis (staging), treatment planning, and delivery. Those readers who are not familiar with computer-aided diagnosis (CAD) systems may find Coiera's [8] recent text *Guide to Health Informatics* (part 8) useful.

9.2 Intelligent Systems in Cancer Diagnosis

Cancer may be diagnosed using a number of approaches: physical examination, blood test, x-ray, ultrasound (US), computer tomography (CT), magnetic resonance imaging (MRI), etc., and confirmed by biopsy. A staging system (e.g., the tumor metasteses nodes (TMN) system [9,10]) is used to indicate the extent of solid organ tumors. To indicate the problems and progress in this area, we have decided to restrict our survey to what we regard as typical exemplars of this field. We have chosen to highlight research into intelligent CAD systems designed to support clinicians and radiologists in the diagnosis of prostate and breast cancer.

9.2.1 Prostate

The prostate is a small gland that forms part of the male reproductive system. Situated just below the bladder, its function is to provide seminal fluid used to transport and protect semen from naturally occurring acids present in the vagina. The prostate is made up of smooth muscle, spongy tissue, and tiny ducts and glands. Because the prostate lies in front of the rectum, an initial diagnosis to check for swelling and abnormality is performed by digital rectal examination (DRE). DRE, once the principal method of diagnosis, has now been replaced by testing for serum prostate-specific antigen (PSA) and transrectal ultrasound (TRUS) guided prostate biopsy as the main method of diagnosis [11]. PSA is produced by the epithelial cells of normal, hyperplastic, and carcinomatious prostatic tissue. There is a positive correlation between the level of PSA and the spread of extraprostatic tumors [12–14]. The early detection or screening for prostate cancer using the PSA test is a major application area for ANN. Table 9.2 compares a number of intelligent CAD systems with respect to their ability to predict the outcome of a biopsy (ground truth). A commercially available system for computer-assisted diagnosis of prostate cancer based on parameters like tPSA, %fPSA, PSAD, etc., is marketed under the trade name Prost Asure. Babaian et al. [17] reported a sensitivity and specificity of 93% and 81% using Prost Asure compared with 74% and 80% using univariate analysis. Although ANN is the most common technique, other approaches to diagnosis based on PSA have used fuzzy systems [26]. Further details of techniques for diagnosis of prostate cancer may be found in a recently published survey by Zhu et al. [27].

Features derived from TRUS [18,24,28] and MRI [29–34] have also been used as inputs to ANN for predicting the pathologic stage of the disease (e.g., organ-confinement, capsular penetration (CP), and seminal vesicle invasion (SVI)). These approaches rely on an accurate automatic segmentation of the prostate organ, which is a very difficult image-processing task. Nevertheless, some promising recent results have been achieved using two-dimensional (2-D) [35] and 2-D–three-dimensional (3-D) hybrid [15] active shape models (ASM) (Figure 9.1).

9.2.2 Breast Cancer Diagnosis

Mammography is the most commonly used method of screening for breast cancer and it is reported that these achieve a mortality reduction of up to 30% [36,37]. In 1999–2000, the U.K. NHS breast screening programme contacted 1.5 million women between the ages of 50 and 64; 75.6% of them responded and of them, about 5% were referred for additional assessment. The current NHS breast screening programme requires a single-medio-lateral oblique x-ray view (however, this is often enhanced by an additional craniocaudal view) to be "read" by an expert radiographer, who reports on abnormal image features [38] such as circumscribed and stellate lesions and calcifications. Under such workloads, CAD systems [39,40] play an important role in both training radiologists [41] and directing their attention toward suspicious lesions (prompting). Over the past 15 years, a number of researchers have developed intelligent CAD systems that are able to detect abnormalities automatically. The performance of some of these systems is summarized in

TABLE 9.2

Published Research on the Use of ANN in the Detection of Prostate Cancer

Reference	Year	Cases	PSA(ng/mL)	Sensitivity (%)	Specificity (%)
Snow et al. [16]	1994	1787	>4.0	87	(overall accuracy)
Babaian et al. [17]	1998	225	<4.0	93	81
Ronco and	1999	442	N/A	82.81	79.64
Fernandez [18]				80.09[a]	(overall accuracy)
Finne et al. [19]	2000	656	4.0–10.0	95	33
				90	46
Babaian et al. [20]	2000	151[b]	2.5–4.0	92	62
Djavan et al. [21]	2002	272	2.5–4.0	95	59
		974	4.0–10.0	95	67
Stephan et al. [22]	2002	928	2.0–4.0	90	63
			4.1–10	90	57
			10.1–20.0	90	46
Stephan et al. [23]	2002	1188	2.0–4.0	90	38
			4.1–10.0	90	62
			2.0–10.0	90	62
			10.1–20.0	90	53
			2.0–20.0	90	61
Porter et al. [24]	2002	319	N/A	95	21[c]
Remzi et al. [25]	2003	820	4.0–10	95	68

[a] Average of 3 ANN models.
[b] In test data set.
[c] Median specificity, five test data sets using cross validation.
Source: Adapted from Zhu, Y., *Segmentation and Staging of Prostate Cancer from MRI Images*, PhD
            Thesis, School of Computing Sciences, University of East Anglia, U.K., 2005.

Table 9.3. A majority of the early approaches use first-generation backpropagation ANN [46,49,51,54], but more recently researchers have turned their attention to other more optimal network structures. For example, Sahiner et al. [48] investigated the classification of regions of interest (ROI) on mammograms using a convolution neural network. Lo et al. [55] evaluated three neural network models in the recognition of medical image patterns associated with breast and lung cancer. There are also several hybrid approaches. For example, Papadopoulos et al. [44] presented an approach using rule-based and neural network subsystems and Zhang et al. [42] investigated a neural genetic algorithm (GA) for feature selection in conjunction with neural and statistical classifiers.

Other research in breast cancer diagnosis considers input features derived from histological data. Pendharker et al. [45] used association rules to study associations of different female hormones with the occurrence of breast cancer within a knowledge discovery and data-mining (KDD) paradigm. They benchmark approaches using data envelope analysis (DEA) and ANN against a standard parametric Fisher's linear discriminant analysis technique and their results show that DEA and ANN outperform the discriminant analysis approach. Moore et al. [56] also published benchmarks for Bayesian, ANN,

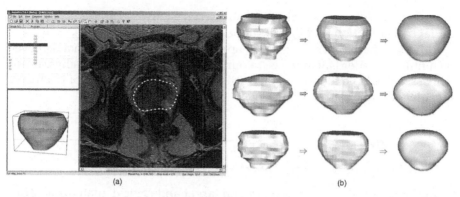

FIGURE 9.1

(a) A screen shot of the "AnnoPro" tool for manual annotation of prostate MR data. (b) Examples of 3-D prostate surfaces construction. From left-hand side to right-hand side: manual annotations on 2-D slices, verified manual annotations with 3-D view, and 3-D prostate surfaces constructed by the automatic surface completion technique. (From Zhu, Y., *Segmentation and Staging of Prostate Cancer from MRI Images*, PhD Thesis, School of Computing Sciences, University of East Anglia, U.K., 2005.)

TABLE 9.3

Published Research on the Use of ANN in the Detection of Breast Cancer

Reference	Year	Approach	Features	Cases	Sensitivity	Specificity
Zhang et al. [42]	2005	Hybrid[a]	Image	322[b]	85%[c]	
Arbach et al. [43]	2003	ANN	Image	160	92.3% (A_z)	
Papadopoulos et al. [44]	2002	MLP	Image	60[d]	82.5% (A_z)	
Papadopoulos et al. [44]	2002	Hybrid[e]	Image	60[d]	91.2% (A_z)	
Pendharker et al. [45]	1999	KDD	Misc	479	66%[c]	
Fogel et al. [46]	1997	—	Image	112	0.95 ($A_z = 89.82\%$)	0.62
Setiono and Liu [47]	1997	KDD	Histolic	579	95.73%[c]	
Sahiner et al. [48]	1996	CNN	Image		87% (Az)	
Zheng et al. [49]	1996	MFNN	Image	30	90.1%	71%
Baker et al. [50]	1995	—	Image	206	0.95	0.62
Fogel et al. [51]	1995	—	Histolic	283	98.05%[c]	

[a] ANN + GA.
[b] DDSM data set [53].
[c] Overall accuracy.
[d] MIAS data set [52].
[e] ANN + RBS.

and linear regression models. Abbass [57] investigated an approach named memetic pareto artificial neural network (MPANN), and Setiono et al. [47,58,59] presented a system for extracting rules from neural networks. Both report results using the Wisconsin breast cancer dataset [60], comprising 579 cases of 9 measurements taken from fine needle aspirates from human breast tissues corresponding to cytological characteristics of a benign or of a malignant pattern. Sehgal et al. [61] used a hybrid technique combining radial basis functions (RBF) and support vector machines (SVM).

Although the performance of ISs appears to be better than that achieved by a human reader [43], none is, to our knowledge, currently in clinical use. Abbass [57] believe the reason for this lies in human perception of intelligent machines as a threat.

9.3 Intelligent Systems in Treatment Planning

In radiotherapy, the volumes of several target and critical anatomical structures are defined to produce a 3-D radiotherapy treatment plan; these include gross tumor volume (GTV), clinical target volume (CTV), internal target volume (ITV), and planning target volume (PTV). They are defined by the International Atomic Energy Agency (IAEA) as follows (Figure 9.2 shows how they are related to one another [62]):

GTV. "The Gross Tumor Volume (GTV) is the gross palpable or visible/ demonstrable extent and location of malignant growth" [63]. The GTV is usually based on information obtained from a combination of imaging modalities (CT, MRI, ultrasound, etc.), diagnostic modalities (pathology and histological reports, etc.), and clinical examination.

CTV. "The clinical target volume (CTV) is the tissue volume that contains a demonstrable GTV and subclinical microscopic malignant disease that has to be eliminated. This volume thus has to be treated

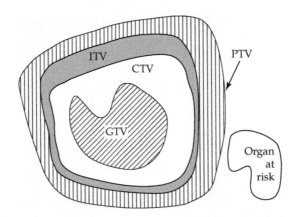

FIGURE 9.2
Graphical representation of the volumes of interest as defined in ICRU Reports No. 50 and 62. (From Parker, W. *Clinical Treatment Planning in External Beam Radiotherapy*, DMRP syllabus on Medical Physics, Chapter 7; http://www-naweb.iaea.org/nahu/dmrp/pdf_files/ Chapter 7.pdf.)

adequately to achieve the aim of therapy, cure, or palliation" [63]. The CTV often includes the area directly surrounding the GTV, which may contain microscopic disease and other areas considered to be at risk and requiring treatment (e.g., positive lymph nodes). The CTV is usually stated as a fixed or variable margin around the GTV, but in some cases it is the same as the GTV (e.g., prostate boost to the gland only).

ITV. The ITV consists of the CTV plus an internal margin. The internal margin is designed to take into account the variations in the size and position of the CTV relative to the patient's reference frame (usually defined by the bony anatomy); that is, variations due to organ motions resulting from breathing and bladder or rectal contents [64].

PTV. "The planning target volume (PTV) is a geometrical concept, and it is defined to select appropriate beam arrangements, taking into consideration the net effect of all possible geometrical variations, to ensure that the prescribed dose is actually absorbed in the CTV" [63]. "The PTV includes the internal target margin and an additional margin for set-up uncertainties, machine tolerances and intra-treatment variations" [64]. The PTV is linked to the reference frame of the treatment machine and is often described as the CTV plus a fixed or variable margin (e.g., PTV = CTV + 10 mm). Usually a single PTV is used to encompass one or several CTVs to be targeted by a group of fields. The PTV depends on the precision of such tools as immobilization devices and lasers, but does not include a margin for the dosimetric characteristics of the radiation beam (i.e., penumbral areas and buildup region), as these will require an additional margin during treatment planning and shielding design.

OAR. Organs at risk (OAR) is another important concept. The OAR is an organ whose sensitivity to radiation is such that the dose received from a treatment plan may be significant compared with its tolerance, possibly requiring a change in the beam arrangement or a change in the dose. Organs with a radiation tolerance that depends on the fractionation scheme should be outlined completely to prevent biasing during treatment plan evaluation.

To improve the accuracy and precision of the treatment and reduce the potential risk on the surrounding healthy tissue, treatment planning needs to consider the uncertainty caused by systematic and random setup errors. The setup errors are normally derived by comparing any individual electronic portal image (EPI) in the beam's eye view (BEV) with its corresponding reference image, normally a digitally reconstructed radiograph (DRR) generated by the treatment planning system (TPS) [65]. Various correction strategies have been considered to reduce the setup error using portal images. A good review on the details can be found in [66] among these, Shalev and Gluhchev [67–69] suggested applying a correction of magnitude,

$k\mu_i$ where μ_i is the cumulative mean setup error and k is a coefficient (≤ 1) whenever the measured displacement exceeds a given action level. In their study, Shalev and Gluhchev propose that for a measured displacement the correction to be applied is modified according to the following:

$$k = \frac{\Sigma^2_{setup}}{\left(\Sigma^2_{setup} + \sigma^2_{setup}\right)}$$

where Σ^2_{setup} is the sum of setup errors and σ^2_{setup} is the variance. Later they developed this approach in the form of an adaptive rule that makes the use of accumulated data to determine the required correction $-k\mu_i$, such that

$$k = \frac{n\Sigma^2_{setup}}{\left(n\Sigma^2_{setup} + \sigma^2_{setup}\right)}$$

where n is the total number of measurements. Following any correction, Σ^2_{setup} is recalculated and fed back into subsequent calculations to produce an adapted value of k. Wong et al. [70] also adopted this adaptive approach, but the correction is made by adjusting the multileaf collimator (MLC) rather than the patient setup.

In the treatment planning process, ISs are also employed in the calculation of dose distribution. At this stage, corrections are often required for contour irregularities caused by oblique beam incidence and the patient's surface. In addition, some irradiated tissues, such as lung and bone, have densities that differ considerably from that of water, therefore, corrections for tissue heterogeneities are also required. Isodose distributions in patients are determined by one of the two radically different approaches:

- Correction-based algorithms
- Model-based algorithms

Between the two approaches, correction-based algorithms are applied for conventional treatment techniques. A radiation beam striking an irregular or sloping patient surface produces an isodose distribution that differs from the standard distributions obtained on flat surfaces with a normal beam incidence. Different organ densities also account for the isodose distribution shift. Correction-based algorithms compensate for irregular patient contours and oblique beam incidence as well as for inhomogeneities to account for varying electron densities of organs. The correction is done by applying wedges, bolus, or compensators in the path of the beam. However, for new and sophisticated treatments such as 3-D conformal radiotherapy (3-DCRT) and intensity modulated radiotherapy (IMRT) [71,72], radical corrections are required, which often make the situation problematic. Model-based algorithms obviate the correction problem by modeling the dose distributions from first principles and accounting for all geometrical and physical characteristics of the particular patient being treated. Among various model-based algorithms, Monte Carlo methods are considered the most promising, as they use well-established probability distributions governing the individual interactions of photons and electrons with the patient and their transport through the patient. The distribution parameters are normally learnt from a large number of histories to reduce stochastic or random uncertainties to acceptable levels. Following the advances in computer technology, Monte Carlo calculation times are expected to be reduced to acceptable levels and it is anticipated that this will make Monte Carlo methods the standard approach to radiotherapy treatment planning [62].

Another IS application in treatment planning is dose optimization. Dose optimization routines including inverse planning may be integrated within the TPS with varying degrees of complexity. Algorithms can adaptively modify beam weights and geometry or calculate modulated beam intensities to satisfy the dose constraint criteria for critical organs. Criteria are set up based on the dose-volume histogram (DVH), using information obtained from CT, MRI, or other digital imaging modalities. These routines can use a predefined beam geometry corresponding to a specific anatomy (e.g., prostate) to shorten calculation times [62].

The TPS also incorporates a capability for 3-D reconstruction and DRR generation. These processes would normally need an automatic segmentation technique, as manual segmentation of CT and MRI slices is inherently time-consuming. ANN has been adopted in various medical segmentation applications over the years. Glass and Reddick [73] introduced a hybrid ANN in segmentation and classification of dynamic contrast-enhanced MRI image although Davis et al. [74] used ANN for x-ray image segmentation.

9.4 Intelligent Systems in Treatment Delivery

When the tumor is subject to motion with a magnitude >2 mm, the ITV margin can put the surrounding healthy tissue at an intolerable higher risk. To reduce the ITV margin or hopefully eliminate it, various approaches have been adopted aimed at tracking the target at the treatment delivery stage. These include 3-DCRT, IMRT, respiratory gated radiotherapy (RGRT), and image guided radiotherapy (IGRT). These approaches extensively employ IS techniques.

3-DCRT starts with the 3-D reconstructed model of the volume of interest (VOI) and DRRs provided by the TPS. The TPS then develops complex plans to deliver highly "conformed" radiation while sparing adjacent healthy tissue. IMRT is an advanced form of 3-DCRT that uses dynamic multileaf collimators (DMLCs), consisting of up to 120 computer-controlled movable "leaves" that conform the radiation beam to the shape of the tumor from any angle while protecting normal adjacent tissue as much as possible. For a fixed gantry position, the opening formed by each pair of opposing MLC leaves is swept across the target volume under computer control with the radiation beam turned on to produce the desired fluence map [75,76].

The research work of Mackie et al. [77] has led to tomotherapy, an advanced form of IMRT, carried out with more than one radiation beam to achieve a uniform dose distribution inside the target volume and a dose as low as possible in surrounding healthy tissues. Further details of tomotherapy can be found in the research paper by Mackie et al. [77]. Another delivery method, intensity modulated arc therapy (IMAT), considered as an alternative to tomotherapy has also been proposed. Yu [78] utilized continuous gantry motion as in conventional arc therapy, but unlike conventional arc therapy, the field shape that is conformed with the MLC changes during gantry rotation. Arbitrary 2-D beam intensity distributions at different beam angles are delivered with multiple superimposing arcs. Earl et al. [79] described an automated inverse treatment planning algorithm capable of producing efficient IMAT treatment plans, called direct aperture optimization (DAO), which can be used to generate inverse treatment plans for IMAT. In contrast to traditional inverse planning techniques where the relative weights of a series of pencil beams are optimized, DAO optimizes the leaf positions and weights of the apertures in the plan. This technique allows any delivery constraints to be enforced during the optimization, eliminating the need for a leaf-sequencing step.

Respiratory gating is implemented by integrating beam control software into the treatment plan instead of the entire area. A physical window is defined at the planning stage and during treatment images are acquired and radiation beam is only switched on when the tumor lies within the window. Respiratory gating is often combined with IMRT to deliver precise doses of radiation to tumor sites. Respiratory gating software focuses radiation on the tumor and spares more healthy tissue. This is particularly useful for tumors near the lungs, where radiation-induced scarring can impair future breathing [80,81].

Respiratory gating is currently a mainstream approach to tackle the problem caused by respiratory motion. However, due to the fact that the beam is only turned on for a relatively shorter period, the treatment period is inevitably prolonged, to ensure that the tumor receives the prescribed dose. In addition, due to lack of constant irradiation, the treatment result may be compromised.

Varian's real-time position management (RPM) respiratory gating system (Figure 9.3) is capable of accurate patient tracking using lightweight retro-reflective markers for real-time respiratory monitoring. This system also employs IS, by way of a predictive filter that automatically withholds treatment delivery when respiration deviates from the normal pattern [82].

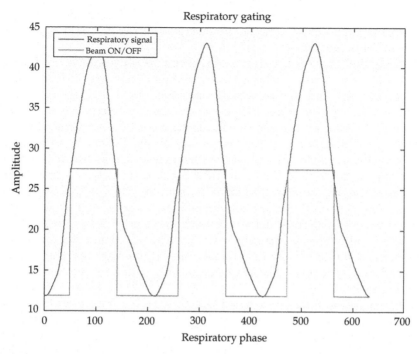

FIGURE 9.3
Schematic of Varian's real time position management (RPM) respiratory gating system.

A program of research carried out at the Harvard Medical School and Massachusetts General Hospital, aimed at accurate tumor motion tracking, also relies on IS. Among them, Neicu et al. [82] introduced synchronized moving aperture radiotherapy (SMART). Here, tumor motion, represented by a sequence of 3-D marker position data and tumor apertures are acquired during the treatment simulation/planning stage. The average tumor trajectory (ATT) is then derived. Using the ATT, tumor motion can be incorporated into IMRT MLC leaf sequence at treatment delivery stage, although the respiratory surrogate or the markers are monitored and used to synchronize the treatment with tumor motion. The beam moves according to the ATT and is turned off when the tumor position is different from the ATT; treatment is resumed when the radiation beam motion and tumor motion are resynchronized.

In their work, Jiang et al. [83] modeled the phase of intrafraction respiratory motion, as this can potentially compromise the result of IMRT. A series of experiments is carried out using a tumor motion simulator that moves sinusoidally to mimic the lung motion. They conclude that the variation in breathing phase at the beginning of dose delivery results in a maximum variation around the mean dose of greater than 30% for one field and reduced to 8% for five field within a single fraction, but after

30 fractions, the variation reduces to less than 1–2%. However, this study did not address the potential damage to the surrounding normal tissue caused by the intrafraction motion.

Tracking the position of a tumor in real time is only one part of the solution to improve accuracy. The treatment system latencies in dose delivery also need to be taken into consideration, which include image acquisition, image processing, communication delays, control system processing, inductance within the motor, mechanical damping, etc. This problem has been addressed by Sharp et al. [84]. Various motion prediction algorithms have been investigated to tackle the systems latencies, such as linear prediction, linear extrapolation, and ANN. In their work they adopt an approach to predicting the dynamic system latency using an ANN and Kalman filter. The parameters for these are established by training the system before each treatment fraction is delivered. Berbeco et al. [85] have designed an integrated radiotherapy imaging system with two diagnostic x-ray imaging sources mounted to the linear accelerator (Linac) gantry. The system is designed to obtain accurate 3-D positional information of a moving tumor in real time.

Research in radiotherapy carried out to date has rarely addressed the problem caused by intrafraction shape variation, which is a major issue surrounding the treatment of certain types of cancer (e.g., prostate). To track the position and the shape variation of the target, several other methods have been investigated. IS are applied here both to track the targets position and to model the shape variation. Recent developments in cone beam imaging (CBI) enable real-time feedback of local image evidence to be realized and hence provide the possibility of a fully implemented IGRT. A series of publications on CBI is found mainly subject to the research work carried out by Jaffray et al. [86–89].

Statistical shape models are able to learn from prior information and fit new shape examples using models of local image evidence. As they have the ability to adapt themselves to fit new data based on a training set of exemplars, these may be considered as applications of IS. Statistical shape models are valuable tools in organ/tumor segmentation and tracking as they are capable of imposing geometrical constraints on the shape and integrating local image evidence [90]. They are extremely useful in IGRT where large variations of target motion and shape are concerned. The incorporation of local image evidence provides a means of feedback control for fine-tuning in treatment planning and dose calculation.

Cootes et al. [91–93] developed a statistical model of shape and appearance called the ASM and active appearance model (AAM). Landmark points are marked on every image in the training set, and a statistical model is computed to describe the shape and texture variation in the image. A best match can be found between the synthesized model image and target image by calculating the appropriate parameters. The statistical models have been successfully applied in a number of medical applications [94]. They have also furthered their research in statistical models for medical image analysis and computer vision [95,96]. Another shape modeling approach known as active contour models (ACM) was first proposed by Kass et al. [97]. To date, their journal article [98] has been cited in 1315 publications comprising a variety of forms, namely, snakes, deformable templates, dynamic contours, and physics-based models. Statistical snakes are a mechanism for bringing a certain degree of prior knowledge, to bear on low-level image interpretation. Typical examples are the inclusion of image noise models [99], however, shape priors can only be modeled indirectly due to the underlying physics-based representation, for example, by using modes of vibration [100]. Models of more specific classes of shapes demand some use of hard constraints, for which the deformable template is created. When applied dynamically to temporal image sequences, the model is defined as a dynamic contour [101]. There are many examples of medical applications of ACM, some focusing on radiotherapy treatment planning (RTP) [102]. Recently, Derraz et al. [103] have introduced ACM into medical image segmentation. ASM and ACM can be used selectively on local image regions, rather than the entire image, thereby enabling images to be processed at the full video rate, 50–60 Hz, making real time tracking possible [101]. Both ASM [104,105] and ACM (in the guise of "deformable models") [106–108] have been extended to 3-D volumetric applications.

Using the improved image contrast and higher resolution on soft tissue provided by the CBI technique together with statistical shape models to accurately track tumor motion, a full implementation of IGRT is now feasible, leading to the concept of adaptive radiotherapy (ART) [62]. In this process, intra- and interfraction tumor shape and position can be accurately tracked, and together with the feedback of local image information, the dose delivery for treatment fractions can also be adaptively modified to compensate for inaccuracies in dose delivery [62].

The tracking methods mentioned earlier rely on prerecorded training samples to extract either motion or shape model templates. Radiotherapy simulators provide an efficient tool for both treatment simulation and the recording of training samples for intelligent dose delivery systems. A radiotherapy simulator consists of a diagnostic x-ray tube mounted on a rotating gantry, simulating geometries identical to those found on megavoltage therapy machines. The simulator enjoys the same degrees of freedom as a megavoltage machine, but instead of providing a megavoltage beam for treatment, it provides a diagnostic quality x-ray beam suitable for imaging [62]. This makes the simulator specifically useful for intensity profile–based tracking schemes such as ASM, AAM, and ACM.

Another highly intelligent example in radiotherapy is the CyberKnife radiosurgery system. Developed in the mid-1990s, CyberKnife delivers the dose with a miniature Linac mounted on a robotic arm. This system has the capability of online target imaging and automatic adjustment of the radiation beam direction to compensate for short-term target motion. As such it can be interpreted as an excellent example of an IS application in robot control [62].

9.5 Conclusions and Future Research Directions

This chapter has attempted to provide an insight into the role of ISs in cancer treatment Intelligence has been interpreted somewhat loosely in an attempt to include as wide a range of approaches as possible. Clearly in any safety critical system there is a need for caution and rigorous testing before any technique is put into routine use. However, there is clear evidence that research undertaken within the past 20 years is beginning to have an impact in terms of commercial systems bringing real benefits for patients.

Owing to the potential for the delivery of a higher radiation dose by IMRT, studies focusing on accurately delivering and verifying prescribed treatments form an important part of future research in this area. The MAESTRO project [109] recently funded by the European Union aims to address these issues by developing better approaches for monitoring patient setup and intrafraction anatomical variations together with a range of technologies for improved dosimetry.

List of Abbreviations

3-DCRT	Three-dimensional conformal radiotherapy
AAM	Active appearance model
ACM	Active contour model
AI	Artificial intelligence
ANN	Artificial neural network
ART	Adaptive radiotherapy
ASM	Active shape model
ATT	Average tumor trajectory
BEV	Beam's eye view
CAD	Computer-aided diagnosis
CBI	Cone beam imaging
CT	Computer tomography
CTV	Clinic target volume
DAO	Direct aperture optimization
DEA	Data envelope analysis
DMLC	Dynamic multileaf collimator
DRE	Digital rectal examination
DRR	Digitally reconstructed radiograph
DVH	Dose-volume histogram
EPI	Electronic portal image
GA	Genetic algorithm
GTV	Gross target volume
IAEA	International atomic energy agency

ICRU	International commission on radiation units and measurements
IGRT	Image guided radiotherapy
IMAT	Intensity modulated arc therapy
IMRT	Intensity modulated radiotherapy
IS	Intelligent systems
ITV	Internal target volume
KDD	Knowledge discovery and datamining
Linac	Linear accelerator
MLC	Multileaf collimator
MLP	Multilayer perceptron
MPANN	Memetic pareto artificial neural network
MRI	Magnetic resonance imaging
NHS	National health service
OAR	Organ at risk
PSA	Prostate-specific antigen
PTV	Planning target volume
RBF	Radial basis function
RGRT	Respiratory gated radiotherapy
ROI	Region of interest
SMART	Synchronized moving aperture radiotherapy
SVM	Support vector machine
TMN	Tumor metasteses nodes
TPS	Treatment planning system
TRUS	Transrectal ultrasound
US	Ultrasound
VOI	Volume of interest

References

1. J. McCarthy, M.L. Minsky, N. Rochester, and C.E. Shannon. A proposal for the Dartmouth summer research project on artificial intelligence. http://www.formal.stanford.edu/jmc/history/dartmouth/dartmouth.html, 1955.
2. A. Turing. Computing machinery and intelligence. *Mind*, 59:433–460, 1950; Reprinted in E.A. Feigenbaum and J. Feldman, *Computers and Thought*. McGraw-Hill, New York, 1963.
3. A. Newell. The knowledge level. *Artif. Intell.*, 18(1):87–127, 1982.
4. J.S. Albus. Outline for a theory of intelligence. *IEEE Trans. Syst. Man Cybern.*, 21(3):473–509, 1991.
5. American Cancer Society. Estimated new cancer cases and deaths by sex for all sites US, 2004. http://www.cancer.org/downloads/STT/CAFF–finalPWSecured.pdf, 2005.
6. IOS. Europe against cancer, 1994.
7. Department of Health. The NHS cancer plan. Technical Report, Department of Health, London, 2000.

8. E. Coiera. *Guide to Health Informatics.* 2nd edition, Arnold, London, 2003.
9. C.F. Montain. Value of the new TNM staging system for lung cancer. *Chest,* 96(1):47s–49s, 1989.
10. L.H. Sobin and C. Wittekind. *UICC TNM Classification of Malignant Tumours.* 6th edition, Wiley-Liss, New York, 2002.
11. NCCN. Clinical practice guidelines in oncology: Prostate cancer. http//www. nccn.org/professionals/physician_gls/default.asp, 2005.
12. A.W. Partin, J.K. Yoo, H.B. Carter, J.D. Pearson, D.W. Chan, J.I. Epstein, and P.C. Walsh. The use of prostate specific antigen, clinical stage and gleason score to predict pathological stage in men with prostate cancer. *J. Urol.,* 150:110–114, 1993.
13. The Alberta CPG Group on Prostate Cancer Screening. Guidline for use of PSA and screening for prostate cancer. Technical Report, Alberta Medical Association, 1999.
14. M. Han, P.H. Gann, and W.J. Catalona. Prostate-specific antigen and screening for prostate cancer. *Med. Clin. North Am.,* 88:245–265, 2004.
15. Y. Zhu. *Segmentation and Staging of Prostate Cancer from MRI Images.* PhD thesis, School of Computing Sciences, University of East Anglia, U.K., October 2005.
16. P.B. Snow, D.S. Smith, and W.J. Catalona. Artificial neural networks in the diagnosis and prognosis of prostate cancer: A pilot study. *J. Urol.,* 152(5 Pt. 2):1923–1926, 1994.
17. R.J. Babaian, H.A. Fritsche, Z. Zhang, K.H. Madyastha, and S.D. Barnhill. Evaluation of prostate index in the detection of prostate cancer: A preliminary report. *Urology,* 51(1):132–136, 1998.
18. A.L. Ronco and R. Fernandez. Improving ultrasonographic diagnosis of prostate cancer with neural networks. *Ultrasound Med. Biol.,* 25(5):729–733, 1999.
19. P. Finne, R. Finne, A. Auvinen, H. Juusela, J. Aro, L. Määttänen, M. Hakamac, S. Rannikko, T.L.J. Tammela, and U.H. Stenman. Predicting the outcome of prostate biopsy in screen-postive men by a multiplayer perceptron network. *Urology,* 56(3):418–422, 2000.
20. R.J. Babaian, H.A. Fritsche, A. Ayala, V. Bhadkamkar, D.A. Johnston, W. Naccarato, and Z. Zhang. Performance of a neural network in detecting prostate cancer in the prostate-specific antigen range of 2.5 to 4.0 ng/ml. *Urology,* 56(6):1000–1006, 2000.
21. B. Djavan, M. Remzi, A. Zlotta, C. Seitz, P. Snow, and M. Marberger. Novel artificial neural network for early detection of prostate cancer. *J. Clin. Oncol.,* 20(4):921–929, 2002.
22. C. Stephan, K. Jung, H. Cammann, B. Vogel, B. Brux, G. Kristiansen, B. Rudolph, S. Hauptmann, M. Lein, D. Schnorr, P. Sinha, and S.A. Loening. An artificial neural network considerably improves the diagnostic power of percent free prostate-specific antigen in prostate cancer diagnosis: Results of a 5-year investigation. *Int. J. Cancer,* 99(3):466–473, 2002.
23. C. Stephan, H. Cammann, A. Semjonow, E.P. Diamandis, L.F. Wymenga, M. Lein, P. Sinha, S.A. Loening, and K. Jung. Multicenter evaluation of an artificial neural network to increase the prostate cancer detection rate and reduce unnecessary biopsies. *Clin. Chem.,* 48(8):1279–1287, 2002.
24. C.R. Porter, C. O'Donnell, E.D. Crawford, E.J. Gamito, B. Sentizimary, A. de Rosilia, and A. Tewari. Predicting the outcome of prostate biopsy in a racially diverse population: A prospective study. *Urology,* 60(5):831–835, 2002.

25. M. Remzi, T. Anagnostou, V. Ravery, A. Zlotta, C. Stephan, M. Marberger, and B. Djavan. An artificial neural network to predict the outcome of repeat prostate biopsies. *Urology*, 62(3):456–460, 2003.
26. I. Saritas, N. Allahverdi, and I.U. Sert. A fuzzy expert system design for diagnosis of prostate cancer. In *CompSysTech '03: Proceedings of the 4th International Conference on Computer Systems and Technologies*. B. Rachey and A. Smrikarov, editors, ACM Press, New York, 2003, pp. 345–351.
27. Y. Zhu, R. Zwiggelaar, and S. Williams. Computer technology in detection and staging of prostate carcinoma: A review. *Med.l Image Anal*, 10(2):178–199, 2006.
28. C.R. Porter and E.D. Crawford. Combining artificial neural networks and transrectal ultrasound in the diagnosis of prostate cancer. *Oncology*, 17(10):1395–1399 and 1403–1406, 2003.
29. A. Tewari and P. Narayan. Novel staging tool for localized prostate cancer: A pilot study using genetic adaptive neural networks. *J. Urol.*, 160(2):430–436, 1998.
30. E.J. Gamito, N.N. Stone, T.J. Batuello, and E.D. Crawford. Use of artificial neural networks in the clinical staging of prostate cancer: Implications for prostate brachytherapy. *Tech. Urol.*, 6(2):60–63, 2000.
31. J.T. Batuello, E.J. Gamito, E.D. Crawford, M. Hun, A.W. Partin, D.G. McLoad, and C. O'Donnell. An artificial neural network model for the assessment of lymph node spread in patients with clinically localized prostate cancer. *Urology*, 57(3):481–485, 2001.
32. M. Han, P.B. Snow, J.M. Brandt, and A.W. Partin. Evaluation of artificial neural networks for the prediction of pathologic stage in prostate carcinoma. *Cancer*, 91(8 Suppl.):1661–1666, 2001.
33. A.M. Ziada, T.C. Lisle, P.B. Snow, R.F. Levine, G. Miller, and E.D. Crawford. Impact of different variables on the outcome of patients with clinically confined prostate carcinoma: Prediction of pathologic stage and biochemical failure using an artificial neural network. *Cancer*, 91(8 Suppl.):1653–1660, 2001.
34. A.R. Zlotta, M. Remzi, P.B. Snow, C.C. Schulman, M. Marberger, and B. Djavan. An artificial neural network for prostate cancer staging when serum prostate specific antigen is 10 ng/ml or less. *J. Urol.*, 169(5):1724–1728, 2003.
35. D. Shen, Y. Zhan, and C. Davatzikos. Segmentation prostate boundaries from ultrasound images using statistical shape model. *IEEE Trans. Med. Imaging*, 22(4):539–551, 2003.
36. E. van den Akker-van Marle, H. de Koning, R. Boer, and P. van der Maas. Reduction in breast cancer mortality due to the introduction of mass screening in The Netherlands: Comparison with the United Kingdom. *J. Med. Screen*, 6(1):30–34, 1999.
37. H. Wan, R. Karesen, A. Hervik, and S.O. Thoresen. Mammography screening in Norway: Results from the first screening round in four counties and the cost-effectiveness of a modeled nationwide screening. *Cancer Causes Control*, 12(1):39–45, 2001.
38. E. Claridge and J. Richer. Characterization of mammographic lesions. In *Proceedings of 2nd International Workshop on Digital Mammography*. S.M. Astley, D.R. Dance and A.Y. Cairns, editors, York, England, July 1994, pp. 241–250.
39. S.M. Astley. Evaluation of computer-aided detection (cad) prompting techniques for mamography. *Br. J. Radiol.*, 78:S20–S25, 2004.
40. S.M. Astley. Computer-based detection and prompting of mamographic abnormalities. *Br. J. Radiol.*, 77:S194–S200, 2004.

41. H. Sumsion, A.G. Davies, G.J.S. Parkin, and A.R. Cowen. Application of a computed radiography database to a mammography reporting environment. In *Digital Mammography*. A.G. Gale and S. Astley, editors, Elsevier Science BV, Amsterdam, 1994, pp. 379–386.

42. P. Zhang, B. Verma, and K. Kumar. Neural vs. statistical classifier in conjunction with genetic algorithm based feature selection. *Pattern Recogn. Lett.*, 26(7):909–919, 2005.

43. L. Arbach, L. Bennett, J.M. Reinhardt, and G. Fallouh. Mammogram breast mass classification with backpropagation neural network. *IEEE Canadian Conference on Electrical and Computer Engineering*, volume 3, Montreal, May 2003, pp. 1441–1444. http://ieeexplore.ieee.org/xpl/freeabs_all.jsp?isnumber= 27522&arnumber=1226168&count=161&index=0.

44. A. Papadopoulos, D.I. Fotiadis, and A. Likas. An automatic microcalcification detection system based on a hybrid neural network classifier. *Artif. Intell. Med.*, 25:149–167, 2002.

45. P.C. Pendharker, J.A. Rodger, G.J. Yaverbaum, N. Herman, and M. Benner. Association, statistical, mathematical and neural approaches for mining breast cancer patterns. *Expert Sys. Appl.*, 17:223–232, 1999.

46. D.B. Fogel, E.C. Wasson, E.M. Boughton, and V.W. Porto. A step toward computer-assisted mammography using evolutionary programming and neural networks. *Cancer Lett.*, 119:93–97, 1997.

47. R. Setiono and H. Liu. NeuroLinear: From neural networks to oblique decision rules. *Neurocomputing*, 17:1–24, 1997.

48. B. Sahiner, H.-P. Chen, N. Petrick, D. Wei, M.A. Helvie, and D.D. Adler. Classification of mass and normal breast tissue: A convolution neural network classifier with spatial domain and texture images. *IEEE Trans. Med. Imaging*, 15(5):598–610, 1996.

49. B. Zheng, W. Qian, and L.P. Clarke. Digital mammography: Mixed feature neural network with spectralentropy decision for detection of microcalcifications. *IEEE Trans. Med. Imaging*, 15(5):589–597, 1996.

50. J.A. Baker, P.J. Kornguth, J.Y. Lo, M.E. Whilliford, and C.E. Floyd. Breast cancer, prediction with artificial neural networks based on BIRADS standard lexicon. *Radiology*, 196:817–822, 1995.

51. D.B. Fogel, E.C. Wasson III, and E.M. Boughton. Evolving neural networks for detecting breast cancer. *Cancer Lett.*, 96:49–53, 1995.

52. J. Suckling, J. Parker, D. Dance, S. Astley, I. Hutt, C. Boggis, et al. The mammographic images analysis society digital mammogram database. *Exerpta Medica*, 1069:375–378, 1994.

53. M. Heath, K.W. Bowyer, D. Kopans, et al. Current status of the digital database for screening mammography. In *Digital Mammography*. N. Karssemeijer, editor, Kluwer Academic, Dordrecht, 1998, pp. 457–460. http://books.google. co.uk/books?id=mKgOAAAACAAJ&dq=Digital+Mammography.

54. C.E. Floyd, J.Y. Lo, A.J. Yun, and D.C. Sullivan. Prediction of breast cancer malignancy using an artificial neural network. *Cancer*, 74:2944–2998, 1994.

55. S.-C.B. Lo, J.-S.J. Lin, M.T. Freeman, and S.K. Mun. Application of artificial neural networks to medical image pattern recognition: Detection of clustered microcalcifications on mammograms and lung cancer on chest radiographs. *J. VLSI Signal Process. Syst.*, 18(3):263–274, 1998.

56. A. Moore and A. Hoang. A performance assessment of Bayesian networks as a predictor of breast cancer survival. In *Second International Workshop on Intelligent*

Systems Design and Application. A. Abraham, B. Nath, M. Sambandham and P. Saratchandran, editors, Dynamic Publishers, Inc., Atlanta, GA, 2002, pp. 3–8.

57. H. Abbass. An evolutionary artificial neural network approach for breast cancer diagnosis. *Artif. Intell. Med.*, 25(3):265–281, 2002.

58. R. Setiono. Generating concise and accurate classification rules for breast cancer diagnosis. *Artif. Intell. Med.*, 18:205–219, 2000.

59. R. Setiono. Extracting rules from pruned neural networks for breast cancer diagnosis. *Artif. Intell. Med.*, 8:37–51, 1997.

60. P.M. Murphy and D.W. Aha. Uc1 repository of machine learning databases. machine-readable data repository. Technical Report, University of California, Department of Information and Computer Science, Irvine, CA, 1992.

61. M.S.B. Sehgal, I. Gondal, and L. Dooley. Support vector machine and generalized regression neural network based classification fusion models for cancer diagnosis. In *Proceedings of 4th International Conference on Hybrid Intelligent Systems (HIS'04)*, IEEE Computer Society, Washington, D.C., 2004, pp. 49–54.

62. Division of Human Health International Atomic Energy Agency. Dosimetry and medical radiation physics tutorial. http://www-naweb.iaea.org/nahu/dmrp. Accessed November 2005.

63. International Commission on Radiation Units and Measurements. Prescribing, recording and reporting photon beam therapy. ICRU Report 50, ICRU, Bethesda, MD, 1993.

64. International Commission on Radiation Units and Measurements. Prescribing, recording and reporting photon beam therapy. ICRU Report 62. (Supplement to ICRU Report 50), ICRU, Bethesda, MD, 1999.

65. British Institute of Radiology. *Geometrical Uncertainties in Radiotherapy: Defining the Planning Target Volume*. Working Party of The British Institute of Radiology, 2003, pp. 44–45.

66. S. Webb. Motion effects in (intensity modulated) radiation therapy: A review. *Phys. Med. Biol.*, 51:R403–R425, 2006.

67. S. Shalev and G. Gluhchev. Decision rules for treatment field set-up displacements. *Proceedings of 3rd International Workshop of Electronic Imaging*, San Francisco, CA, October 1994.

68. S. Shalev and G. Gluhchev. Intervention correction of patient set-up using portal imaging: A comparison of decision rules. *Int. J. Radiat. Oncol. Biol. Phys.*, 32:216, 1995.

69. G. Gluhchev. The magnitude of treatment field set-up parameter correction in radiation therapy. *Radiother. Oncol.*, 48:79–82, 1998.

70. J. Wong, D. Yan F. Vicini, et al.. Adaptive modification of treatment planning to minimize the deleterious effects of treatment set-up errors. *Int. J. Radiat. Oncol. Biol. Phys.*, 38:197–206, 1997.

71. T. Bortfeld. IMRT: A review and preview. *Phys. Med. Biol.*, 51:R363–R379, 2006.

72. A. Ahnesjo, B. Hårdmark, U. Isacsson, and A. Montelius. The IMRT information process-mastering the degrees of freedom in external beam therapy. *Phys. Med. Biol.*, 51:R381–R482, 2006.

73. J.O. Glass and W.E. Reddick. Hybrid artificial neural network segmentation and classification of dynamic contrast-enhanced mr imaging (demri) of osteosarcoma. *Magn. Resonance Imag.*, 16(9):1075–1083, 1998.

74. D. Davis, L.Y. Su, and B. Sharp. Neural networks for x-ray image segmentation. *Proceedings of 12th International Conference on Control Systems and Computer Science*, Romania, 1999.

75. Radiation Oncology Palo Alto Medical Foundation. Three-dimensional conformal radiotherapy. http://www.pamf.org/radonc/tech/3d.html. Accessed November 2005.
76. Radiation Oncology Palo Alto Medical Foundation. Intensity modulated radiation therapy (IMRT). http://www.pamf.org/radonc/tech/imrt.html. Accessed November 2005.
77. T.R. Mackie, T. Holmes, S. Swerdloff, P. Reckwerdt, J.O. Deasy, J. Yang, B. Paliwal, and T. Kinsella. Tomotherapy: A new concept for the delivery of dynamic conformal radiotherapy. *Med. Phys.*, 20(Nov–Dec):1709–1719, 1993.
78. C.X. Yu. Intensity-modulated arc therapy with dynamic multi-leaf collimation: An alternative to tomotherapy. *Phys. Med. Biol.*, 40:1435–1449, 1995.
79. M.A. Earl, D.M. Shepard, S. Naqvi, X.A. Li, and C.X. Yu. Inverse planning for intensity-modulated arc therapy using direct aperture optimisation. *Phys. Med. Biol.*, 48:1075–1089, 2003.
80. Stanford Hospital & Clinics, Stanford Cancer Centre. Respiratory gating. http://cancer.stanfordhospital.com/forPatients/services/radiationTherapy/respiratoryGating/default. Accessed November 2005.
81. Treatment Delivery Varian Medical Systems, Radiation Therapy. RPM respiratory gating. http://www.varian.com/orad/prd057.html. Accessed November 2005.
82. T. Neicu, H. Shirato, Y. Seppenwoolde, and S.B. Jiang. Synchronized moving aperture radiation therapy (smart): Average tumour trajectory for lung patients. *Phys. Med. Biol.*, 48:587–598, 2003.
83. S.B. Jiang, C. Pope, K.M. Al Jarrah, J.H. Kung, T. Bortfeld, and G.T.Y. Chen. An experimental investigation on intra-fractional organ motion effects in lung imrt treatment. *Phys. Med. Biol.*, 48:1773–1784, 2003.
84. G.C. Sharp, S.B. Jiang, S. Shimizu, and H. Shirato. Prediction of respiratory tumour motion for real-time image-guided radiotherapy. *Phys Med. Biol.*, 49:425–440, 2004.
85. R.I. Berbeco, S.B. Jiang, G.C. Sharp, G.T.Y. Chen, H. Mostafavi, and H. Shirato. Integrated radiotherapy imaging system (iris): Design considerations of tumour tracking with linac gantry-mounted diagnostic x-ray systems with flat-panel detectors. *Phys. Med. Biol.*, 49:243–255, 2004.
86. D.A. Jaffray, J.H. Siewerdsen, J.W. Wong, and A.A. Martinez. Flat-panel cone-beam computed tomography for image-guided radiation therapy. *Int. J. Radiat. Oncol. Biol. Phys.*, 53(5):1337–1349, 2002.
87. D.A. Jaffray and J.H. Siewerdsen. Cone-beam computed tomography with a flat-panel imager: Initial performance characterization. *Med. Phys.*, 27(6):1311–1323, 2000.
88. J.H. Siewerdsen and D.A. Jaffray. Optimization of x-ray imaging geometry (with specific application to flat-panel cone-beam computed tomography). *Med. Phys.*, 27(8):1903–1914, 2000.
89. J.H. Siewerdsen and D.A. Jaffray. Cone-beam computed tomography with a flat-panel imager: Effects of image lag. *Med. Phys.*, 26(12):2635–2647, 1999.
90. B. Fisher. 2-D deformable template models: A review. http://homepages.inf.ed.ac.uk/rbf/CVonline/LOCAL_COPIES/ZHONG1/node1.html, 1999. Accessed November 2005.
91. T.F. Cootes, C.J. Taylor, D.H. Cooper, and J. Graham. Active shape models their training and application. *Proceedings of Computer Vision and Image Understanding 1995*, U.K., 1995.

92. T.F. Cootes, G.J. Edwards, and C.J. Taylor. Active appearance models. *IEEE Trans. Pattern Anal. Mach Intell.*, 23(6):681–685, 2001.

93. T.F. Cootes and C.J. Taylor. Constrained active appearance model. *Proceedings of ICCV 2001*, U.K., 2001.

94. T.F. Cootes, A. Hill, C.J. Taylor, and J. Haslam. The use of active shape models for locating structures in medical images. *Image Vision Comput.*, 12(6):355–366, 1994.

95. T.F. Cootes and C.J. Taylor. Statistical models of appearance for medical image analysis and computer vision. *Proceedings of SPIE Medical Imaging 2001*, U.K., 2001.

96. T.F. Cootes and P. Kittipanyangam. Comparing variations on the active appearance model algorithm. *Proceedings of BMVC 2002*, U.K., 2002.

97. M. Kass, A. Witkin, and D. Terzopoulos. Snakes: Active contour models. *Proceedings of 1st International Conference on Computer Vision*, London, June 8–11, 1987, pp. 259–268. http://www.visionbib.com/bibliography/journal/icc.html#ICCV87.

98. M. Kass, A. Witkin, and D. Terzopoulos. Snakes—active contour models. *Int. J. Comput. Vision*, 1(4):321–331, 1987.

99. C. Chesnaud, P. Refregier, and V. Boulet. Statistical region snake-based segmentation adapted to different physical noise models. *IEEE Trans. Pattern Anal. Mach Intell.*, 21:1145–1157, 1999.

100. C. Nikou, G. Bueno, F. Heitz, and J.-P. Armspach. A joint physics-based deformable model for multimodel brain image analysis. *IEEE Trans. Med. Imaging*, 20(10):1026–1037, 2001.

101. A. Blake and M. Isard. *Active Contours; The Application of Techniques from Graphics, Vision, Control Theory and Statistics to Visual Tracking of Shapes in Motion*. Springer, New York, 1998.

102. S.D. Fenster and J.R. Kender. Sectored snakes: Evaluating learned-energy segmentations. *IEEE Trans. PAMI*, 23(9):1028–1034, 2001.

103. F. Derraz, A.B. Belkaid, M. Beladgham, and M. Khelif. Application of active contour models in medical image segmentation. *Proceedings of International Conference on Information Technology: Coding and Computing (ITCC'04)*, Australia, 2004.

104. T. Cootes, C. Beeston, G. Edwards, and C. Taylor. Unified framework for atlas matching using active appearance models. *Inform. Process. Med. Imaging*, 1999.

105. R.H. Davies, C.J. Twining, T.F. Cootes, J.C. Waterton, and C.J. Taylor. 3D statistical shape models using direct optimisation of description length. *Proc. ECCV'2002*, 3:3–20, 2002.

106. L.D. Cohen and I. Cohen. Finite-element methods for active contour models and balloons for 2-D and 3-D images. *IEEE Trans. Pattern Anal. Mach Intell.*, 15:1131–1147, 1993.

107. D. Metaxas. Deformable models for segmentation, 3D shape and motion estimation and recognition. *Proc. 10th Br. Mach. Vision Conf.*, 1:1–11, 1999. Invited paper.

108. F. Weichert, M. Wawro, and C. Wilke. A 3D computer graphics approach to brachytherapy planning. *Int. J. Cardiovasc. Imaging*, 20:173–182, 2004.

109. J. Barthe. Methods and advanced equipment for simulation and treatment in radio oncology. http://www.maestro-research.org/index.htm, 2005.

10

Fuzzy Systems and Deformable Models

Gloria Bueno

CONTENTS

10.1 Introduction

Fuzzy systems and deformable models are well-known examples of artificial intelligence (AI) system. This chapter describes the segmentation of medical images with these AI models. Three major application areas are treated: the detection of lesions on mammography images, the detection of radiotherapy relevant organs on computerized tomography (CT) images of the pelvic area, and the three-dimensional (3-D) modeling of brain structures.

With medical imaging playing an increasingly prominent role in the diagnosis and treatment of disease, the medical image analysis community has become preoccupied with the challenging problem of extracting, with the assistance of computers, clinically useful information about anatomic structures imaged through CT, MR, PET, and other modalities [1]. Both fuzzy and deformable models have played an important role aiming to solve this problem.

The impetus for using fuzzy models came from control theory. Many important medical and industrial applications based on fuzzy theory have made their way into the marketplace in the recent years [2,3]. Section 10.2 contains a brief introduction to the basic ideas underlying fuzzy pattern recognition, tailored to illustrate fuzzy clustering segmentation applied to medical images. Sections 10.4.1 and 10.4.2 contain case studies involving unsupervised and supervised segmentation of medical images with two spatial dimensions, two-dimensional (2-D) images, in applications such as mammography and radiotherapy treatment planning.

Moreover, the challenge is also to extract boundary elements belonging to the same structure and integrate them into a coherent and consistent model of the structure. Traditional low-level image-processing techniques, which consider only local information, can make incorrect assumptions during this integration process and generate infeasible object boundaries. To solve this problem, deformable models have been presented [4].

Deformable models offer a powerful approach to accommodate the significant variability of biological structures over time and across different individuals. They are able to segment, match, and track images of anatomic structures by exploiting (bottom-up) constraints derived from the image data together with (top–down) *a priori* knowledge about the location, size, and shape of these structures [4]. Section 10.3 contains a brief introduction to the mathematical foundations of deformable models. Section 10.4.1 illustrates the application of a geometrical deformable model for segmenting region of interest (ROI) in radiotherapy treatment planning. The geometrical deformable model is combined with a fuzzy C-means clustering (FCMC) segmentation based on the theory introduced in Section 10.2. Section 10.3 explains the application of a physics-based statistical deformable model for modeling anatomical structures with three spatial dimensions (3-D images) in applications such as brain analysis.

Finally, the conclusions and future research direction are drawn in Section 10.5.

10.2   Fuzzy C-Means Clustering

Fuzzy set theory provides a powerful mathematical tool for modeling the human ability to reach conclusions when information is imprecise and incomplete. This is sometimes the case of medical images that can exhibit various levels of noise and low-contrast densities resulting in ill-defined shapes [1,5].

Fuzzy sets are a generalization of conventional set theory introduced by Zadeh [6] in 1965 as a mathematical way to represent uncertainties. Fuzzy set theory applied to image segmentation is a fuzzy partition of the image data, $I(x, y) = \{x_1, x_2, \ldots, x_n\}$, into c fuzzy subsets or c specified classes. That is, $\{x_m (i, j): \rightarrow [0, 1]: m = 1, \ldots, c\}$, which replaces a crisp membership function, divides the image into c regions by means of a nonfuzzy segmentation process. Thus, image data is clustered into the c different classes by using an unsupervised FCMC algorithm. This is achieved by computing a measure of membership, called fuzzy membership, at each pixel [7]. The fuzzy membership function, constrained to be between 0 and 1, reflects the degree of similarity between the data value at that location and the centroid of its class. Thus, a high membership value near 1 means that the data value (pixel intensity) at that location is close to the centroid for that particular class. The FCMC algorithms are then formulated as the minimization of the squared error with respect to the membership functions, U, and the set of centroids, $\{V\} = \{v_1, v_2, \ldots, v_n\}$.

$$J(U, V : X) = \sum_{i=1}^{c}\sum_{k=1}^{n} (u_{ik})^m \, \| x_k - v_i \| \qquad (10.1)$$

where $u_{ik} = u_i(x_k)$ is the membership of x_k in class i, and $m \geq 1$ is a weighting exponent of each fuzzy membership.

The FCMC objective function, Equation 10.1, is minimized when high membership values are assigned to pixels whose intensities are close to the centroid for its particular class and low membership values are assigned when the pixel intensity is far from the centroid. Taking the first derivatives of Equation 10.1 with respect to u_{ik} and v_k, setting those equations to zero yields necessary conditions for Equation 10.1 to be minimized. Then u_{ik} and v_k are defined as

$$u_{ik} = \left[\sum_{j=1}^{c} \left(\frac{\| X_k - V_i \|}{\| X_k - V_j \|} \right)^{\frac{2}{m-1}} \right]^{-1}, \quad v_i = \frac{\sum_{k=1}^{n}(u_{ik})^m x_k}{\sum_{k=1}^{n}(u_{ik})^m} \quad \forall \, i, k \qquad (10.2)$$

Iterating through these conditions leads to a grouped coordinate descent scheme for minimizing the objective function. The stopping criterion is determined for each iteration by $E_i < \varepsilon$, where

$$E_i = \sum_{i=1}^{c} \| v_{i, t+1} - v_{i,t} \| \quad \forall \, t \qquad (10.3)$$

The resulting fuzzy segmentation can be converted to a hard or crisp segmentation by assigning each pixel solely to the class that has the highest membership value for that pixel. This is known as a maximum membership segmentation. Once the method has converged a matrix, U, with the membership or degree to which every pixel is similar to all of the c, different classes are obtained. Every pixel is assigned the class for which the maximum membership is found. That is, if $\max(u_{ik}) = u_{ik}$, then x_k is assigned the label associated with the class i. Some results of this FCMC algorithm are shown in Section 10.4. Further modifications of the FCMC algorithm may be done by introducing more terms to the objective function (Equation 10.1), for example, to cope with noisy images [7].

The result of the FCMC algorithm may be quite variable according to the number of selected clusters, c, and the position of the centroids, $\{V\}$. To apply the FCMC algorithm to the problem of interest within the research projects carried out by the Universidad de Castilla-La Mancha group (UCLM-ISA group), a proper configuration of both c and $\{V\}$ has been found from analysis of image histograms; this is illustrated in Section 10.4.

10.3 Deformable Models

The mathematical foundations of deformable models represent the confluence of geometry, physics, and approximation theory. Geometry serves to represent object shape, physics imposes constraints on how the shape may vary over space and time, and optimal approximation theory provides the formal mechanisms for fitting the models to measure data. The physical interpretation views deformable models as elastic bodies that respond to applied force and constraints.

The deformable model that has attracted the most attention to date is, the active contour model (ACM), well-known as "snakes," presented by Kass et al. [8]. The mathematical basis present in snake models is similar to many deformable models.

10.3.1 Energy Minimizing Deformable Models

Geometrically, a snake is a parametric contour embedded in the image plane $(x, y) \in \mathbb{R}^2$. The contour is represented as a curve, $v(s) = (x(s), y(s))$, where x and y are the coordinate functions and $s \in [0, 1]$ is the parametric domain. The curve evolves up to the ROI, subject to constraints from a given image $I(x, y)$. Initially a curve is set around the ROI by minimizing an energy functional, the curve moves along a direction normal to itself aiming to stop at the boundary of the ROI. The energy functional is defined as

$$E_{snake}(s) = \int_0^1 [E_{internal}(v(s)) + E_{ext\_potential}(v(s))]ds \qquad (10.4)$$

The first term $E_{internal}$ represents the internal energy of the spline curve due to the mechanical properties of the contour, stretching, and bending [8]. It is a sum of two components: the elasticity and rigidity energy.

$$E_{internal}(s) = \left(\frac{\alpha(s)}{2}|v_s(s)|^2 + \frac{\beta(s)}{2}|v_{ss}(s)|^2\right) \qquad (10.5)$$

where $\alpha(s)$ controls the tension of the contour, whereas $\beta(s)$ controls its rigidity. Thus, these functions determine how the snake stretches or bends at any point s of the spline curve.

The second term couples the snake to the image:

$$E_{ext\_potential}(s) = P(v(s)) \tag{10.6}$$

where $P(v(s))$ denotes a scalar potential function defined on the image plane. It is responsible for attracting the contour toward the object in the image (external energy). It can be expressed as a weighted combination of energy functionals. To apply snakes to images, external potentials are designed whose local minima coincide with intensity extrema, edges, and other image features of interest. For example, the contour will be attracted to intensity edges in an image by choosing a potential $P(v(s)) = -c|\nabla[G_\sigma * I(x, y)]|$, where c controls the magnitude of the potential, ∇ is the gradient operator, and $G_\sigma * I(x, y)$ denotes the image convolved with a Gaussian smoothing filter.

In accordance with the calculus of variations, the contour $v(s)$ that minimizes the energy of Equation 10.4 must satisfy the Euler–Lagrange equation, such that:

$$-\frac{\partial}{\partial s}\left(\frac{\alpha(s)}{2}|v_s(s)|^2\right) + \frac{\partial^2}{\partial s^2}\left(\frac{\beta(s)}{2}|v_{ss}(s)|^2\right) + \nabla P(v(s, t)) = 0 \tag{10.7}$$

This vector-valued partial differential equation expresses the balance of internal and external forces when the contour rests at equilibrium.

Traditional snake models are limited in several aspects such as sensitivity to the initial contours, nonfree parameters, and not handling changes in the topology of the shape. In other words, when considering more than one object in the image, for instance, for an initial prediction of $v(s)$ surrounding all of them, it is not possible to detect all the objects; special topology-handling procedures must be added. Some techniques have been proposed to solve these drawbacks. These techniques are based on information fusion, dealing with ACM in addition to region properties [5,9] and curvature-driven flows [10–13].

10.3.2 Dynamic Deformable Models

A potential approach to computing the local minima of a functional such as Equation 10.4 is to construct a dynamic system that is governed by the functional and allow the system to evolve to equilibrium. The system may be constructed by applying the principles of Lagrangian mechanics. This leads to dynamic deformable models that unify the description of shape and motion, making it possible to quantify not only static shape, but also shape evolution through time [4].

A simple example is a dynamic snake that can be represented by introducing a time-varying contour $v(s, t) = (x(s, t), y(s, t))$ with a mass density $\mu(s)$ and a damping density $\gamma(s)$. The Lagrange equations of motion for a snake with the internal and external energies given by Equation 10.4 are as follows:

$$\mu \frac{\partial^2 v}{\partial t^2} + \gamma \frac{\partial v}{\partial t} - \frac{\partial}{\partial s}\left(\frac{\alpha(s)}{2}|v_s(s)|^2\right) + \frac{\partial^2}{\partial s^2}\left(\frac{\beta(s)}{2}|v_{ss}(s)|^2\right) + \nabla P(v(s,t)) = 0 \quad (10.8)$$

The first two terms on the left-hand side of this partial differential equation represent inertial and damping forces. Referring to Equation 10.7, the remaining terms represent the internal stretching and bending forces, whereas the right-hand side represents the external forces. Equilibrium is achieved when the internal and external forces balance and the contour comes to rest, (i.e., $\partial^2 v/\partial t^2 = \partial v/\partial t = 0$), which yields the equilibrium condition (Equation 10.7).

10.3.2.1 Discretization and Numerical Simulation

To compute a minimum energy solution numerically, it is necessary to discretize the energy (Equation 10.4). The usual approach is to represent the continuous geometric model v in terms of linear combinations of local- or global-support basis functions. Finite elements, finite differences, and geometric splines [4] are local representation methods, whereas Fourier bases are global representation methods. The continuous model $v(s)$ is represented in discrete form by a vector $\mathbf{U}$ of shape parameters associated with the basis functions. The discrete form of energies such as the energy (Equation 10.4) for the snake may be written as

$$E(U) = \frac{1}{2}\mathbf{U}^T\mathbf{K}\mathbf{U} + P(\mathbf{U}) \qquad (10.9)$$

where $\mathbf{K}$ is called the stiffness matrix and $P(\mathbf{U})$ is the discrete version of the external potential. The minimum energy results are obtained from setting the gradient of Equation 10.9 to 0, which is equivalent to solving the set of algebraic equations.

$$E(U) = \frac{1}{2}\mathbf{U}^T\mathbf{K}\mathbf{U} = -\nabla P = \mathbf{F} \qquad (10.10)$$

where $\mathbf{F}$ is the generalized external force vector. The discretized version of the Lagrangian dynamics equation may be written as a set of second-order ordinary differential equations for $\mathbf{U}(t)$:

$$\mathbf{M\ddot{U}} + \mathbf{C\dot{U}} + \mathbf{KU} = \mathbf{F} \qquad (10.11)$$

This is the governing equation of the physical model, which is characterized by its mass matrix $\mathbf{M}$, its stiffness matrix $\mathbf{K}$, and its damping matrix $\mathbf{C}$. The time derivatives in Equation 10.8 are approximated by finite differences and explicit or implicit numerical time-integration methods that are applied to simulate the resulting system of ordinary differential equations in the shape parameters **U**. This model is explained in Section 10.3 for the purpose of 3-D statistical modeling.

10.3.3 Geodesic Active Contour Model

Recently, there has been an increasing interest in level set segmentation methods. Level set, introduced in [11], involves solving the active contour (AC) minimization (Equation 10.4) by the computation of minimal distances curve. Thereby, the AC evolves following the geometric heat flow equation.

Consider a particular class of snake models in which the rigidity coefficient is set to zero, that is, $\beta = 0$. Two main reasons motivate this selection: (i) It allows the derivation of the relation between these energy-based active contours and geometric curve evolution and (ii) the regularization effect on the geodesic active contours comes from curvature-based curve flows, obtained only from the other terms in Equation 10.4. The latter provides the means to achieve smooth curves in the proposed approach without having the high-order smoothness given by $\beta \neq 0$ in energy-based approaches. Moreover, this smoothness component in Equation 10.4 appears to minimize the total squared curvature. It is possible to prove that the curvature flow used in the geodesic model decreases the total curvature. The use of the curvature-driven curve motions as smoothing terms was proved to be very efficient in previous literature [11,13]. Hence curve smoothing obtained with $\beta = 0$, has only the first regularization term, leading to a reduction of Equation 10.4 to

$$E_{\text{geo}}(s) = \int_0^1 \left(\frac{\alpha(s)}{2} |v_s(s)|^2 \right) ds - \int_0^1 c |\nabla I[v(s)]| \, ds \qquad (10.12)$$

Observe that, by minimizing functional, Equation 10.12, we are trying to locate the curve at the points of maxima $|\nabla I|$ (acting as edge detector) while keeping a certain smoothness in the curve (object boundary). This is actually the goal in general formulation (Equation 10.4) as well.

It is possible to extend Equation 10.12, generalizing the edge detector part in the following way: Let $\{g : [0, \infty [\rightarrow \mathbf{R}^+\}$ be a strictly decreasing function. Hence, we can replace $-|\nabla I|$ with $g|\nabla I|^2$, obtaining a general energy functional given by

$$E_{\text{geo}}(s) = \int_0^1 \left(\frac{\alpha(s)}{2} |v_s(s)|^2 \right) ds + \int_0^1 c \, g \left(|\nabla I[v(s)]| \right)^2 ds \qquad (10.13)$$

The solution of the particular energy snake model of Equation 10.13 is given by a geodesic curve in a Rieman space induced from the image $I(x, y)$, where a geodesic curve is a local minimal distance path between given points. To show this, the classical Maupertuis' principle from dynamical systems together with the Fermat's principle is used [13,14]. By assuming

$E_{internal} = E_{ext\_potential}$, that is, $E_0 = 0$, it is possible to reduce the minimization of Equation 10.4 to the following form:

$$\min_{v(s)} = \int_0^1 g\left(|\nabla I(x, y)| (v(s)) \cdot |v_s(s)|\, ds \right. \tag{10.14}$$

where g is a function of the image gradient used for the stopping criterion. By using Euler–Lagrange, and defining an embedding function of the curve $v(s)$, $\psi(t, s)$, this is an implicit representation of $v(s)$. Assuming that $v(s)$ is a level set of a function $\psi(t, s) : [0, a] \times [0, b] \rightarrow \Re$, the following equation for curve/surface evolution is derived:

$$\frac{\partial \psi}{\partial t} = g(\nabla I)(C + K)|\nabla \psi| \tag{10.15}$$

where C is a positive real constant and K is the Euclidian curvature in the direction of the normal, $\vec{N}$.

To summarize the force (C + K) acts as the internal force in the classical energy-based snake model, smoothness being provided by the curvature part of the flow. The heat-flow ($K\vec{N}$) is the regularization curvature flow that replaces the second-order smoothness term in Equation 10.4. The external-image-dependent force is given by the stopping function, $g(\nabla I)$. Thus, $g(\nabla I)$ stops the evolving curve when it arrives at the object's boundaries.

The advantage of using a level set representation is that the algorithm can handle changes in the topology of the shape as the surface evolves in time and it is less sensitive to the initialization. However, there are also some drawbacks in terms of efficiency and convergence [12]. It is also nonfree parameters, ψ_t is dependent on the time step Δt and the spatial resolution. The proposed algorithm explained in Section 10.4.2 provides an improved stopping criterion for the active curve evolution where the equilibrium is guaranteed.

10.3.4 Statistical Physics-Based Deformable Models

In some cases, it may be desirable to simultaneously outline several structures physically connected to each other. This section presents a probabilistic physical deformable model for the representation of such multiple anatomical structures. The physical model is characterized by its mass matrix **M**, its stiffness matrix **K**, and its damping matrix **C**; and its governing equation may be written as Equation 10.11, where **U** represents the nodal displacements of an initial mesh $\mathbf{X}_0$. $\mathbf{X}_0$ is a set of 3-D points sampled on a spherical surface, following a quadrilateral cylinder topology [15]. Thus, the model nodes are stacked in vector:

$$\mathbf{X}_0 = \left[x_1^0, y_1^0, z_1^0, x_{N'N'}^0, y_{N'N'}^0, z_{N'N}^0 \right] \tag{10.16}$$

where N designates the number of points in the direction of the geographical longitude and N' is the number of points in the direction of the geographical latitude of the sphere.

The image force vector **F** is based on the Euclidean distance between the mesh nodes and their nearest contour points [15–17].

Because Equation 10.11 is of order $3NN'$, where NN' is the total number of nodes of the spherical mesh, it is solved in a subspace corresponding to the truncated vibration modes of the deformable structure [15,17], using the following change of basis:

$$\mathbf{U} = \mathbf{\Phi}\tilde{\mathbf{U}} = \sum_i \tilde{u}_i \phi_i \tag{10.17}$$

where $\mathbf{\Phi}$ is a matrix and $\tilde{\mathbf{U}}$ is a vector, ϕ_i is the ith column of $\mathbf{\Phi}$, and $\tilde{u}_i$ is the ith scalar component of vector $\tilde{\mathbf{U}}$. By choosing $\mathbf{\Phi}$ as the matrix whose columns are the eigenvectors of the eigenproblem,

$$\mathbf{K}\phi_i = \omega_i^2 \mathbf{M}\phi_i \tag{10.18}$$

and using the standard Rayleigh hypothesis [15], matrices **K**, **M**, and **C** are simultaneously diagonalized:

$$\begin{cases} \mathbf{\Phi}^{\mathsf{T}}\mathbf{M}\mathbf{\Phi} = \mathbf{I} \\ \mathbf{\Phi}^{\mathsf{T}}\mathbf{M}\mathbf{\Phi} = \mathbf{\Omega}^2 \end{cases} \tag{10.19}$$

where $\mathbf{\Omega}^2$ is the diagonal matrix whose elements are the eigenvalues ω_i^2 and **I** the identity matrix. Substituting Equation 10.17 and premultiplying by $\mathbf{\Phi}^{\mathsf{T}}$ yields

$$\ddot{\tilde{\mathbf{U}}} + \tilde{\mathbf{C}}\dot{\tilde{\mathbf{U}}} + \mathbf{\Omega}^2\tilde{\mathbf{U}} = \tilde{\mathbf{F}} \tag{10.20}$$

where $\tilde{\mathbf{C}} = \mathbf{\Phi}^{\mathsf{T}}\mathbf{C}\mathbf{\Phi}$ and $\tilde{\mathbf{F}} = \mathbf{\Phi}^{\mathsf{T}}\mathbf{F}$.

In many computer vision applications [17], when the initial and the final states are known, it is assumed that a constant load **F** is applied to the object. Then, the model equation is called the equilibrium governing equation and corresponds to the static problem:

$$\mathbf{KU} = \mathbf{F} \tag{10.21}$$

In the new basis, Equation 10.21 is simplified to $3NN'$ scalar equations:

$$\omega_i^2 \tilde{u}_i = \tilde{f}_i \tag{10.22}$$

In Equation 10.22, ω_i designates the ith eigenvalue, the scalar $\tilde{u}_i$ is the amplitude of the corresponding vibration mode (corresponding to eigenvector ϕ_i). Equation 10.22 indicates that instead of computing the displacements vector **U** from Equation 10.21, its decomposition may be computed in terms of the vibration modes of the original mesh. The number of vibration modes retained in the object description is chosen to obtain a compact but adequately accurate representation. A typical *a priori* value covering many types of standard deformations is the quarter of the number of degrees of freedom in the system [15] (i.e., 25% of the modes are kept). Although a high-resolution description of the surfaces is not provided, this truncated representation provides a satisfactory compromise between accuracy and complexity of the representation. The spherical model is initialized around the structures of interest [43]. The vibration amplitudes are explicitly computed by Equation 10.22, where rigid body modes ($\omega_i = 0$) are discarded and the nodal displacements may be recovered using Equation 10.17. The physical representation $X(\tilde{U})$ is finally given by applying the deformations to the initial spherical mesh:

$$X(\tilde{U}) = X_0 + \Phi \tilde{U} \tag{10.23}$$

Thus, this parameterization is applied for the different segmented objects in the training set and their statistical learning is performed. For each image $i = 1, \ldots, n$ in the training set, a vector a_i containing the M_s lowest frequency vibration modes describing the S different anatomical structures is created.

$$a_i = \left(\tilde{U}_i^1, \tilde{U}_i^2, \ldots, \tilde{U}_i^s \right)^T \tag{10.24}$$

Where

$$\tilde{U}_i^s = \left(u_1^s, u_2^s, \ldots, u_{M_s}^s \right)_i \tag{10.25}$$

Random vector a is statistically constrained by retaining the most significant variation modes in its Karhunen-Loéve (KL) transform:

$$a = \bar{a} + Pb \tag{10.26}$$

where $\bar{a}$ is the average vector of vibration amplitudes of the structures belonging to the training set, P is the matrix whose columns are the eigenvectors of the covariance matrix $\Gamma = [(a - \bar{a})^T (a - \bar{a})]$ and $b_i = P^T (a_i - \bar{a})$ are the coordinates of $(a - \bar{a})$ in the eigenvector basis.

The deformable multiobject model is finally parameterized by the m most significant statistical deformation modes stacked in vector b. By modifying b, the different objects are deformed in conjunction [43], according to the anatomical variability observed in the training set.

Given a set of S initial spherical meshes, X_{INIT}, corresponding to the structures described by the joint model:

$$X_{INIT} = \begin{pmatrix} X_0^1 \\ \vdots \\ X_0^s \end{pmatrix} \tag{10.27}$$

the statistical deformable model $X(a)$ is thus represented by

$$X(a) = X_{INIT} + \overline{\Phi}\, a \tag{10.28}$$

Combining Equations 10.26 and 10.28, we have

$$X(b) = X_{INIT} + \overline{\Phi}\bar{a} + \overline{\Phi}pb \qquad (10.29)$$

where

$$\overline{\Phi} = \begin{bmatrix} \Phi_1 & 0 & \vdots & 0 \\ 0 & \Phi_2 & \vdots & 0 \\ \vdots & \vdots & \ddots & \vdots \\ 0 & 0 & \vdots & \Phi_s \end{bmatrix} \qquad (10.30)$$

$$P = \begin{pmatrix} P_1 \\ \vdots \\ P_s \end{pmatrix}, \bar{a} = \begin{pmatrix} \bar{a}_1 \\ \vdots \\ \bar{a}_s \end{pmatrix} \qquad (10.31)$$

In Equation 10.30, the columns of any $3NN' \times 3M_s$ matrix Φ_s are the eigenvectors of the spherical mesh describing the surface s. The spatial relation between the different structures, as well as the anatomical variability observed in the training set can therefore be compactly described by a limited number of parameters (typically $m \simeq 10$, corresponding to a compression ratio of about 10,000:1).

The statistically learned deformable model represents the relative location of different anatomical surfaces and it is able to accommodate their significant variability across different individuals. The surfaces of each anatomical structure are parameterized by the amplitudes of the vibration modes of a deformable spherical mesh. For a given image in the training set, a vector containing the largest vibration modes describing the different deformable surfaces is created. This random vector is statistically constrained by retaining the most significant variation modes of its Karhunen-Loéve expansion on the training population. By these means, the conjunction of surfaces is deformed according to the anatomical variability observed in the training set. The joint model has been applied to isolate the brain and segment the different structures included in the model; it is illustrated in Section 10.4.3.

10.4 Medical Image Analysis with Deformable Models

Although originally developed for application to problems in computer vision and computer graphics, the potential of deformable models for use in medical image analysis was quickly realized. Dynamic models are valuable for medical image analysis, since most anatomical structures are deformable and continually undergo nonrigid motion *in vivo*. Moreover, dynamic models exhibit intuitively meaningful physical behaviors. They have been applied to images generated by imaging modalities as varied as x-ray, CT, angiography, MR, and ultrasound. 2-D and 3-D deformable models have been used to segment, visualize, track, and quantify a variety of anatomic structures ranging in scale from the macroscopic to the microscopic [4].

10.4.1 Lesion Detection on Mammography Images

Breast cancer continues to be an important health problem among women population. Early detection is the only way to improve breast cancer prognosis and to significantly reduce women mortality. Therefore, computer-aided mammography is an important and challenging task in automated diagnosis. It is by using computer-aided diagnosis (CAD) systems that radiologists can improve their ability to detect and classify mammography lesions. A project carried on at the UCLM-ISA group has been prompted by this need to develop CAD systems for automatic analysis of breast lesions. In this study, a FCMC algorithm is used to get a preliminary tissue classification to undergo analysis of the breast lesions. Experimental results are given on different mammograms with various densities and abnormalities. A qualitative validation shows the success of the FCMC algorithm for a preliminary segmentation of mass lesions.

Therefore, early detection by mammography analysis is the key to improvement in breast cancer prognosis and application of a proper treatment. Recently, computer-aided mammography screening (CAD systems) has received great attention because of its speed and consistency [18,19]. CAD schemes could provide help and be an efficient tool for radiologist [20]. Although there are several studies on computer-aided mammography, automated interpretation of mammogram lesions still remains very difficult [21]. The major reasons are as follows:

- The size of the images.
- The very low contrast of the regions of interest within the breast tissue.
- The dense tissues may cause suspicious areas to be almost invisible.
- The skin thickening may also cause suspicious areas to be almost invisible.
- The dense tissues may be easily misinterpreted as calcifications.
- The misinterpreted calcifications may yield a high false-positive rate.

To deal with these problems, many methods for automated digital mammography processing have been studied [22–27]. These methods are based on

- Mathematical morphology to analyze features related to tissue lesions.
- Multiresolution image processing to detect micro calcifications.

- Analysis of local edge orientation and texture features to detect spiculated masses.
- Basic analysis of shape properties, including compactness; moments; Fourier descriptors; and intensity, changes to identify tumors.

In this study, the usefulness of texture analysis has been investigated by applying fuzzy logic theory. Several mammogram images were randomly chosen from a generic database [28], with different background tissue, various classes of abnormalities, and their severity. The mammography images shown here are as follows:

a. G: fatty-glandular tissue, ASYM: asymmetry, M: malignant

b. G: fatty-glandular tissue, CALC: calcification, M: malignant

c. G: fatty-glandular tissue, CALC: calcification, B: benign

d. F: fatty tissue, CIRC: well-defined/circumscribed masses, B: benign

e. G: fatty-glandular tissue, CIRC: well-defined/circumscribed masses, M: malignant

f. G: fatty-glandular tissue, MISC: ill-defined masses, M: malignant

g. F: fatty tissue, CIRC: well-defined/circumscribed masses, B: benign

As mentioned earlier, the results of the FCMC algorithm may be quite variable according to the number of selected clusters, c, and the position of the centroids, V. Moreover, proper selection will improve accuracy and convergence of the algorithm. The aim is to locate the mammogram tissues and the lesions by using a proper configuration of both c and V. Figure 10.1 shows the result of the unsupervised algorithm using $c = 15$ clusters, where the initial centroids have been chosen randomly. Figure 10.2 presents the result of the supervised algorithm (in the sense of setting previously the number of clusters c), where the initial estimation of the centroids has been chosen based on a histogram analysis. The number of clusters are $c = 6$ corresponding to the peaks of the histogram. The peaks are found automatically by looking for sequences of pixels in the histogram that follow a peak-like pattern. To avoid false peaks because of the noise, a five-pixel peak pattern is employed. It is possible to see how the algorithm has segmented the different tissues within the mammograms and some selected masses.

This technique has been integrated within a CAD system developed by the group, together with other image-processing tools for mammography image analysis. It has been shown that the FCMC is suitable for mass lesion segmentation. Thus, 50 images with mass lesions were segmented and quantitatively evaluated by experts. The whole segmentation process is automatic and consists of three steps: selection of the mammography tissue, unsupervised FCMC followed by a threshold processing. The threshold is

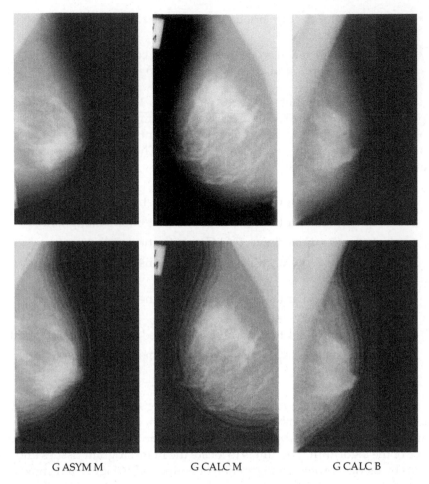

| G ASYM M | G CALC M | G CALC B |

FIGURE 10.1
Unsupervised FCMC segmentation. First row: original images; second row: unsupervised FCMC with $c = 15$ clusters.

automatically set according to both, the gray level and size of the detected regions. The method gives satisfactory results and is quite efficient since it takes less than 1 min to segment the ROI on a 1024 × 1024 image (on a Pentium® 4 CPU 2.66 GHz 512 MB RAM). This process is illustrated in Figure 10.3 with a F CIRC B mass. Figures 10.4 and 10.5 show the original image, the supervised FCMC, and the final segmentation of the mass lesions on a G CIRC M and a G MISC M mammography images.

10.4.2 Segmentation for Radiotherapy Treatment Planning

CT imaging is one of the most widely used radiographic techniques. When a patient undergoes a CT scan, a sequence of 2-D image slices is generated, each of which represents one cross section image of the 3-D human body.

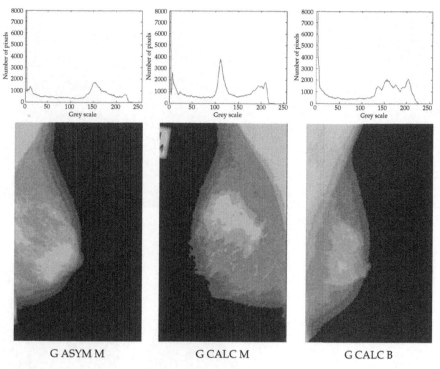

G ASYM M   G CALC M   G CALC B

FIGURE 10.2
FCMC segmentation. First row: histogram of the original images of Figure 10.1; second row: FCMC with $c = 6$ clusters.

One of the main tasks for diagnosis and treatment of diseases is the segmentation of organs of interest, which is the case in radiotherapy treatment planning (RTP) [29,30].

Automatically identifying organs from CT image series is challenging; this is also the case for CTs of the pelvic area, which are the images we are working with. This is due to the following: (1) the images usually have blurred edges and low contrast due to the partial volume effects resulting from spatial averaging, patient movement, and reconstruction artifacts; (2) the fact that different organs may have very similar gray levels, which create an additional difficulty for segmentation between adjacent organs. As a result, general gray-level-dependent image segmentation techniques for edge detection and image classification may not clearly separate overlapped organ regions and correctly extract the desired organs; (3) the administration of contrast media and different machine set up conditions, the same organ may exhibit different gray-level values in different cases or in different image slices of the same case. Such gray-level variations are sometimes significant, making it impossible to apply a simple thresholding technique; and (4) the anatomical structures of an organ in different image slices may be different. Even for the same slice position, its shape may vary significantly from one patient to the other.

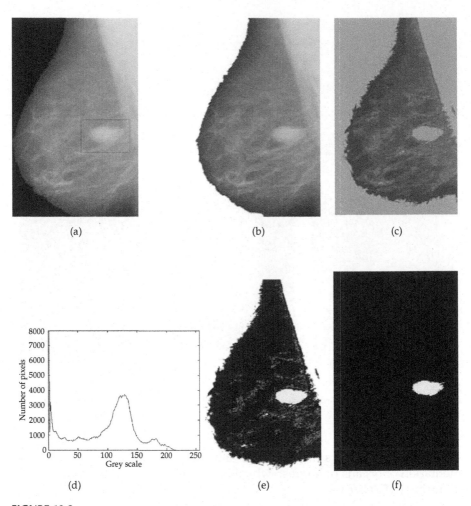

FIGURE 10.3
Benign circumscribed mass lesion segmentation on a fatty tissue. (a) Original image with the ROI selected by a clinician, (b) selection of the mammography tissue, (c) unsupervised FCMC segmentation with five clusters based on the histogram, (d) histogram of the original image, (e) segmentation after applying an intensity threshold of 190 to (c), (f) final segmentation after applying a bilevel threshold to the size of the segmented regions in (e).

The traditional methods to identify and delineate organs are generally done manually; this is time-consuming and a tedious process, besides the results of which are also dependent on the operator. There is a need for CT automatic image segmentation algorithms; to this end different methods have been illustrated in the literature based on edge detection [10,31], texture analysis, deformable models [32], and neural networks [33,34]. They show good results for a few cases, but some problems remain—such as poor boundary segmentation, or overfragmented ROI, or they are still dependent on manual intervention. In this work, carried on under MAESTRO's project

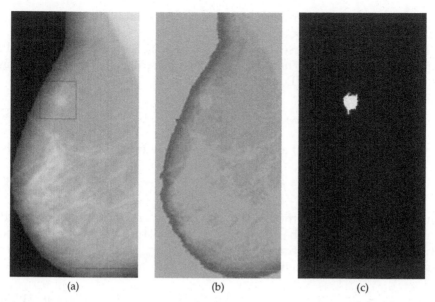

FIGURE 10.4
Malignant mass lesion segmentation on a fatty-glandular tissue. (a) Original image; (b) FCMC five clusters; and (c) segmentation, $T = 185$.

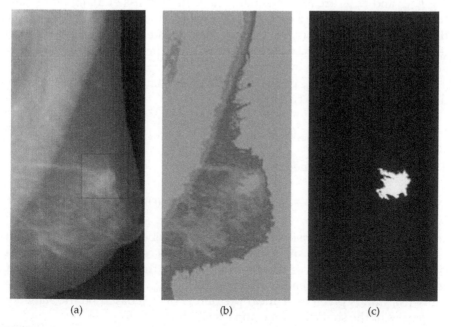

FIGURE 10.5
Malignant ill-defined mass lesion segmentation on a fatty-glandular tissue. (a) Original image; (b) FCMC four clusters; and (c) segmentation, $T = 162$.

framework [35], a hybrid method based on the combination of fuzzy and deformable models has been implemented.

The proposed method uses the model described in Section 10.3, based on the Equations 10.4 and 10.12. Thus, the $E_{internal}$ has been calculated in the same way as the classical model [8] (Equation 10.5) and the $E_{ext\_potential}$ is expressed as

$$E_{ext\_pontential}(s) = E_{images}(s) + E_{external}(s) \qquad (10.32)$$

where E_{images} is expressed as a weighted combination of energy functionals:

$$E_{images}(s) = w_{line} E_{line} + w_{edge} E_{edge} + w_{term} E_{term} \qquad (10.33)$$

The purpose is to attract the snake to lines, edges, and terminations depending on the highlighted characteristics of the structure under consideration. This is achieved by adjusting the weights, w_{line}, w_{edge}, and w_{term} that provide a wide range of snake behavior. The three energy functionals are defined as

$$E_{line} = I_{FCMC}(x, y), \qquad E_{edge} = |\nabla[G_\sigma * I(x, y)]|, \qquad E_{term} = K(x, y) \quad (10.34)$$

where $G_\sigma(x, y)$ is a Gaussian of standard deviation σ, $\sigma = 0.5$ has been used, and $K(x, y)$ is the curvature of the lines in a smoothed image. Thus, $K(x, y)$ is used to find terminations of line segments and corners. Both terms defining E_{edge} and E_{term} are also calculated as done by Kass et al. [8]. The image functional E_{line} is usually defined as the image intensity itself; here we use the FCMC image, $I_{FCMC}(x, y)$. The aim is to create a stronger potential by highlighting ROI edges, where the snake will be attracted toward these edges.

The last term, $E_{external}$, comes from external constraints and it has been defined by a sum of two components:

$$E_{external}(s) = E_{distance}(v(s)) + E_{pressure}(v(s)) \qquad (10.35)$$

The functionals are defined as

$$E_{distance}(v(s)) = \|\mathbf{x_g} - v_k\|, \qquad E_{pressure}(v(s)) = -\frac{\rho}{2} P(I(u)) \qquad (10.36)$$

where $\mathbf{x_g}$ is the fuzzy subset of the class $\mathbf{g}$ as described in Section 10.2, $E_{distance}$ is the distance from the snake points to the ROIs center of gravity enclosed by the initial spline curve (snake region). Thus, the $E_{distance}$ directs the AC toward a user-defined feature. Finally, the $E_{pressure}$ improves the snake stability [36], and is given by a linear pressure, $P(I(x, y))$ based on the statistical properties of the snake region μ and σ such that:

$$P(I(x, y)) = 1 - \frac{|I(x, y) - \mu|}{k\sigma} \qquad (10.37)$$

Thus, these energy functionals are derived for the geodesic AC model following Equation 10.12. To this end, the FCMC is used for the stopping criterion.

$$\begin{cases} g(v(s)) = 1.0 & \text{if } v(s) \in x_j, \ J = ROI \\ g(v(s)) = \dfrac{1}{1 + |\nabla[G_\sigma * I(x, y)]|} & \text{otherwise} \end{cases} \tag{10.38}$$

Therefore, in addition to the criteria of minima energy, the FCMC cluster membership is also taken into account for active contour model convergence. Thus, among the entire minima energy points candidate for final boundary, the preferred ones are those belonging to the corresponding cluster. This cluster is automatically set according to both: the gray level and size of the anatomical structure under consideration.

Thus, a five-peak model was applied to automatically find the maximum and minimum values of the histogram and thereby the set of centroids, c. To achieve an appropriate clustering, including all therapy relevant regions without loosing information, $\{V\} \in [5–15]$, clusters were considered. Figures 10.6b through 10.9b show the results of the FCMC applied to a 2-D CT data set of the

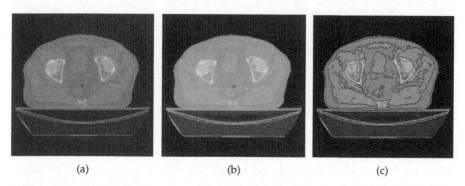

| (a) | (b) | (c) |

FIGURE 10.6
Fuzzy-geodesic segmentation (view four of the sample CT set).

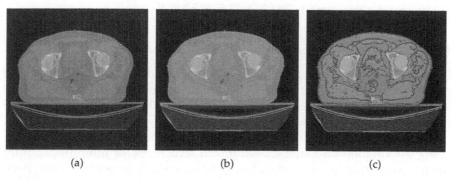

| (a) | (b) | (c) |

FIGURE 10.7
Fuzzy-geodesic segmentation (view five of the sample CT set).

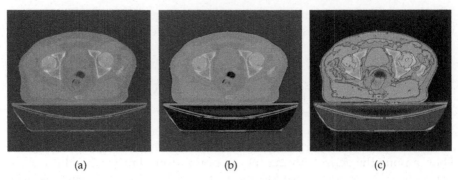

FIGURE 10.8
Fuzzy-geodesic segmentation (view six of the sample CT set).

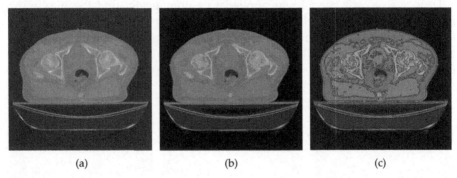

FIGURE 10.9
Fuzzy-geodesic segmentation (view nine of the sample CT set).

human pelvic area. The FCMC is used for the segmentation with the geodesic active contour models as explained before. Figures 10.6c through 10.9c show the results of this fuzzy geodesic deformable model applied to the 2-D CT data set. The results are quite promising, since all the ROI for a RTP of this zone are detected with very low computational time, that is, average of 30 s in a 2-D 512 × 512 image on a Pentium® 4 CPU 2.66 GHz 512 MB RAM.

10.4.3 Three-Dimensional Brain Structures Modeling

Deformable models have principally been used to describe and characterize pathological shape deformations [37], to register single modal images [38], to label and segment different anatomical structures [39,40], or to track temporal structure deformations [15].

In this study, a 3-D statistical physics-based deformable model (SDM) carrying information on multiple anatomical structures for multimodal brain image processing is presented. Several anatomical structures have been taken into consideration, such as the head (skull and scalp), brain, ventricles, and cerebellum [41]. The goal is to describe the spatial relations among these anatomical structures as well as the biological shape variations observed over a representative population of individuals.

In the proposed approach, the surfaces of the anatomical structures of interest are first extracted from a training set of 3-D MRI. To this end, a 24-patients training set is aligned in the same reference coordinate system [42] and segmented using semi-automatic segmentation algorithms [43,44].

The initial model is a sphere mesh initialized around the segmented structure of interest, where the initial spherical mesh is superimposed on the structures to be parameterized. The vibration amplitudes are explicitly computed by Equation 10.22, where rigid body modes $w_i = 0$ are discarded and the nodal displacements may be recovered using Equation 10.17.

The physical representation, $\mathbf{X}(\tilde{\mathbf{U}})$, is finally given by applying the deformations to the initial spherical mesh (Equation 10.23), where a multiplanar view of the deformable models at equilibrium (25% of the vibration modes are kept) is presented.

Thus, the different surfaces of a particular patient are hierarchically described in terms of vibrations of an initial spherical mesh. The next step applies the mentioned parameterization to each patient of the training set and performs statistical learning for the anatomical structures taken into consideration (Equations 10.26–10.29). A set of joint deformations of the model anatomical structures may be obtained by modifying vector $\mathbf{b}$ (Equation 10.26).

The model may be used as a simplified anatomical representation of the images belonging to the training set. If the training set is representative enough of a population, the model may also be used to analyze images of patients not belonging to the training set. To this end, the 24 structures of our database were carefully selected, with the aid of an expert neurologist. Additionally, the database is conceived in such a way that it can be incrementally augmented by new elements. The segmentation of the different structures included in the model is described here as an application of the joint statistical model.

Before presenting this application, note that the equation describing the configuration of the statistical model (Equation 10.29), may be separated into S coupled equations describing the different anatomical parts of the model.

The next step involves the determination of the statistical model parameters $\mathbf{b}$ that best describe the segmented head surface:

$$X_1(\mathbf{b}) = X_0^1 + \Phi_1\bar{\mathbf{a}}_1 + \Phi_1 P_1 \mathbf{b} = X_1(\tilde{U}^1) \tag{10.39}$$

The Equation 10.39 is overconstrained: there are $3NN'$ equations (the head surface coordinates $\mathbf{X}_1$) and m unknowns (the components of $\mathbf{b}$). The overconstrained system is solved by standard least squares, yielding the following pseudoinverse solution for the deformation parameter $\mathbf{b}$:

$$\mathbf{b}^* = [(\Phi_1 P_1)^T \Phi_1 P_1]^{-1} (\Phi_1 P_1)^T [X_1(\tilde{U}^1) - X_0^1 - \Phi_1\bar{\mathbf{a}}_1] \tag{10.40}$$

The other patient anatomical structures surfaces (brain, etc.) are then recovered by introducing the estimated parameter $\mathbf{b}^*$ in Equation 10.40, describing the other parts ($s = 2, 3, 4$) of the statistical model:

$$\mathbf{X}_2(\mathbf{b}^*) = \mathbf{X}_0^2 + \Phi_2 \bar{\mathbf{a}}_2 + \Phi_2 \mathbf{P}_2 \mathbf{b}^*$$

$$\vdots \quad \vdots \quad \vdots \qquad\qquad\qquad (10.41)$$

$$\mathbf{X}_s(\mathbf{b}^*) = \mathbf{X}_0^s + \Phi_s \bar{\mathbf{a}}_s + \Phi_s \mathbf{P}_s \mathbf{b}^*$$

Equation 10.41 provides a prediction of the location of the considered surfaces, obtained by exploiting the spatial relationships between the head and the other anatomical structures, coded in the learned statistical representation. This initial prediction may be further refined by standard iterative refinement algorithms such as energy-based approaches [8], iterative closest point techniques [45], or gray-level profile matching [46].

Some results of the segmentation may be seen in Ref. 43, where quantitative validation of the brain isolation was done by using statistical approaches [41]. The sensitivity, specificity, and accuracy of the method were used as quality measures obtained and 93% of agreement against manual segmentation. The computational time required to (1) parameterize the head surface, (2) estimate the statistical deformation parameters $\mathbf{b}^*$, and (3) predict one structure of interest is, about 5 min cpu time on a standard PC Pentium® 4 CPU 2.66 GHz 512 MB RAM for a 128^3 image volume.

10.5   Conclusions and Further Work

This chapter has described fuzzy and deformable models within the framework of AI Systems applied to some of the most prevalent diseases in our society, that is, cancer (breast and prostate) and neurology.

A FCMC algorithm has been presented for tissue classification on mammogram images. Tissue classification is a necessary step in many medical imaging applications, including the detection of pathology and general CAD. Owing to the nature of mammograms, fuzzy logic appears to be an appropriate choice to process mammogram images. The advantages of FCMC are that it is automatic and robust to work for different types of tissues. The results show that the FCMC algorithm is a good tissue classifier. Further analysis is being carried out to categorize the different breast lesions. To this end, different classification algorithms are being evaluated with a view to being integrated into a prototype intelligent mammography workstation as a computer-aided system for screening mammograms.

The advantage of using fuzzy theory and deformable models by means of a fuzzy geodesic model has been also investigated for segmenting anatomical structures in CT images for RTP. The segmentation approach is based on a geometrical deformable model evolving constrained to a fuzzy intensity image based on a FCMC algorithm. Moreover, the fuzzy reasoning in addition to statistical information is also used for final contour convergence. The model aimed to address some of the drawbacks found in traditional snake models. In particular, it was shown to be independent of initialization, free parameters, and could also preserve the topology. The analysis shows good results yielding gains in reproducibility, efficiency, and time. Further analysis is being performed for all the therapeutic ROI in the pelvic area, to enable their representation using a multiobject physical model.

Finally, a physically based 3-D statistical deformable model embedding information on the spatial relationships and anatomical variability of multiple anatomical structures, as observed over a representative training population has been presented. The model has been used to describe different brain structures (head, brain surface, ventricles, and cerebellum). Preliminary applications of the statistical deformable model included the automatic segmentation of the intracranial cavity (brain isolation). Quantitative validation has shown that a 24-patients trained model was able to provide automatic accurate brain isolations on individuals not belonging to the training set. The major advantage of statistical models is that they naturally introduce *a priori* statistical knowledge that provides useful constraints for ill-posed image-processing tasks such as image segmentation. Consequently, they are less affected by noise, missing data, or outliers. As an example, the statistical deformable model was applied to the isolation of the brain structure from postoperative images, in which missing anatomical structures lead standard voxel-based techniques to erroneous segmentations. The statistical deformable model presented in this chapter may be considered as a first step toward the development of a general-purpose probabilistic anatomical atlas of any part of the human body, for 3-D segmentation, labeling, registration, and pathology characterization.

References

1. Duncan J.S. and Ayache N., Medical image analysis: Progress over two decades and the challenges ahead, *IEEE Trans. PAMI*, 22, 85–106, 2000.
2. Nguyen H.T. and Sugeno M., *Fuzzy Systems: Modeling and Control*, Kluwer, Norwell, MA, 1998.
3. Klir G. and Yuan B., *Fuzzy Sets and Fuzzy Logic Theory and Applications*, Prentice Hall, Englewood Cliffs, NJ, 1995.
4. McInerney T. and Terzopoulos D., Deformable models in medical image analysis: A survey, *Med. Image Anal.*, 2(1), 91–108, 1996.
5. Solaiman B., Debon R., Pipelier F., Cauvin J.-M., and Roux C., Information fusion: Application to data and model fusion for ultrasound image segmentation, *IEEE Trans. Biomed. Eng.*, 46(10), 1171–1175, 1999.
6. Zadeh L.A., Fuzzy sets, *Inf. Control*, 8, 338–335, 1965.
7. Mohamed N., A modified fuzzy C-means algorithm for bias field estimation and segmentation of MRI data, *IEEE Trans. Med. Imag.*, 21(3), 193–200, 2002.

8. Kass M., Witkin A., and Terzopoulos D., Snakes: Active contour models, *Int. J. Comput. Vis.*, 14(26), 321–331, 1998.
9. Ray N., Havlicek J., Acton S.T., and Pattichis M., Active contour segmentation guided by AM-FM dominant componente analysis, *IEEE Int. Conf. Image Process.*, 1, pp. 78–81, 2001.
10. Yu Z. and Bajaj C., "Image segmentation using gradient vector diffusion and region merging, *IEEE International Conference on Pattern Recognition*, Quebec, Canada, (11–15) August, 2, pp. 941–944, 2002.
11. Malladi R., Sethian J.A., and Vemuri B.C., Shape modeling with front propagation: A level set approach, *IEEE Trans. PAMI*, 17, 158–175, 1995.
12. Wang H. and Ghosh B., Geometric active deformable models in shape modeling, *IEEE Trans. Image Process.*, 9(2), 302–308, 2000.
13. Caselles V., Kimmel R., and Sapiro G., Geodesic active contours, *Int. J. Comput. Vis.*, 22(1), 61–79, 1997.
14. Dubrovin B.A., Fomenko A.T., and Novikov S.P., *Modern Geometry, Methods and Applications I, Springer-Verlag*, Heidelberg, 1984.
15. Nastar C. and Ayache N., Frequency-based nonrigid motion analysis: Application to four-dimensional medical images, *IEEE Trans. Pattern Anal. Machine Intell.*, 18, 1069–1079, 1996.
16. Borgefors G., On digital distance tranforms in three dimensions, *Comput. Vision Image Understanding*, 64(3), 368–376, 1996.
17. Pentland A. and Sclaroff S., Closed-form solutions for physically based shape modeling and recognition, *IEEE Trans. Pattern Anal. Machine Intell.*, 13, 730–742, 1991.
18. Taylor P., Champness J., Given-Wilson R., Johnston K., and Potts H., Impact of computer-aided detection prompts on the sensitivity and specificity of screening mammography, *Health Technol. Assess.*, 9(6), 1–58, 2005.
19. Alberdi E., Povykalo A., Strigini L., and Ayton P., Effects of incorrect computer-aided detection (CAD) output on human decision-making in mammography, *Acad. Radiol.*, 11(8), 909–918, 2004.
20. Jiang Y., Nishikawa R., Schmidt R., Toledano A., and Doi K., Potencial of computer-aided diagnosis to reduce variability in radiologist interpretations of mammograms depicting microcalcifications, *Radiology*, 220, 787–794, 2001.
21. Petrick N., Sahiner B., Chan H., Helvie M., Paquerault S., and Hadjiiski L., Breast cancer detection: Evaluation of a mass-detection algorithm for computer-aided diagnosis-experience in 263 patients, *Radiology*, 224, 217–224, 2002.
22. Cheng H.D., Chen C.H., and Chiu H.H., Fuzzy homogeneity approach to image thresholding and segmentation, *Inform. Sci.: An Int. J.*, 98(1–4), 237–262, 1997.
23. Mendez A.J., Tahoces P.G., Lado M.J., and Soute M., Computer-aided diagnosis: Texture features to discriminate between malignant masses and normal breast tissue in digitized mammograms, *CAR'97 Comput. Aided Radiol. Int. Conf.*, 1, 342–346, 1997.
24. Pham D.L. and Prince J.L., Adaptive fuzzy segmentation of magnetic resonance images, *IEEE Trans. Med. Imag.*, 18(9), 737–752, 1999.
25. Polakowski W.E., Cournoyer D.A., Rogers S.K., DeSimio M.P., Ruck D.W., Hoffmeister J.W., Raines R.A., Computer-aided breast cancer detection and diagnosis of masses using difference of gaussians and derivative-based feature saliency, *IEEE Trans. Med. Imag.*, 16(6), 811–819, 1997.
26. Kobatake H. et al., Computerized detection of malignant tumors on digital mammograms, *IEEE Trans. Med. Imag.*, 18(5), 369–378, 1999.

27. Ferrari R.J. et al., Analysis of asymmetry in mammograms via directional filtering with gabor wavelets, *IEEE Trans. Med. Imag.*, 20(9), 953–964, 2001.

28. Mammography Image Analysis Society, *Digital Mammogram Database*, Manchester, U.K., 1994.

29. Haas O., *Radiotherapy Treatment Planning. New System Approaches*, Springer-Verlag, Heidelberg, 1998.

30. Purdy J.A., 3D treatment planning and intensity-modulated radiation therapy, *Oncology*, 13, 155–168, 1999.

31. Bueno G., Fisher M., Burnham K., and Haas O., Automatic segmentation of clinical structures for RTP: Evaluation of a morphological approach, *MIAU Int. Conf. U.K.*, 22, 73–76, 2001.

32. Tsagaan B., Shimizu A., Kobatake H., Miyakawa K., and Hanzawa Y., Segmentation of kidney by using a deformable model, *Proc. Int. Conf. Image Process.*, 3, 1059–1062, 2001.

33. Koss J.E, Newman F.D., Johnson T.K., and Kirch D.L., Abdominal organ segmentation using texture transform and a Hopfield neural network, *IEEE Trans. Med. Imag.*, 18(7), 640–648, 1999.

34. Lee C., Chung P., and Tsai H., Identifying multiple abdominal organs from CT image series using a multimodule contextual neural network and spatial fuzzy rules, *IEEE Trans. Inf. Technol. Biomed.*, 7(3), 208–217, 2003.

35. Barthe J., Methods and advanced equipment for simulation and treatment in radio oncology, http://www.maestro-research.org/index.htm, 2005.

36. Ivins J. and Porrill J., Active region models for segmenting medical images, *IEEE Trans. Image Process.*, 227–231, 1994.

37. Martin J., Pentland A., Sclaro S., and Kikinis R., Characterization of neuropathological shape deformations, *IEEE Trans. Pattern Anal. Machine Intell.*, 20(2), 97–112, 1998.

38. Gee J., Reivich M., and Bajcsy R., Elastically deforming 3D atlas to match anatomical brain images, *J. Comput. Assist. Tomogr.*, 17(2), 225–236, 1993.

39. Zeng X., Staib L., Schultz R., and Duncan J., Segmentation and measurement of the cortex from 3D MR Images using coupled-surfaces propagation, *IEEE Trans. Med. Imag.*, 18(10), 927–937, 1999.

40. Thompson P., MacDonald D., Mega M., Holmes C., Evans A., and Toga A., Detection and mapping of abnormal brain structure with a probabilistic atlas of cortical surfaces, *J. Comput. Assist. Tomogr.*, 21(4), 567–581, 1997.

41. Bueno G., Heitz F., and Armspach J.P., 3D segmentation of anatomical structures in MR images on large data sets, *Magn. Reson. Imag.*, 19, 73–88, 2001.

42. Nikou C., Armspach J.P., Heitz F., Namer I.J., and Grucker D., MR/MR and MR/SPECT registration of brain images by fast stochastic optimization of robust voxel similarity measures, *Neuroimage*, 8(1), 30–43, 1998.

43. Bueno G., Nikou C., Heitz F., and Armspach J.P., Construction of a 3D physically-based multi-object deformable model, *ICIP 2000, Proc. IEEE Int. Conf. Image Process.*, 268–271, 2001.

44. Nikou C., Bueno G., Heitz F., and Armspach J.P., A joint physics-based statistical deformable model for multimodal brain image analysis, *IEEE Trans. Med. Imag.*, 20(10), 1026–1037, 2001.

45. Declerck J., Subsol G., Thirion J.P., and Ayache N., Automatic retrieval of anatomical structures in 3D medical images, *Technical Report 2485, INRIA*, 1995.

46. Kelemen A., Szekely G., and Gerig G., Elastic model-based segmentation of 3D neuroradiological data sets, *IEEE Trans. Med. Imag.*, 18(10), 828–839, 1999.

11

Texture Analysis and Classification Techniques for Cancer Diagnosis

James Shuttleworth, Alison Todman, Raouf Naguib,
Robert Newman, and Mark K. Bennett

CONTENTS

A common application of automated systems in medicine is image analysis. Such systems have been applied to the investigation of skin lesions and tissue biopsies, analysis of various forms of medical imaging, and many less patient-centered applications such as pill counting.

In many cases, morphological analysis of image content can be difficult or unsuitable. Images without clear delineation of components make it difficult to automatically measure morphological features, as do images in which there is a large variation in the shape of the components.

One form of image analysis that has proved to be useful in these cases is texture analysis, especially in the analysis of digitized tissue slides.

The texture of an image, as perceived by a human, is a very subjective property making it very difficult to assess accurately. Words such as fine, course, smooth, rough, regular, or irregular are commonly used to describe texture, but these terms are subjective. Texture analysis is a way to extract reproducible values for features that can describe an image in terms of "coarseness," "smoothness," etc.

There are many texture analysis techniques, ranging from the purely statistical to those that incorporate morphological information in the form of texture elements or texels.

Automatic classification based on image features is a common image analysis problem. Classification is a matter of analyzing the measurable features and finding a way to use them to predict class membership. The classes might be types of tumor, prognosis, or severity. In trivial cases, there is a strong and direct correlation between one of the features and the classification, and it is a simple matter to scale the feature to create a good predictor. Unfortunately, trivial cases in real-world applications are rare and we must rely on more complex methods based on multiple features to predict classification.

This chapter provides an introduction to texture analysis and statistical techniques for image classification, followed by a detailed example of a system for automated staging of dysplasia in colorectal tissue that combines multiscale color texture analysis with discriminant analysis.

11.1 Texture Analysis

Morphological analysis allows us to extract features from images that have shapes that are either delineated by a user or found using preprocessing. In many situations, this shape-based approach can be useful in the analysis of an image. However, there are many types of images that do not have clearly delineated components, making morphological analysis difficult or impossible.

Before examining the medical case, first consider a general example of texture in images. Imagine that a method for estimating the number of trees in an image is implemented using morphological analysis. First, assuming that the images will all contain trees with green leaves, we may ignore all pixels that do not have a high green component. This would leave us with (hopefully) grass and tree pixels only. Each separate area of green could then be taken to be a possible tree, and morphological analysis could be used to decide whether each blob of green is tree-shaped enough to be counted as a tree. This algorithm may work perfectly well when the trees are separate and not seen against a green background (such as grass), but beyond that, it falls apart.

Now imagine that we had data on the texture of trees—we could use this to select pixels likely to contain trees and make an estimate on tree numbers from the area that they cover. The texture analysis approach is not going to be confused by partially occluded trees, trees against grass or clumps of trees that touch. In fact, if the data we have on tree textures are detailed enough, we could possibly detect different types of trees or spot those that need attention—a pine and a sycamore would have different textural properties, as would a tree getting plenty of water and one that has drooping leaves.

Texture is often described as "rough," "smooth," "regular," etc., but such terms are not enough to accurately describe most textures—for example, the difference in the texture of an image of a sycamore tree and a pine tree (Figure 11.1). We could say that the sycamore has a "rougher" texture than the pine, or maybe "finer" is a better description. Adding more tree types would make this even more difficult: How can we describe textures accurately, such that the descriptions can be used later to distinguish between trees of different types?

Apart from the inability of such descriptions to describe texture in anything but the broadest of terms, these descriptions are subjective. One person might say that one image is twice as rough as another, whereas another person might say that it is three times as rough, or even that it is not rougher, but finer or courser.

Pine   Sycamore

FIGURE 11.1
Tree textures.

Because of the inaccurate and subjective nature of such descriptions, it is necessary to have a more formal and reproducible method for analyzing texture. Many different techniques have been developed to analyze and describe texture in images, but they can all be classified as statistical, structural, or spectral techniques.

The following sections discuss each class of texture analysis along with a description of some of the more common techniques. For a more thorough review of texture analysis, see Haralick (1986) and more recent publications such as Tuceryan and Jain (1998) and Singh and Singh (2002).

11.1.1 Statistical Texture Analysis

Statistical texture analysis "generates parameters to characterize the stochastic properties of the spatial distribution of gray levels in an image" (Haralick, 1986). In other words, statistical texture analysis is the analysis of the intensity and location of pixels in an image.

Additionally, there are simple techniques that ignore even the spatial distribution of gray levels and analyze only the gray-levels.

11.1.1.1 Histogram Texture Analysis

Many of the techniques that incorporate no spatial information are based on histogram statistics such as simple measures of mean, median, or skew. These first-order histogram features are very efficient in terms of processing speed and memory, but do not correlate well with human perception of texture. For example, examine Figure 11.2. The first image is that of sharp noise, what we would call a rough texture. Images 2, 3, and 4 are generated by repeatedly applying a Gaussian filter to produce progressively "smoother" images. They are placed in order of the skewness feature. Skewness, in this case, was calculated as the histogram mean–median. There appears to be a correlation between skewness and "smoothness"—greater skew seems to indicate a smoother image.

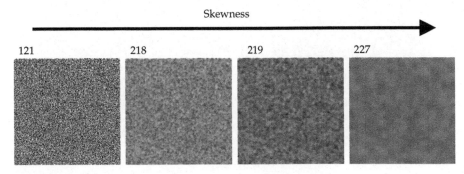

FIGURE 11.2
Skewness illustration.

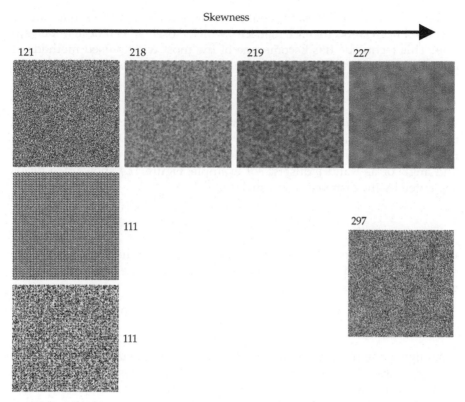

FIGURE 11.3
Skewness does not necessarily correlate with the human subjective idea of 'smoothness'.

However, this is not always the case. Figure 11.3 shows that an image that would not be described as smooth can still have a high skewness measurement. It also illustrates how texture features that ignore spatial data completely cannot differentiate between textures with equivalent histograms but different textures.

This poor correlation between human texture perception and first-order statistics arises because human texture discrimination additionally uses second-order features and, in some cases, higher order features (Julesz, 1962).

11.1.1.2 Co-Occurrence Matrices

Co-occurrence matrix-based texture features are second-order methods that incorporate spatial information. Co-occurrence matrices contain the number of pixels of any pair of intensities separated by a given displacement vector as described in detail in the following sections. From this information, more reliable texture features can be extracted that are dependent on pixel organization, as well as intensity distribution, unlike the histogram-based techniques described in Section 11.1.1.1.

Haralick et al. (1973) suggested a set of textural features that can be extracted from co-occurrence matrices, which closely match human perception. This technique has become one of the most widely used methods of texture analysis.

11.1.1.2.1 Distance Vectors

A co-occurrence matrix contains the number of occurrences of pixels with two given intensities separated by a distance vector. The distance vector maps a source pixel to a target pixel and can be described either as a distance and angle or as x and y offsets. For example, Figure 11.4 shows two pixels separated by the distance vector such that

$$\theta = 117.57° \quad d = 2.24 \text{ pixels}$$

where θ is the angle (measured from the vertical) from the first pixel a to the second pixel b, and d is the distance between them. The same relationship can be expressed using offsets as

$$x\text{offset} = 2 \quad y\text{offset} = 1$$

Although either representation can be used, because distances between pixels are discrete, it is simpler to select offsets than angles and distances. To find an angle and distance that point to exactly one pixel, we must first choose that pixel and calculate the distance vector, whereas any two integers used as offsets will always point to precisely one pixel. Choosing arbitrary angles and distances usually results in the target of the relationship being a weighted average of up to four pixels. For example, if we choose $\theta = 40°, d = 3$, then the target of the relationship lies 1.93 pixels to the right of the source pixel and

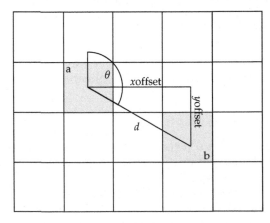

FIGURE 11.4
Two pixels and the distance vector between them.

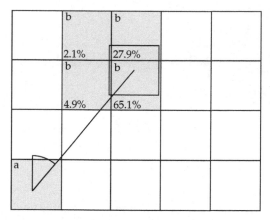

FIGURE 11.5
Calculating the intensity of a target pixel that covers more than one pixel in the image.

2.30 pixels above it (see Figure 11.5) and the intensity of that pixel must be calculated as 4.9% of the pixel 1 to the right and 2 above the source, 65% of the pixel 2 to the right and 2 above the source, 2.1% of the pixel 1 to the right and 3 above the source, and 27.9% of the pixel 2 to the right and 3 above the source. Hence, the offset approach is adopted throughout the remainder of this chapter.

11.1.1.2.2 Constructing the Matrix

A co-occurrence matrix, P, generated from an image I will be square with sides of length O, where O is the number of gray levels in I. The contents of the matrix are defined by

$$P(m,n) = |((i,j),(k,l)) \in S \wedge I(i,j) = m \wedge I(k,l) = n|$$

where S is the set of pairs in I that fit the distance vector. For example, in an image where there are 5 pairs of pixels separated by the given distance vector such that the source pixel has intensity 1 and the target pixel has intensity 10

$$P(1,10) = 5$$

Figure 11.6 shows (on the left-hand side) the intensities of an image 4 × 4 pixels in size with four possible intensities. Using the distance vector,

$$xoffset = 1, \quad yoffset = 0$$

the table on the right-hand side of Figure 11.6 shows the resulting co-occurrence matrix. Start pixel intensities are shown down the left-hand side of the matrix and target pixel intensities are along the top. From the matrix, it

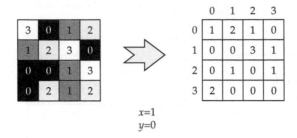

$x=1$
$y=0$

FIGURE 11.6
Example of co-occurrence matrix calculation.

can be seen that two pairs of pixels fitting the distance vector exist such that the source pixel has intensity 3 and the target pixel has intensity 0.

11.1.1.2.3 Normalization

The sum of the elements in a co-occurrence matrix is equal to the number of pixel pairs that fit the distance vector in the image. Hence, images of different sizes could have different values for measured texture features, although they may have the same texture. To prevent this, co-occurrence matrices should be normalized.

Normalization is achieved simply by dividing each cell in the matrix by the total number of valid pixel pairs. The resulting normalized gray-level co-occurrence matrix then contains probabilities rather than absolute values, that is, if the cell $(5,5)$ contains the value 0.5, then for a given pixel of intensity 5, the pixel related to it by the distance vector has a 50% chance of also having intensity 5.

11.1.1.2.4 Binning

Images with a higher brightness resolution (i.e., more possible intensities or gray-levels) produce sparses co-occurrence matrices. Because more densely populated matrices tend to produce more reliable texture measurements, binning can be employed to "shrink" the matrix. Binning is a simple procedure in which the total range of values is divided by some smaller amount—the required number of bins. The matrix dimensions will then be defined by the number of bins and each intensity in the image will fall into one of these bins. For example, an image with 256 possible intensities may be used to populate a co-occurrence matrix of length 8. When a pixel of intensity 0–31 is encountered, it is translated into the co-occurrence matrix as 0, pixels of intensity 32–63 will be translated to 1, and so on.

11.1.1.2.5 Obtaining Features

Co-occurrence matrices do not directly provide any measure of texture that can easily be used for classification. Texture information is rather derived from statistical relationships such as entropy, correlation, or homogeneity, calculated from the values in the co-occurrence matrix.

The arrangement of co-occurrence matrices allows some reasoning to be applied to what the values and their locations indicate about the source image. Contrast, for example, is a measure of pixel dissimilarity. High-contrast images tend to yield co-occurrence matrices in which the majority of values lie away from the diagonal running from $(0,0)$ to (n,n) for a matrix using n bins, because along this line run the pixel pairs in which both pixels have the same value.

A list of commonly used textural features extracted from co-occurrence matrices is as follows:

Entropy: $$\sum_i \sum_j P(i,j) \log P(i,j)$$

Contrast: $$\sum_i \sum_j (i-j)^2 P(i,j)$$

Correlation: $$\sum_i \sum_j \frac{(i-u_x)(j-u_y) P(i,j)}{\sigma_x \sigma_y}$$

Homogeneity: $$\sum_i \sum_j \frac{P(i,j)}{1+|i-j|}$$

Dissimilarity: $$\sum_i \sum_j P(i,j) |i-j|$$

Angular second moment (ASM): $$\sum_i \sum_j P(i,j)^2$$

Energy: $$\sqrt{\mathrm{ASM}}$$

Horizontal mean (μ_x): $$\sum_i \sum_j i P(i,j)$$

Vertical mean (μ_y): $$\sum_i \sum_j j P(i,j)$$

Horizontal variance (σ_x^2): $$\sum_i \sum_j P(i,j)(i-\mu_x)^2$$

Vertical variance (σ_y^2): $$\sum_i \sum_j P(i,j)(j-\mu_y)^2$$

Horizontal standard deviation (σ_x): $\sqrt{\sigma_x^2}$

Vertical standard deviation (σ_y): $\sqrt{\sigma_y^2}$

Where P is a normalized co-occurrence matrix and $P(i,j)$ is the probability of the source pixel having value i and the target pixel having value j for any pixel pair fitting the distance vector used to populate the matrix.

11.1.1.2.6 Directionality

The general method for calculating co-occurrence matrices is direction or rotation dependant, that is, the distance vector works only in one direction, and rotating the image is likely to change the results in the co-occurrence matrix. Rotational dependence can be significantly reduced (Shuttleworth et al., 2002a) by calculating co-occurrence matrices for the given distance vector and for the vector rotated by 90°, 180°, and 270°. Transposing a co-occurrence matrix results in the matrix that would be calculated using the same distance vector rotated by 180°, and so it is possible to calculate two instead of four matrices and use a computationally cheaper transposition function to determine their opposites. As an example of the effect of rotational dependence, Figure 11.7 shows the response of homogeneity calculated from co-occurrence matrices generated from the image in Figure 11.8. Note that the second value of homogeneity varies much less than the direction-dependant one. The original image (shown in Figure 11.8) is a 100 × 100 pixel image of tissue taken from a colon tumor. The image has been rotated by 360° in 30° steps and a central region of 70 × 70 pixels extracted—the largest square subregion that can be extracted at all rotations.

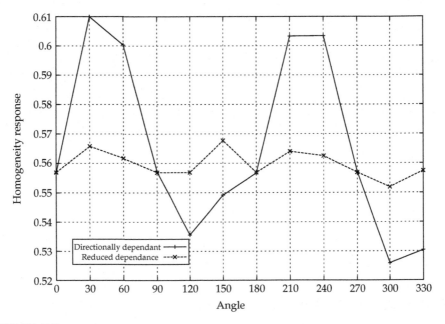

FIGURE 11.7
Homogeneity response from a rotated image using direction-dependant and reduced-dependence techniques.

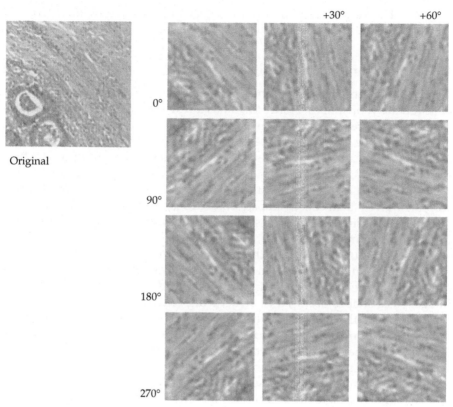

FIGURE 11.8
Colon tumor tissue image rotated at 30° intervals.

11.1.1.2.7 Advanced Co-Occurrence Matrix Techniques

Co-occurrence matrices have been used in texture analysis for more than 50 years, and many advances have been made within this period. A brief discussion of these is also beyond the scope of this chapter, but interested readers may wish to start with a general multiscale co-occurrence matrix technique (Metzler et al., 2000), dual-scale color texture analysis for microscopy image analysis (Shuttleworth et al., 2002b), or techniques to examine co-occurrence matrices before features are extracted from them (Walker et al., 1995; Zucker and Terzopoulos, 1980).

11.1.1.3 Gradient Analysis

Gradient analysis uses the local contrast of an image to assess the coarseness of texture. Local contrast can be easily computed across an image using convolution kernels such as Roberts, Prewitt, or Sobel. These kernels detect "edges" in an image—areas where light and dark meet. For any particular edge, the greater the contrast between either side, the greater the response to the filter. The average response for the image or image region then gives an indication of the coarseness of the image or region.

11.1.2 Structural Texture Analysis

Structural texture analysis is concerned with shape. Texture can be thought of as a pattern of shapes. Figure 11.9b has a texture that could be defined as a black square repeated and translated across the image. More complicated textures require more complicated descriptions; however, there are a number of techniques that have been developed to extract this structural texture from images.

One approach to structural texture analysis defines texture in terms of texture primitives (also called subpatterns, texons, or texels). Once these have been identified in the image, rules for repetition that model the texture found in the original image are created.

A similar, but simpler, method of structural texture analysis uses mathematical morphology to determine the repetitiveness of shapes in an image. Rather than attempting to discover dominant texels within an image, basic shapes are used as structuring elements in an erosion operation. Counting the remaining pixels gives an indication of the occurrence of shapes similar to the structuring element in the texture of the image. A feature vector may be composed of number of pixels found in the image after erosion for each selected structuring element.

Mathematical morphology usually relies on a binary image being the source, although there are techniques for gray-scale morphology. A major limiting factor is the reliance on selecting a set of structuring elements that

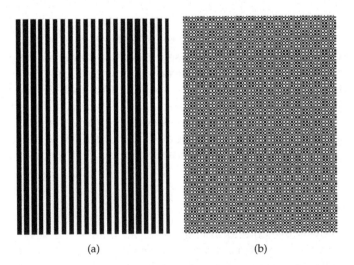

(a)   (b)

FIGURE 11.9
(a) and (b) have the same ratio of light and dark pixels, but the texture of the two images is qualitatively different.

are good for all images, rather than extracting shape information directly from the image.

It has already been noted that spatial information is important in determining image texture and indeed these methods rely heavily on shape. It should be noted, however, that intensity is important in gray-level texture analysis and in these techniques; shape is used almost to the exclusion of variations in intensity.

11.1.3 Spectral Texture Analysis

The statistical and spatial texture analysis methods discussed earlier are all applied to the spatial domain of an image, that is, they process two-dimensional matrices of intensities.

However, texture is characterized by repeating shapes and patterns and variations in local and global intensities, which can also be represented in terms of frequency information. It is possible, therefore, to analyze texture using spectral texture analysis techniques.

There are many different applications of spectral analysis in texture analysis, and a detailed discussion of all of them is beyond the scope of this book. Instead, we present the key features of the main spectral texture analysis methods: Fourier analysis, wavelets, and Gabor filters.

Fourier analysis is the use of the Fourier transform to translate an image to and from the spectral domain. The Fourier transform decomposes an image into a summation of horizontal and vertical cosine functions such that the interference between the combined functions reproduces the original image. The resulting collection of functions is usually represented as another image as shown in Figure 11.10. In the first transform, the composition is quite simple because there is only one shape and it is smooth. In the transform, the intensity shows the strength of a component, the distance from the center shows its frequency, and angle about the center is the orientation of the function. In the first example, there are large symmetrical low-frequency components and not others. In the second example, we have many high-frequency components at all angles because of the sharpened edges of the circle.

It is possible to translate an image into the frequency domain as well as reverse the process with an inverse Fourier transform. This allows modifications to be made to the frequency components of images before returning to the spatial domain.

There are many ways to apply the Fourier transform in texture analysis. It is possible to extract measurements directly from the frequency domain or even to apply operations such as masking before returning to the spatial domain. Features extracted from the frequency domain can be selected for particular frequencies by measuring the strength of components in a ring around the center at a given distance. Orientation can be measured by measuring the combined strength of components in a given "wedge" from the frequency domain. Variations and other texture applications of Fourier analysis include a "warping" tolerant method of texture analysis for natural images (Hsu et al., 1993).

Just as Fourier analysis allows the analysis of frequency and orientation in an image, wavelet analysis also allows the analysis of scale. In other words, wavelets allow the image to be analyzed based on scale as well as frequency. Fourier analysis makes use of sine and cosine basis functions, whereas wavelet analysis uses finite basis functions that are better suited to high-frequency data—in the case of image analysis, noisy images, or images with edges. This also means that it is possible to analyze local texture, which is not possible with Fourier analysis because all the resulting components act globally. Detailed information on wavelet-based texture analysis can be found in Scheunders et al. (1998) and Tuceryan and Jain (1998).

It is often necessary to analyze local texture, a job for which Fourier analysis is not well suited, and wavelets might not be an alternative. In this case, it

FIGURE 11.10
Example of Fourier transforms.

is possible to use Fourier analysis to examine texture through a window that moves across the image, just as a convolution kernel is applied repeatedly with its center over each pixel in turn. Fourier transforms are affected by discontinuities between opposing image edges however, it is necessary to assume that the image is repeated indefinitely in either direction, just as the component functions are repeated. In a full-sized image, these effects can be fairly problematic but in a small window repeated across the image, the effects can make this approach unusable. The Gabor transform reduces the effect of window edges by using a soft-edged Gaussian window. See Tuceryan and Jain (1998) for further details.

11.1.4 Color Texture Analysis

All of the methods of texture analysis discussed earlier have been based on gray-scale images, that is, the images being analyzed had only one channel: luminance. In many cases, there is information in the patterns of distribution of other properties of the image, such as hue or individual color components.

Many color models have been developed, mostly based on either color mixing (such as red, green, and blue—RGB) or measurements of chrominance and luminance (such as hue, saturation, and brightness—HSB). Whichever color model is used, the total information in the image remains the same, but the way in which the individual pixels are expressed changes.

Applying texture analysis to color images can be as simple as selecting one or more channels from an appropriate model or models. Research suggests that the selection of color model can have an effect on the outcome of texture analysis (Singh et al., 2002; Drimbarean and Whelan, 2001).

11.2 Classification

Classification is often an important stage in image understanding. Feature analysis usually produces quantitative measurements of particular characteristics of an image, but does not necessarily provide any indication of how these features can be used to reason about the image.

For example, in Section 11.1, we discussed different ways of extracting texture features from an image, most of which can measure more than one feature. To use these features to decide whether a given image contains sycamore trees or pine trees, we need to know which feature or combination of features vary with tree type, and the way in which they vary. In the case of the simple sycamore or pine example, there may be one obvious feature that always results in a value close to 0 for a sycamore and a value close to 1 for a pine, making classification a trivial process. In a more complex situation, for example, with 5–10 types of trees, it is unlikely that a single feature will correlate well with a tree type. One way of dealing with this is to

i. Use statistical analysis to first find the combination of features that vary with class membership and

ii. Use a classifier that combines these features to predict class membership.

Unsupervised grouping is possible, although in most cases it is based on a set of given classes rather than those that appear to exist after analysis. In these cases, it is necessary to have some prior knowledge to select features that can be used for classification. This is often in the form of a set of images that have already been classified by a field expert.

The information presented in the following sections should be considered a starting point only—interested readers should consult more specialized texts of Breiman (1984), Schürmann (1996), and Raudys (2001).

11.2.1 Problems with Human Classification

Human classification of images is often a time-consuming task, which makes automated classification of images desirable. In many areas of medicine, the number of images that often need to be examined visually is immense, and automated classification is desirable. Quality control often requires visual examination of products for cracks, fissures, omissions, defects, or tone variance. Analysis of tissue, fluids, or films is a necessary, but time-consuming part of many areas of medicine.

However, the human visual system suffers from fatigue when throughput is high and there is frequent variability between individual assessments of an image. In some areas, producing a "gold standard" against which to measure the success of an automated system is therefore difficult, for example, the examination of dysplasia in histopathological images (Bosman, 2001; Eaden et al., 2001; Coppola and Karl, 1999).

In these cases, it can be difficult to produce a reliable known-good set of training images, and it is often necessary to attempt to emulate a single domain expert's classification to produce a reproducible and reliable classification system.

11.2.2 Selecting Features

Feature selection is the process by which a potentially large set of features is reduced to the smallest set of features that can be used to classify the given training set with minimum classification error.

To determine the best set of features, it is necessary to define a criterion function. The result of this function determines how "good" a set of features is. This is almost always tied to the classification step. For example, let F be the complete set of features, and F_s a subset of those features. $C(F_s)$ gives the classification accuracy when applied to F_s, where 1 is 100% accurate and 0 is 0% accurate. A sensible function for feature selection would then simply be $C(F_s)$, and the set of features that is selected would be the set with the maximum value of $C(F_s)$.

In simple cases (where F is small), it is possible to simply try every combination of the elements of F. As the size of F increases, however, the cost of attempting classification for every combination goes up. For example, with 3 features, there are only 9 combinations, but with 4 features there are 15 combinations, and with just 10 features, there are 1023 possible combinations of features. Because of this, many feature selection techniques use more complex methods than an exhaustive search to select features, and are more closely tied to the classification step.

Because of this close relationship between feature selection and classification, feature selection is discussed with each associated classification algorithm, whereas concepts that apply to most or all of the feature selection and classification algorithms are discussed here.

11.2.2.1 Feature Reduction

In many cases where a system for classification is based on measured features, features are extracted without any prior knowledge of how well they will correlate with class. In these cases, it is common to extract many features—perhaps all measurable features. With such a large set of features, it can be helpful to reduce the set to a more manageable size, especially if the classification technique does not have a feature selection step. Without actually using classification, it is sometimes possible to reduce the feature set by removing features that correlate with one another—features that essentially measure the same thing, although possibly on different scales.

Factor analysis or principal components analysis (PCA) can be used to combine such features into a single feature that represents either most of the variability of the removed features or all of the variability of the removed features. This also has the added benefit that all the features in the new set will be orthogonal, that is, they will be independent features that measure different properties. Detailed explanation of factor analysis and PCA is beyond the scope of this chapter, and there are many texts devoted to the subject. See Kline (1993) for a simple introduction to the subject.

Multidimensional scaling (MDS) can be used in a similar manner to PCA, but allows more control over the dimensionality of the resulting feature set.

Other techniques useful for analyzing and manipulating sets of features include multiple regression and multiple correspondence analysis.

11.2.2.2 Multivariate Normality

All of the reduction techniques discussed earlier, and many classification techniques, are based on the multivariate general linear hypothesis, and thus generally require variables to be—or be close to—"multivariate normal." The key requirements for multivariate normality are

- *Skewness (a measure of symmetry)*. It must be kept to a minimum, although there are methods to compensate for this (translation) and some classification techniques are somewhat robust in this respect.
- *Kurtosis*. It measures the "peakedness" of a data set. Again, this must be kept to a normal level.
- *Orthogonality*. Its variables must be independent. Some classification techniques, such as discriminant analysis, are not troubled by associated variables. Also referred to as multicollinearity or, if variables are 100% identical, singularity.
- *Homoscedasticity*. Its variability is roughly equivalent for all variables. Many techniques will be unreliable if this is not true, that is, if the data are heteroscedastic.

For data sets that do not fit the requirements of a particular classification technique, there are a few ways to analyze the data. If the problem is only with collinearity, then feature reduction using any of the techniques described earlier will remove the offending variables, or techniques that select features that progressively account for remaining variance, such as discriminant analysis, can be used.

Skewness or kurtosis can be removed using translation—rescaling the data to make it fit the normal model. Nonlinearity might suggest the use of a nonlinear classification method or translation of the data.

11.2.3 Classifying Feature Vectors

Having a set of features and a number of cases with predetermined classifications, the next step is to attempt to use this information to classify new cases automatically based on their feature vectors.

This could be quite simple—one of the features may correlate perfectly with class membership. However, in most useful applications, this is not the case. Just as there is the possibility that determining class membership is trivial, there is also a possibility that the extracted features have absolutely no correlation with class membership, either individually or as parameters to some tailored classification function. Luckily, most cases fall between these extremes, but the problem of determining which features, combined in which fashion, provide the best predictor of classification is not trivial.

The methods described in later sections are the most commonly used methods of automated classification. We will describe each method, but first we will discuss some common points.

Because we are concerned with image classification based on a set of image analysis features, we refer to the set of features used to classify an image as a feature vector, and the partition of the classification scale to which the image belongs as the class or classification. The methods discussed in later sections are not image analysis tools, they are statistical tools, and so the nomenclature in much of the associated literature will be different. Feature vectors are sets of independent variables, from which the dependant variable (class) is to be inferred.

Most methods for automated classification can be divided into two distinct parts or phases. The first phase is investigatory and generally needs to be applied only once to any problem—assuming, of course, that the end result is a usable classification system. This first phase is the discovery of a classifier: a combination of features that can be used to predict class membership. To construct the classifier, it is usually necessary to have a set of cases with known classifications. This set of cases is called the learning or training set and it is to this set that the classifier is molded such that given a feature vector, the classifier predicts the correct class.

The second phase is classification. In this phase, the classifier is fed feature vectors from new (i.e., not in the training set) cases to predict classification. In almost all applications of classifiers, a second set of feature vectors from known cases is presented to the classifier to determine its efficacy before any "real-life" cases are presented. This is necessary to ascertain the reliability of the classifier or to show that the classification determination phase was erroneous.

Most linear classification methods result in a weighted sum of some or all of the features, which, unlike the classifier discovery phase are computationally cheap.

Not all methods fit this division perfectly. For example, artificial neural networks (ANNs) have a less distinct partition between phases—the network is usually trained by presenting cases to be classified, at first without any previous information, and feedback on the correctness of its decision is given to the network. Each new case presented to the network allows it to adjust its internal state to predict membership better. Although ANNs are usually trained fairly thoroughly before being used on "real" data, in many cases the network will be allowed to evolve even when it is being used by eliciting user feedback, especially on difficult cases.

11.2.3.1 Regression

Unless there is a nonlinear relationship between features and classification, linear regression provides a way to predict classification based on measured features.

There is a whole host of variations on this method, but first it is important to understand the most basic case.

The simplest case is a two-variable regression, where the values of one variable are estimated from the other, that is, we have one independent variable and one dependant variable. If we call the independent variable x, and the dependant variable y, then the regression equation is of the form

$$y' = bx + a$$

where b is the regression coefficient, a the regression constant, and y' the predicted value for y.

This is linear regression and assumes that the relationship between the variables is linear, which is why the equation is that of a straight line, with b being the slope and a being the offset.

Of course in many cases, there is no simple correlation between a single independent variable and the dependant variable we are trying to predict. In such cases, we must use multiple regression. The regression equation would then have multiple partial coefficients. The equation would now be of the form

$$y' = a + b_1 x_1 + b_2 x_2 + b_3 x_3 + \cdots + b_n x_n$$

for n independent variables. Each independent variable has its own partial coefficient ($b_1 \ldots b_n$), thus back into image analysis terms, each feature in the vector is weighted separately and summed to decide on a class. Because we are generally concerned with discrete classes, we could either determine, from known cases, the maximum and minimum values for y' for each class or scale the coefficients such that y' is between 1 and 5, for example, for a five-class analysis.

Figure 11.11 shows an example plot of the data in a simple two-variable regression. The solid line shows the independent variable whereas the crosses show the dependant variable. For simplicity and to make the plot easy to read and understand, in this case, we have used $y = x$ for the independent variable, resulting in a straight line. The regression function is shown as a dashed line

$$y' = 70 + 2$$

thus, the coefficient is 2 and the constant is 70.

With real-life data, the relationship between the independents and the dependant is unlikely to be exactly the same as the regression equation, just as the crosses do not exactly fit the dashed line in Figure 11.11. The discrepancies between y (the observed values of the dependant) and y' (the predicted values of the dependant) are called the residuals. Thus, for case n, the residual is $y_n - y'_n$. The accuracy (the closeness of fit between actual and predicted results) can be measured as the sum of the residuals, $\Sigma_n (y_n - y'_n)$, which should be as close to 0 as possible. However, because the standard deviation may differ with different data sets, this simple summing of residuals does not provide a good measure for comparison. Instead, it is common to use a Pearson product moment correlation (Pearson's correlation) between y and y', called the "multiple correlation coefficient" (noted as R rather than r to show that it is not just a correlation between the dependant and an independent variable). The equation is

$$R = \frac{\Sigma yy' - \frac{\Sigma y \Sigma y'}{n}}{\sqrt{\left(\Sigma y^2 - \frac{(\Sigma y)^2}{n}\right)\left(\Sigma y'^2 - \frac{(\Sigma y')^2}{n}\right)}}$$

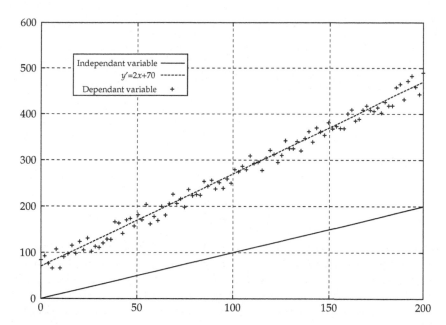

FIGURE 11.11
Example of regression.

R will be a value between 0 and 1, where 0 shows no correlation and 1 shows complete correlation. Although Pearson's correlation can also show negative correlation (values between 0 and −1), the regression equation is constructed such that this is not possible.

This is the second phase of classification (see Section 11.2.3)—having a regression equation, which is a function of the feature vector for the given case. With the knowledge of how to apply it and its correlation with the cases in the training set, we can begin to classify new cases.

This is the result of the regression. Regression itself is the process of producing this result. Actually finding the regression constant and the coefficients for each feature or independent variable can be considered as a separate problem to that of applying them and reasoning about their efficacy, and has more than one solution. Here, we will briefly discuss the most commonly used regression method: least square fitting.

Selected values for the regression constant and regression coefficient should give the least possible distance between observed values and the line that is described to predict them. Although it is possible, in most cases the real distance between the points and the regression line is not used—for simplicity, only the vertical discrepancies are counted, which are easier and computationally cheaper to determine and differ only slightly from the orthogonal distance to the line.

The squares of the vertical distances are used. If we define the function $f(i)$ to be the square of the vertical distance between the observed and predicted values of the point *i*, we get

$$f(x) = (y_x - (a + bx))^2$$

for a two-variable regression, or

$$f(x) = (y_x - (a + b_1x + b_2x + b_3x + \cdots + b_nx))^2$$

for a more general regression with *n* independent variables.

The sum of the squares:

$$R^2 = \sum f(x)$$

thus, R^2 (not related to Pearson's correlation mentioned earlier) is the sum of the squares of vertical deviations. Determining the values of the constant and coefficients is now just a matter of recursively minimizing the value of R^2. There are many sources for details of the algebraic manipulations required to do this, and even algorithms and source code, such as Weisstein (2004).

Other types of regression include polynomial regression, which results in a polynomial regression equation and many major and minor variations.

11.2.3.2 Artificial Neural Networks

ANNs are becoming an increasingly popular choice for classification tasks, especially in cases where it is difficult to make choices about which features will be useful for classification. In addition to classification, ANNs are also used for forecasting and modeling of complex systems, signature verification, voice recognition, and many other areas where algorithmic solutions do not exist or are difficult to find, and in situations where humans can determine an answer fairly easily but automated systems struggle.

By creating (not necessarily or commonly physically) a network of interconnected nodes that behave in a manner similar to the neurons in an animal brain, an ANN can be trained to analyze complex data.

Each node, or artificial neuron, may have many inputs, which are either inputs for the network as a whole or the outputs of another node. Each node may also have many outputs, which will become either inputs to other nodes or outputs for the network. Figure 11.12 shows a simple network with nodes and connections labeled. The leftmost layer is called the input layer and the rightmost is the output layer. Layers between the two are called hidden layers. In Figure 11.12, there is an input layer with three nodes, a hidden layer with two nodes, and an output layer with two nodes.

The behavior of an individual node is deceptively simple: if the combination of its inputs is below a certain threshold, it does nothing, but if the combination of its inputs exceeds the threshold, it "fires," that is, its output

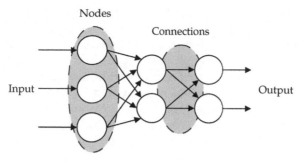

FIGURE 11.12
A simple artificial neural network.

becomes active. The combination of inputs is a simple weighted sum, so some inputs will have more effect than others, and the weights may be negative, allowing some inputs to be inhibitory. More formally, for the inputs $x_1 \ldots x_N$ to a node with N inputs with weights $w_1 \ldots w_N$, the combined input value (also called activation) is

$$a = \sum_{n=1}^{N} w_n x_n$$

It is common for outputs to be binary (0 if the neuron is not firing and 1 if it is) and although this is not absolutely necessary, using any other system (e.g., floats between 0 and 1 or unlimited values) does not provide any extra benefit because input weights are already used to amplify the effect of some inputs and diminish others.

If we call the threshold for our node t, then the node output, o is expressed as

$$o = \begin{cases} 1 & a > t \\ 0 & a < t \end{cases}$$

Although each node has a different threshold and a different set of input weights, all nodes function in this simple manner. It is the values of the weights, thresholds in the nodes, and the connections between them that allow ANNs to accomplish complex tasks with such simple components—the whole is greater than the sum of its parts.

As with regression and most classification methods, determining values for the variables in the classifier is a much more complex task than applying the resulting classifier. In the case of regression, there was a clear distinction between the exploratory phase and the application, but with ANNs, the values for the internal variables of the nodes (threshold and weights) are determined by "teaching" the network. When the teaching has reached a stage where the network can classify cases to the required level of accuracy, the network itself becomes the classifier. The learning might not stop here as it is possible to allow the network to continue to learn in a real-world environment by soliciting feedback from users.

An ANN that is to be used for classification will have an input for each feature in the feature vector and an output representing each possible classification. Classification is performed by feeding the values of measurable features for a case to the network inputs and by reading the network outputs. In a two-class system, there would be two outputs, one of which should fire for each case. If a case is presented that causes both or neither output neurons to fire, the network has encountered a case for which its training did not prepare it. In these cases, it would be necessary to fall back on human classification, which could also be used to further the training of the ANN.

In a supervised learning model, training of the network involves the adjustment of weights and thresholds based on an attempted classification. The adjustment of weights and thresholds will either reduce the likelihood of getting the same result with the given input if the answer is incorrect or reinforce current values if a correct result is obtained. Changes to weights are controlled by the learning rate—a parameter that states the allowable size for weight changes.

A complete discussion of ANN theory is not presented here and interested readers are encouraged to read some of the many texts on the subject. Some common types of neural networks and learning techniques are as follows:

Feedforward architecture. This is a common variety of neural network topology in which connections are always forward between one layer and the next from input to output. Figure 11.12 is an example of this topology. Feedback architectures allow connections in the same layer.

Backpropagation. Backpropagation networks are the most common type of ANNs. These are multilayered feedforward networks in which feedback is used to change the weights backward through the network, output layer first. Forwardpropagation performs the same sequential weight adjustment in reverse—input layer to output layer.

Hebbian learning. This is a learning method in which the weights of connections between two neurons are increased if both neurons are active at the same time.

Widrow–Hoff learning rule. This rule also called the delta rule, is a method of training single layer networks.

ANNs used for other purposes may use unsupervised learning techniques, but for classification against a predefined class system, it is important to mold the network to the desired classifications rather than allow it to attempt to determine inherent classes in the data.

ANNs are popular for image classification because of their adaptability and generality, but for moderately complex networks also it can be difficult to reason about the effect of features on classification.

11.2.3.3 Discriminant Analysis

Strictly being a form of regression, discriminant analysis is a method specifically developed for predicting class membership. There are three types of discriminant analysis.

i. *Direct discriminant analysis.* All variables are entered into the equation at once.

ii. *Hierarchical discriminant analysis.* A predefined ordering is used to enter variables.

iii. *Stepwise discriminant analysis.* Probably the most useful method of discriminant analysis in classification unless there is a reason to predetermine the relative weights of features. Stepwise discriminant analysis adds variables to the classifier one at a time, selecting the most useful features from the remaining set of unused features each time. This is the method discussed here.

As with most techniques, data used in discriminant analysis must be multivariate normal (see Section 11.2.2.2), although some skewness and multicollinearity are tolerated. Singularity and outlying values are more of a problem.

Discriminant analysis produces a discriminant function or set of discriminant functions similar to the regression function produced by linear regression. For a scenario with n used features, $x_1 \ldots x_n$, the discriminant function would be

$$D = a + b_1 x_1 + b_2 x_2 + \cdots + b_n x_n$$

where b_i is the weight applied to feature i and a is some constant. Although this looks exactly like a linear regression function, the value it evaluates to is not just a direct prediction of the dependant variable (the class). Instead, D is engineered to differ as greatly as possible when presented with cases of different classes.

For a two-class system, there will be only one discriminant function and so the mean value of D for each of the classes should be as distant as possible. A dividing point can then be determined, which can be used to determine class membership. Figure 11.13 shows an example of distribution of D for a two-class experiment, and the location of the divider. In this case, there are no misclassifications, although with real-life data, there is often some overlap of the tail ends of the distribution for each class.

For scenarios with multiple classes, there will be more than one discriminant function, which gives a multidimensional result for each case. Plotting the results allows the space to be divided into classes in the same way as the two-class scenario. Figure 11.14 shows an example of a three-class experiment with two discriminant functions (x and y axes) used to predict the

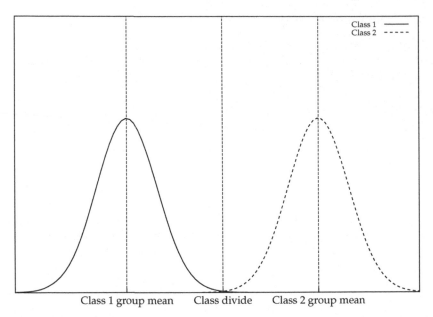

Class 1 ——
Class 2 - - - - -

Class 1 group mean   Class divide   Class 2 group mean

FIGURE 11.13
Example of a discriminant function distribution for a two-class system.

stage of cancer progression in colon tissue based on a number of textural features.

In an experiment with three classes, as in Figure 11.14, there will be two discriminant functions. There are always fewer discriminant functions than classes, although they may not be used always. The significance of the functions is not constant and each consecutive discriminant function in an experiment is less than the previous one. Looking again at our colon tissue classification example, note that the difference in means between classes along the x-axis (the first discriminant function) is greater than that along the y-axis (the second discriminant function). In most cases, only the first two or three functions are necessary for classification, and there is no significant advantage of using more.

In a stepwise discriminant analysis, variables are added to the equation one at a time. At each step, the variable that contributes most to variation between classes is selected. Using Wilk's lambda (Λ) as a test of the difference in means between classes, the significance of the change in Λ, if a variable is added, is calculated using an F test and is called the F "to enter." A threshold is set such that it states the minimum value of F for a variable to be entered into the discriminant function, and at each step the variable with the largest F to enter is selected, unless there are no variables with a value of F above the threshold. Variables that have already been selected also have an associated F value, the F "to remove," which is calculated at each step, and can lead to a variable being removed from the analysis if its F value falls below the threshold.

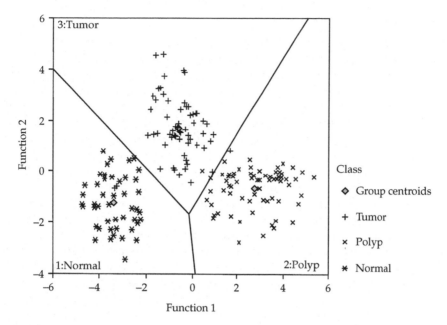

FIGURE 11.14
Example of a three-class discriminant plot showing divisions between classes.

After a number of steps, there will be no variables in the unused set that have a large enough F to be entered and no variables in the used set with an F low enough to be removed.

Further information on discriminant analysis can be found in Cacoullos (1973).

11.2.4 Logistic Regression

Logistic regression is a variation on regression that is intended to predict values for discrete dependant variables such as class. Unlike linear multiple regression, there are no formulaic procedures to arrive at the required coefficients, and so an iterative approach is used to progressively refine the coefficients until they converge on fixed values.

Logistic regression generally produces similar results to discriminant analysis on the same data, although for systems with multiple binary predictors, logistic regression is preferred.

As with the other classification methods mentioned in this section, it is usual for the classifiers to be constructed by statistical software, but the Statistical Package for the Social Sciences (SPSS), one of the most popular statistical analysis packages, can only carry out logistic regression on data sets in which the dependant has only two classes.

11.2.5 Classification Trees

A classification tree is a less statistically rigorous method of classification. Cases are analyzed by testing them against a hierarchical set of questions, until a classification can be made.

As a simple example, imagine a system for assigning people to a relevant youth group based on their age and sex. For example, boys younger than 16 can join the Boy scouts, whereas girls in the same age range can join the Girl guides. Youths between 16 and 18 are offered membership to the local youth club, and people over the age of 18 are told to get a job. The sequence of tests is best described by a tree of questions, with each leaf node being a classification. Figure 11.15 shows the youth group classification tree.

Decisions do not have to have binary results, but it is usually simpler to express decisions with multiple outcomes as a series of binary outcomes. For example, if we wanted to add a playgroup for girls or boys below 8 years, we could alter the first node to be a decision based on age rather than a decision based on the Boolean result of an expression containing age, that is, the first node would simply contain "age," and the three possible outcomes would be "<8," "≥8 and ≤18," and ">18." Instead, it is simpler to add a node before the current first node or between the first node and the "age ≤16" node with the decision based on the expression "age <8."

The strength of classification trees lies in their flexibility. There are no requirements for multivariate normality, and classification trees can cope with scenarios in which classes may have multiple centroids even if another class centroid lies between them.

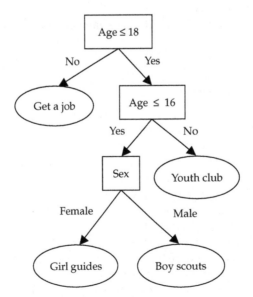

FIGURE 11.15
Decision tree for assigning youths to youth groups.

11.2.6 Other Methods

Many other techniques and variations of those presented earlier have not been discussed here. These have been omitted because they are not straight-forward classification methods. Cluster analysis methods such as hierarchical or k-means clustering are used to determine classes inherent in the data instead of creating a classifier for a known classification scheme. It is a tool for discovering classes, not just for classification.

Nonlinear estimation is a generalization of multiple regression in which the relationship between the independents and the dependant is not linear, and requires the nature of the relationship to be decided beforehand.

11.3 Using Color Texture Analysis to Automatically Determine Dysplastic Severity in Colon Tissue Biopsies

Colorectal cancer is the third most common malignant neoplasm in the world. In the United Kingdom, colon cancer is the second most common cancer-related cause of death, and kills around 17,000 people annually, with approximately 34,000 new cases every year (Quinn et al., 2001). About 60% of patients die within 5 years of diagnosis.

With the continuing research of cancer treatment, the likelihood of survival after diagnosis is increasing. As with most other types of cancer, early diagnosis of colon cancer can drastically increase the chances of successful treatment. This has prompted schemes similar to the U. K. National Health Service pilot scheme for earlier diagnosis of "at-risk" groups. With the number of incidents of colon cancer steadily increasing and the large number of cases likely to be identified by such schemes, the volume of cases investigated by pathologists is almost certain to increase.

A particularly time-consuming task is the classification of colon tissue samples. Currently, each sample must be analyzed individually by trained pathologists under a microscope to determine whether the tissue is normal or abnormal; and in cases of abnormal tissue, it is graded according to severity. Research into automated classification of colon tissue samples using gray-level texture has yielded promising results (Esgiar et al., 1998; Hamilton et al., 1987) and yet, diagnosis in all cases still requires human judgment, which is subject to interobserver (Eaden et al., 2001) and even intraobserver (Coppola and Karl, 1999) variation.

One feature that has been largely overlooked, especially in the area of cytological and histological microscopy image analysis, is color texture. A technique commonly employed worldwide to increase the visual contrast between areas of differing cytological content is dual staining with hematoxylin and eosin. This dual staining procedure highlights cell nuclei blue and cytoplasm pink or red. The information that could be extracted from the pattern of hue and saturation is lost when color information is discarded.

Here, we present the results of our research into the automation of the assessment of dysplasia based on color texture analysis and discriminant analysis.

11.3.1 Image Acquisition

Ninety-two 5 μm slices of colon tissue were taken, showing normal, polyp, and cancerous tumor tissues. Dual staining (using hematoxylin and eosin) was used to increase contrast between areas of tissue with different cytological properties, a commonly used technique to aid visual staging of samples that highlights cell nuclei blue and cytoplasm pink or red. The images were classified by a qualified, experienced gastrointestinal pathologist.

Of these images, 46 were digitized at ×100 objective and the remaining 46 at ×40. Digitization was carried out using a Leitz microscope and a digital camera. From these images, 200 × 200 pixel regions of interest were extracted, with selection carried out by attempting to capture as many visually dissimilar areas as possible. In total, the data set used for the assessment of color texture in colon tissue classification contained 140 images taken at ×40 objective and 175 at ×100. Each of these groups contains images taken from normal colon sections; colon sections taken from colon polyps, and colon sections from tumors. These three classifications represent three stages in colon cancer progression and show an increasing degree of dysplasia from the normal group to the tumor group. The breakdown of these images is shown in Table 11.1.

11.3.2 Texture over Morphology

We have made significant progress in our research by moving away from morphological indicators—the features that almost all other research in this area has concentrated on. This approach suffers from the same problems that clinicians face: morphological features that vary with dysplastic severity are not clearly defined.

In our studies, we have investigated textural features, and in particular, color texture features. Because dysplasia is simply the disorganization of tissue, textural features are an ideal way to measure this disorganization.

Using multiresolution co-occurrence matrix texture analysis applied to color images, we have been able to achieve high levels of predictive accuracy.

TABLE 11.1

Number of Test Images at Each Objective and Classification

Magnification	Normal	Polyp	Carcinoma	Total
×40	41	44	55	140
×100	46	68	61	175
Total	87	112	116	315

11.3.3   Obtaining Low-Frequency Texture Measurements

The simplest way to measure this low-frequency texture, using co-occurrence matrices, is to increase d. Figure 11.16 shows a tissue sample taken from a dysplastic colon polyp. Figure 11.17 shows the regional texture response using the energy metric with $d = 16$ and a window size of 25 extracted from the hue channel of Figure 11.16 using methods for increasing rotational invariance as discussed in Shuttleworth et al. (2002a). Comparing this texture image with the image produced using $d = 2$ (Figure 11.18), we can see that the response to finer details, such as edges, is reduced and the response to larger features, such as areas of differing hue, has increased.

An undesired feature of this approach is the effect of variations in pixel values in regions that, for the purposes of assessing structural abnormalities, would be classed as homogeneous by a human observer. These variations can be seen in Figure 11.17 as gaps and rough edges.

The variance caused by fine detail could be removed by scaling down the image. This idea is attractive for many reasons: high-frequency noise

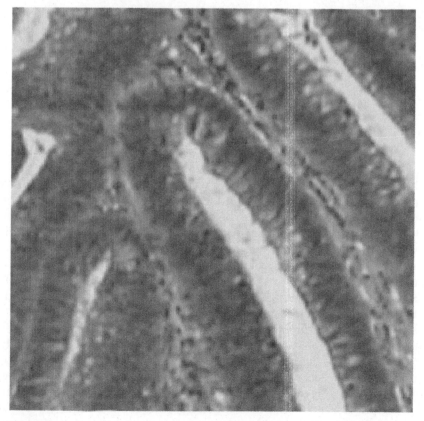

FIGURE 11.16
An example of colon tissue taken from a polyp; $d = 16$.

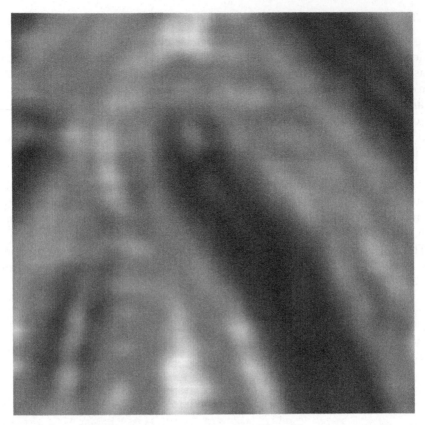

FIGURE 11.17
Texture response created by windowing the feature extraction.

will be lost, smaller values of d can be used, and there are fewer valid pixel pairs, making the extraction of the measurements a much less processor-intensive task. Figure 11.19 shows the energy response of the original image scaled down to 25%. The texture image has been rescaled to match the earlier examples. Unfortunately, the technique has disadvantages as well as advantages. While we lose high-frequency noise, we also lose some image data; the reduction in the number of valid pixel pairs increases calculation speeds but also gives a less reliable result—a 200 × 200 pixel image will have roughly 40,000 valid pairs, whereas a 50 × 50 image has only 2500, which means we are now making measurements with only 6.25% of the original data.

To combat the problems associated with extracting low-frequency texture information as described earlier, we have devised a variation of the co-occurrence matrix technique that combines the favorable elements of both simple approaches. By measuring a weighted average of the source pixel value and its neighbors, and likewise for the target pixel, the technique can operate with large pixel distances, without suffering from the effects of pixel variations in regions that we wish to be treated as homogeneous, and the decrease

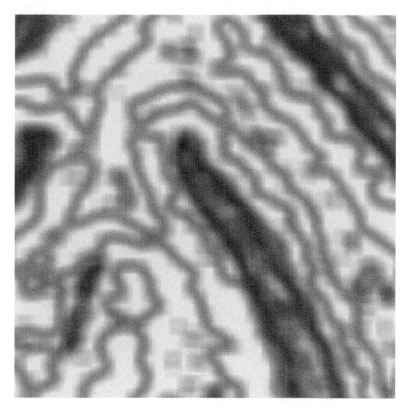

FIGURE 11.18
Using a smaller value of $d = 2$.

in the number of valid pixel pairs is negligible. Figure 11.20 shows the energy response of the new technique applied to the original image. Note that regional variations are minimal, whereas structural features of the original image are clearly visible.

To create the texture image in Figure 11.20, the contribution of neighboring pixels was calculated as a Gaussian function of the distance between each pixel in the neighborhood at the central pixel, with a neighborhood defined as the pixels within an 8-pixel radius of the source or target. To improve execution times, this can be easily implemented simply by convolving the source image with a Gaussian kernel before extracting texture information.

11.3.4 Color Texture

In an earlier study (Shuttleworth et al., 2002a), we have shown that the use of color texture yields a higher level of classification accuracy than is possible using gray-level texture analysis alone. In this work, we have analyzed data from two different color spaces (RGB and HSB). Various texture measurements were extracted from these six channels using co-occurrence matrices such as

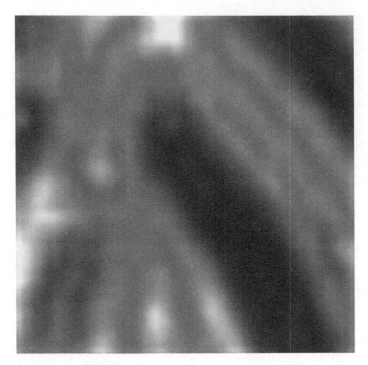

FIGURE 11.19
Extracting texture information on a scaled image.

correlation, entropy, and contrast. These data are then analyzed, using discriminant analysis, to determine which features correlate with classification.
Table 11.2 shows results obtained after classification using gray-level texture
analysis compared to results obtained using all six channels. Accuracy, sensitivity, specificity, positive predictive value (PPV), and negative predictive
value (NPV) are shown for a two-class experiment (normal, abnormal).

11.3.5 Dual Objective Analysis

Because dysplasia is a result of abnormal and prolific tissue growth, it is
exhibited at both the structural and cellular levels. Clinicians assess dysplasia on both of these levels by using multiple microscope objectives. Our
recent research (Shuttleworth et al., 2002b) has shown that by exploiting this
use of scale, it is possible to further increase the accuracy of our classification technique. By adjusting the parameters of the co-occurrence matrix
algorithm to compare pixels at much greater distances than commonly used
and include information from neighboring pixels, we were able to extract
texture measurements relating to structural features of the tissue. Combining the high- and low-frequency data, an even greater classification accuracy
was obtained (Table 11.3). Because our method is now able to mimic changes
in magnification, only the ×100 images were needed.

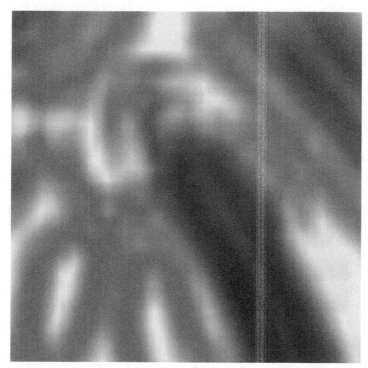

FIGURE 11.20
Extracting texture information after low-pass.

TABLE 11.2

Classification Accuracy Using Gray-Level Texture Analysis and Color
Texture Analysis

Objective	Accuracy (%)	Sensitivity (%)	Specificity (%)	PPV (%)	NPV (%)
Gray-level texture					
×100	94.8	93.8	97.8	99.1	84.9
×40	75.0	76.8	70.7	86.4	55.8
Color texture					
×100	97.1	96.9	97.8	99.2	91.8
×40	94.3	92.9	97.6	98.9	85.1

TABLE 11.3

Classification Accuracy for Combined Fine and Gross Texture

	Accuracy (%)	Sensitivity (%)	Specificity (%)	PPV (%)	NPV (%)
Combined	99.4	100.0	97.8	99.2	100.0

11.3.6 Further Advances

The concepts presented in this chapter are taken from real research. Advances have been made not only in investigating simple classification, but also in color representation of colon images (Shuttleworth et al., 2005a), class-free analysis, and the difficulties of learning histopathological microscopy (Shuttleworth et al., 2005b).

11.4 Conclusion

In this chapter, we have examined a number of algorithms and methods that are routinely applied in medical image analysis and classification research. Texture analysis is particularly useful in areas of medical imaging, where automated delineation of structure is difficult or impossible.

Regardless of the features in question, the feature selection and classification algorithms discussed earlier have many uses. They could be applied to data from patients' records (Arochena et al., 2001) just as effectively as to image data.

In many areas of medicine, the existing classification systems are a result of the difficulties associated with achieving repeatable objective analysis. In colon dysplasia grading, for example, there are problems of interobserver (Eaden et al., 2001) and intraobserver variations (Coppola and Karl, 1999). Research into variation in the analysis of dysplasia (Bosman, 2001) suggests that the imposition of a discrete classification system onto the inherently continuous progression of dysplasia as one of the main causes of classification difficulty, and that the skills used by pathologists are difficult to measure and articulate. Because of this, developing new methods of feature extraction (Shuttleworth et al., 2005a) and combination is an interesting area of current research, examining class-free results and inherent classes.

In the area of colon dysplasia analysis and in the wider field of medical image and data analysis, there have been many successes. However, there are many problems that are yet to be tackled.

References

Arochena, H. E., Todman, A., Naguib, R. N. G., Wheaton, M., and Wallis, M. (2001). Discussion of an algorithm for the prediction of attendance at first invitation from a UK NHS breast screening unit. *Proceedings of the IEEE EMBS UK & RoI Postgraduate Conference on Biomedical Engineering and Medical Physics*, p. 17, Aston University, Birmingham.

Bosman, P. F. T. (2001). Dysplasia classification: Pathology in disgrace? *Journal of Pathology*, 194:143–144.

Breiman, L. (1984). *Classification and Regression Trees.* Kluwer Academic Publishers, Dordrecht.

Cacoullos, T. (Ed.), (1973). *Discriminant Analysis and Applications.* Academic Press, New York.

Coppola, D. and Karl, R. C. (1999). Barrett's esophagus and barrett's-associated neoplasia: Etiology and pathologic features. *Cancer Control, Journal of the Moffit Cancer Center,* 6(1):21–27.

Drimbarean, A. and Whelan, P. F. (2001). Experiments in colour texture analysis. *Pattern Recognition Letters,* 22:1161–1167.

Eaden, J., Abrams, K., McKay, H., Denley, H., and Mayberry, J. (2001). Inter-observer variation between general and specialist gastrointestinal pathologists when grading dysplasia in ulcerative colitis. *Journal of Pathology,* 194:152–157.

Esgiar, A. N., Naguib, R. N. G., Sharif, B. S., Bennet, M. K., and Murray, A. (1998). Microscopic image analysis for quantitative measurement and feature identification of normal and cancerous colonic mucosa. *IEEE Transactions on Information Technology in Biomedicine,* 2(3):197–203.

Hamilton, P. W., Allen, D. C., Watt, P. C. H., Patterson, C. C., and Biggart, J. D. (1987). Classification of normal colorectal mucosa and adenocarcinoma by morphometry. *Histopathology,* 11:901–911.

Haralick, R. M. (1986). Statistical image texture analysis. *Handbook of Pattern Recognition and Image Processing.* Academic Press, New York.

Haralick, R. M., Shanmugam, K., and Dinstein, I. (1973). Textural features for image classification. *TransSMC,* 3(6):610–621.

Hsu, T.-I., Calway, A. D., and Wilson, R. (1993). Texture analysis using the multiresolution Fourier transform. *Proceedings of the 8th Scandinavian Conference on Image Analysis,* pp. 823–830, IAPR.

Julesz, B. (1962). Visual pattern discrimination. *IRE Transactions on Information Theory,* 8(36):84–92.

Kline, P. (1993). *An Easy Guide to Factor Analysis.* Routledge, London.

Metzler, V., Aach, T., Palm, C., and Lehmann, T. (2000). Texture classification of gray-level images by multiscale cross-cooccurrence matrices. In *Proceedings of the International Conference on Pattern Recognition (ICPR),* volume 2, p. 2549. IEEE Computer Society Press, Washington.

Quinn, M., Babb, P., Brock, A., Kirby, L., and Jones, J. (2001). *Cancer trends in England and Wales 1950–1999.* Office for National Statistics, Norwich.

Raudys, S. (2001). *Statistical and Neural Classifiers.* Springer-Verlag, Heidelberg.

Scheunders, P., Livens, S., de Wouwer, G. V., Vautrot, P., and Dyck, D. V. (1998). Wavelet-based texture analysis. *International Journal on Computer Science and Information Management,* 1(2):23–34.

Schürmann, J. (1996). *Pattern Classification: A Unified View of Statistical and Neural Approaches.* Wiley-Interscience, New York.

Shuttleworth, J. K., Todman, A. G., Naguib, R. N. G., and Newman, R. M. (2005a). Enhancing feature extraction from colon microscopy images using colour space rotation. In *Proceedings of the Medical Imaging Understanding and Analysis (MIUA),* pp. 11–14.

Shuttleworth, J. K., Todman, A. G., Naguib, R. N. G., Newman, B. M., and Bennett, M. K. (2002a). Colour texture analysis using co-occurrence matrices for classification of colon cancer images. *Proceedings of the IEEE Canadian Conference on Electrical and Computer Engineering,* Winnipeg, Manitoba.

Shuttleworth, J. K., Todman, A. G., Naguib, R. N. G., Newman, B. M., and Bennett, M. K. (2002b). Multiresolution colour texture analysis for classifying colon cancer images. *Proceedings of the joint 4th Annual International Conference of the EMBS and Annual Fall Meeting of the BMES*, Houston, TX.

Shuttleworth, J. K., Todman, A. G., Naguib, R. N. G., and Norrish, M. (2005b). Learning histopathological microscopy. In *Pattern Recognition and Image Analysis, Proceedings of the Third International Conference on Advances in Pattern Recognition (ICAPR'05)*, volume 3687, *Lecture Notes in Computer Science*. S. Singh, M. Singh, C. Apte and P. Perner (Eds), Springer, Heidelberg.

Singh, M., Markou, M., and Singh, S. (2002). Colour image texture analysis: Dependence on colour spaces. *Proceedings of the 15th International Conference on Pattern Recognition*, Quebec, Canada.

Singh, M. and Singh, S. (2002). Spatial texture analysis: A comparative study. In *Proceedings of the 16th International Conference on Pattern Recognition*, volume 1, p. 10676.

Tuceryan, M. and Jain, A. K. (1998). Texture analysis. In *The Handbook of Pattern Recognition and Computer Vision*. 2nd edition. C. H. Chen and P. S. P. Wang (Eds), World Scientific Publishing Co, Singapore.

Walker, R. F., Jackway, P., and Longstaff, I. D. (1995). Improving co-occurrence matrix feature discrimination. *Proceedings of the 3rd Conference on Digital Image Computing: Techniques and Applications*, 6–8 December, Brisbane, Australia. pp. 643–648.

Weisstein, E. W. (2004). Least squares fitting. *MathWorld—A Wolfram Web Resource*. http://mathworld.wolfram.com/LeastSquaresFitting.html.

Zucker, S. W. and Terzopoulos, D. (1980). Finding structure in co-occurrence matrices for texture analysis. *Computer Graphics and Image Processing*, 12:286–308.

Index